THE GREEN WORLD

THE GREEN WORLD

An Introduction to Plants and People

Richard M. Klein

Department of Botany
University of Vermont, Burlington

HARPER & ROW, PUBLISHERS
New York, Hagerstown, San Francisco, London

Sponsoring Editor: Kyle Wallace
Project Editor: Claudia Kohner
Designer: Emily Harste
Senior Production Manager: Kewal K. Sharma
Compositor: American Book–Stratford Press, Inc.
Printer and Binder: Halliday Lithograph Corporation
Illustrator: Jane R. Ulrich
Art Studio: J&R Technical Services Inc.

Cover art from *Thornton's Temple of Flora* published in 1807.
Illustrated is the bird-of-paradise, *Strelitzia reginae.*
Used by permission of The New York Botanical Garden.

THE GREEN WORLD
An Introduction to Plants and People

Library of Congress Cataloging in Publication Data

Klein, Richard M
      The green world.

      Includes index.
      1. Botany.   2. Botany, Economic.   3. Ethnobotany. I. Title.
QK47.K63   581   78-14875
ISBN 0-06-043703-0

Contents

CHAPTER 8
Plants That Changed History
340

CHAPTER 9
Plants in the Environment
394

Preface

Botany, like any other subject, can be studied from many different points of view. At the college level, botany is usually studied as a biological science in which the structure, function, evolution, and life histories of plants are examined in detail. We not only study plants but are dependent upon them for food, shelter, and even the oxygen we breath. Our daily lives involve activities in which plants and plant products are constantly used. The relationships between plants and people are multidimensional and are frequently so much a part of the fabric of our lives that they rarely enter our conscious thought. We use plant symbolisms in everyday speech in such expressions as "putting down roots" or "knocking on wood." Birth, marriage, and even death are often memorialized with the use of plants. There are religious, economic, historical, political, and many other dimensions of our concern for plants; the evaluation of these dimensions forms the framework of this book.

For centuries, people have lied, cheated, and killed to obtain particular plants or to ensure their control of a source of supply. The exploitation of the Americas provided Europe with the potato, Indian corn, tobacco, quinine, and other plants; and these plants proved to be economically and politically more important than the gold and silver extracted from the Indians. The introduction of sugar and cotton into North America resulted in an economic need for slavery, the historical, economic, and social consequences of which are still with us. The rise and fall of the Venetian city-state, the Western hegemony of China, the "troubles" and famines in Ireland, and the settling of the North American prairies are historical events deriving from our use and misuse of plants. Thus, through an examination of the green world, we can better understand the modern world and the civilizations from which Western culture has evolved.

I believe botany is an intellectual activity in every way comparable to history, philosophy, art, or political science and that there is an intellectual excitement to be obtained by learning what plants are and how they live and grow. Knowing the structure of a cell or a leaf is basic to understanding how plants capture and utilize solar energy, how they move water, and how they produce the chemicals we use. Hence this book includes aspects of formal botanical science, but it is not merely an abbreviated and simplified textbook of general botany; this would be like trying to transmit the excitement of Shakespeare by an abbreviated version of *King Lear*. Essays on botanical topics have been inserted where such information is relevant. These, too, are merely sketched and do not pretend to cover the topic.

Textbook illustrations are included to enhance the text—both visually and pedagogically. To emphasize the continuity of our long association with plants, I have chosen many old woodcuts and etchings instead of putting primary emphasis on modern photographs. Many of these old illustrations are beautiful—and beauty is an important part of our mental picture of a plant. Where possible, diagrams of plant structure are drawn isometrically, since most people visualize things better in three dimensions. It is hoped that such pedagogical considerations will benefit student and instructor alike.

If a textbook is to serve a function beyond that of helping the student pass an examination, it ought to arouse the interest and curiosity of the reader. In a book attempting to cover a very broad range of topics, none can be presented in detail, and there are as many plant-people related topics excluded as are included in this volume. Although the background, interests, and expertise of the instructor are extremely important, ultimate responsibility and intellectual pleasure are self-generated. Each section of this book ends with Additional Readings designed to provide access to subtopics. Some citations are technical; some are beautifully written expositions on aspects of the history, sociology, or botany of a topic; and others blend science and art. Some of the citations deal with current methods of processing and marketing.

This preface is followed by a Classification of the Plant Kingdom, which will serve as an additional aid to the student.

Although my name appears as the author, there have been many collaborators whose expertise, critical judgments, and friendly interest have been invaluable. I want to thank the following persons who read all or part of the manuscript: Drs. M. J. Behan, J. C. Cavender, W. M. Hess, R. W. Hoshaw, R. L. Hulbary, D. T. Klein, R. L. Mansell, C. A. Schroeder, and D. W. Smith. Some collaborators are personally unknown to me, and many have been dead for several centuries; I have, in the words of Isaac Newton, stood on the shoulders of giants. I must, however, assume full responsibility for any factual errors and for the many interpretations and personal opinions found herein. Since the focus of this book is on aspects of plant biology that have affected the lives of people in many civilizations over thousands of years, my own biases and opinions—which are conditioned by my personal life and the history of the Western civilization that has molded me—cannot fail to influence my writing. If you disagree with what I have said, so much the better!

R.M.K.

Classification
of the Plant Kingdom

Taxonomists have developed formal taxonomic hierarchies to facilitate the retrieval of information. Every species, however defined, is categorized in ascending sequence with each species belonging to a genus, every genus in a family (-aceae), every family in an order (-ales), orders subsumed within classes (-phyceae or -opsida), and classes organized into divisions (-phyta). This classification scheme is necessarily a linear arrangement, which may or may not show the relationships, affinities, or lines of descent to other taxa of equal rank; thus, a printed scheme is necessarily artificial. Where affinities are indicated, the systematist uses a phylogenetic tree whose branches are designed to show how one taxon is related to another. Most phylogeneticists tend to agree on the classes and higher categories, but at the order level and below there is considerable difference of opinion. As we learn more about the genetics, physiology and biochemistry, structure and life cycles of plants, classification schemes are appropriately revised.

Taxonomy and systematics are not merely biological stamp collecting, but are vital to evolutionary science. Evolution is, in turn, the base upon which all life sciences rest. The methods and theory of plant taxonomy and plant phylogeny—facinating and challenging sciences in their own right—can be found in many texts and advanced treatises. The scheme presented here utilized the following references.

Ainsworth, G. C., F. K. Sparrow, and A. S. Sussman (eds.). *The Fungi: An Advanced Treatise*, vol. IVab. New York: Academic Press, 1973.

Benson, L. D. *Plant Classification*. Boston: D. C. Heath, 1957.

Buchanan, R. E., and N. E. Gibbons (eds.). *Bergey's Manual of Determinative Bacteriology*, 8th ed. Baltimore: Williams & Wilkins, 1974.

Cronquist, A. *The Evolution and Classification of Flowering Plants*. Boston: Houghton Mifflin, 1968.

Klein, R. M., and A. Cronquist. "A Consideration of the Evolutionary and Taxonomic Significance of Some Biochemical, Micromorphological, and Physiological Characters in the Thallophytes. *Quarterly Review of Biology* 42 (1967):108–296.

Round, F. E. *The Biology of the Algae*. New York: St. Martin's Press, 1965.

Scagel, R. F., et al. *An Evolutionary Survey of the Plant Kingdom*. Belmont, Ca.: Wadsworth, 1965.

Kingdom: Plantae

SUBKINGDOM 1: PROCARYOTA

DIVISION: Schizophyta
    Class: Bacterophyceae (bacteria)
        Pseudomonadales (photo & chem)
        Eubacteriales (*Escherichia*)
        Chlamydobacteriales (*Crenothrix*)
        Caryophanales (*Caryophanon*)
        Spirochaetales (*Treponema*)
        Hypomicrobiales (*Rhodomicrobium*)
        Myxobacteriales (*Chondromyces*)
        Actinomycetales (*Streptomyces*)
        Mycoplasmatales (*Mycoplasma*)
        Rickettsiales (*Rickettsia*)

    Class: Cyanophyceae (blue-green algae)
        Chroococcales (*Gleocapsa*)
        Chamaesiphonales (*Dermacarpa*)
        Oscillatoriales (*Oscillatoria*)
        Nostocales (*Nostoc*)
        Scytonemales (*Tolypothrix*)
        Stigonematales (*Stigonema*)
        Rivulariales (*Gleotricha*)

SUBKINGDOM 2: EUCARYOTA

Cyanidium caldarium

DIVISION: Rhodophyta (red algae)
    Class: Bangiophyceae
        Bangiales (*Porphyra*)
    Class: Floridiophyceae
        Nemalionales (*Nemalion*)
        Gelidiales (*Gelidium*)
        Cryptonemiales (*Corallina*)
        Gigartinales (*Chondrus*)
        Rhodymeniales (*Rhodymenia*)
        Ceramiales (*Callothamnion*)

DIVISION: Chlorophyta (green algae)
    Class: Chlorophyceae
        Volvocales (*Volvox*)
        Tetrasporales (*Tetraspora*)
        Chlorococcales (*Chlorella*)
        Ulotrichales (*Ulothrix*)
        Ulvales (*Ulva*)
        Prasiolales (*Prasiola*)
        Chaetophorales (*Pleurococcus*)
        Oedogoniales (*Oedogonium*)
    Class: Bryopsidophyceae
        Cladophorales (*Cladophora*)
        Acrosiphoniales (*Urospora*)
        Sphaeropleales (*Sphaeroplea*)
        Siphonocladales (*Valonia*)
        Dasycladales (*Acetabularia*)
        Derbesiales (*Halicystis*)
        Codiales (*Bryopsis*)
        Caulerpales (*Halimeda*)

    Class: Conjugatophyceae
        Gonatozygales (*Genicularia*)
        Mesotaeniales (*Mesotaenium*)
        Desmidiales (*Closterium*)
        Zygnemales (*Spirogyra*)
    Class: Charophyceae
        Charales (*Chara*)
    Class: Prasinophyceae
        Pyraminonadales (*Micromonas*)
        Halosphaerales (*Malosphaera*)

DIVISION: Euglenophyta (euglenoid flagellates)
    Class: Euglenophyceae
        Euglenales (*Euglena*)

DIVISION: Pyrrophyta
    Class: Desmophyceae
        Desmiales (*Dinophysis*)
        Peridinales (*Peridinium*)
    Class: Dinophyceae
        Dinophiales (*Gonaulax*)
        Dinotrichiales (*Dinothrix*)
    Class: Cryptophyceae
        Cryptomonadales (*Chilomonas*)

DIVISION: Chrysophyta
    Class: Chloromonadophyceae
        Chloromonadales (*Hornellia*)

Class: Xanthophyceae (yellow-green algae)
    Heterochloridales (*Chloramoeba*)
    Rhizochloridales (*Rhizochloris*)
    Heterocapsales (*Gleochloris*)
    Heterococcales (*Chlorobotrys*)
    Heterotrichales (*Tribonema*)
    Heterosiphonales (*Vaucheria*)
Class: Chrysophyceae (golden algae)
    Chrysomonadales (*Ochromonas*)
    Rhizochrysidales (*Chrysamoeba*)
    Chrysocapsales (*Gleochrysis*)
    Chrysosphaerales (*Stichogloea*)
    Chrysotrichales (*Phaeodermatium*)
    Isochrysidales (*Isochrysis*)
    Prymensiales (*Chrysochromulina*)
Class: Bacillariophyceae (diatoms)
    Centrales (*Cyclotella*)
    Pennales (*Navicula*)

DIVISION: Phaeophyta (brown algae)
Class: Isogeneratae
    Ectocarpales (*Ectocarpus*)
    Sphacelariales (*Halopteris*)
    Cutleriales (*Cutleria*)
    Tilopteridales (*Tilopteris*)
    Dictyotales (*Dictyota*)
Class: Heterogeneratae
    Chordariales (*Mesogloia*)
    Sporochnales (*Carpomitra*)
    Desmarestiales (*Desmarestia*)
Class: Cyclosporae
    Fucales (*Fucus*)

DIVISION: Bryophyta
Class: Musci (mosses)
  Subclass: Sphagnidae
    Sphagnales (*Sphagnum*)
  Subclass: Andreaeidae
    Andreaeales (*Andreaea*)
  Subclass: Bryidae
    Polytrichales (*Polytrichium*)
    Arthrodontales (*Arthrodontum*)
    Haplolepidales (*Fissidens*)
    Diplodelidales (*Funaria*)
  Subclass: Tetraphidae
    Tetraphidales (*Tetraphis*)
Class: Hepaticae (liverworts)
    Sphaerocarpales (*Sphaerocarpos*)
    Marchantiales (*Marchantia*)
    Metzgeriales (*Metzgeria*)
    Calobryales (*Haplomitrium*)
    Jungermanniales (*Bazzania*)
Class: Anthocerotae (hornworts)
    Anthocerotales (*Anthoceros*)

DIVISION: Psilophyta (wisks)
Class: Psilophyceae
    Psilotales (*Psilotum*)
    Psilophytales (fossils: *Rhynia*)

DIVISION: Lycopodophyta (club mosses)
Class: Lycopodiae
    Lycopodiales (*Lycopodium*)
    Selaginellales (*Selaginella*)
    Isoetales (*Isoetes*)
    Lepidendrales (fossils: *Lepidodendrum*)

DIVISION: Equisetophyta (horsetails)
Class: Calamophyceae
    Hyeniales (fossils: *Protohyenia*)
    Pseudoborniales (fossils: *Pseudobornia*)
    Sphenophyllales (fossils: *Sphenophyllum*)
    Calamitales (fossils: *Calamites*)
    Equisetales (*Equisetum*)

DIVISION: Filocophyta (ferns)
Class: Pteropsidae
    Cladoxylales (fossils: *Pseudosporachnus*)
    Protopteridales (fossils: *Protopteridium*)
    Coenopteridales (fossils: *Botryopteris*)
    Ophioglossales (*Ophioglossum*)
    Archaeopteridales (fossils: *Archaeopteris*)
    Filicales (true ferns)
    Marsileales (*Marsilea*)
    Salviniales (*Azolla*)
    Marattiales (*Danaea*)

DIVISION: Gymnospermae
Class: Cycadicea (cycads)
    Lyginopteridales (fossils: seed ferns)
    Caytoniales (fossils: *Caytonia*)
    Bennettitales (fossils: *cycadoides*)
    Cycadales (*Cycas*)
Class: Pinicea (conifers)
    Ginkgoales (*Ginkgo*)
    Cordaitales (fossils: *Callixoxylon*)
    Pinales (pines, etc.)
    Taxales (yews)
Class: Gneticae
    Gnetales (*Gnetum*)
    Ephedrales (*Ephedra*)
    Welwitchniales (*Welwitchia*)

DIVISION: Angiospermae (flowering plants)
Subdivision: Dicotyledonae
Class: Magnoliidae
Magnoliales (nutmeg, laurel)
Piperales (pepper)
Papaverales (poppy)
Nymphaeales (lotus)
Aristolochiales (birthwort)
Class: Hamameliidae
Trochodendrales (*Tetracentrum*)
Hamamelidales (sycamore)
Eucommiales (*Eucommia*)
Urticales (elm, *Cannabis*)
Leitneriales (corkwood)
Myricales (sweetgale)
Fagales (birch, oak)
Juglandales (beech)
Class: Caryophyllales
Caryophyllales (cactus, pinks)
Batales (*Batis*)
Polygonales (rhubarb, buckwheat)
Plumbaginales (sea lavender)
Class: Dilleniidae
Dilleniales (peony)
Theales (tea, *Camellia*)
Malvales (baobab, linden)
Lecythidales (brazilnut)
Sarraceniales (pitcherplant)
Violales (violet, cucumber)
Salicales (willow, poplar)
Capparales (crucifers)
Ericales (heaths, blueberry)
Diapensiales (*Diapensia*)
Ebenales (persimmon)
Primulales (*Primula*)
Class: Rosiidae
Rosales (rose, legumes)
Podostemales (riverweed)
Haloragales (water milfoil)
Myrtales (*Lythrum*)
Proteales (Macadamia nut)
Cornales (dogwood)
Santalales (mistletoe)

Rafflesiales (*Rafflesia*)
Celastrales (holly)
Euphorbiales (*Poinsettia*)
Rhamnales (grape)
Sapindales (orange, maple)
Geraniales (geranium)
Linales (flax)
Polygalales (milkwort)
Umbellales (carrot)
Class: Asteriidae
Gentianales (gentian)
Polemoniales (tomato)
Lamiales (mint)
Schrophulariales (African violet)
Campanulales (Canterbury bells)
Rubiales (coffee)
Dipsacales (fuller's teasel)
Asterales (*Chrysanthemum*)

Subdivision: Monocotyledonae
Class: Alismatidae
Alismatales (arrowhead)
Hydrocharitales (*Elodea*)
Najadales (pondweed)
Triuridales (*Thiuris*)
Class: Commelinidae
Commelinales (spiderwort)
Eriocaulales (pipewort)
Restionales (*Restio*)
Juncales (rush)
Cyperales (grasses, cereals)
Typhales (cattail)
Bromeliales (pineapple)
Zingiberales (ginger, banana)
Class: Arecidae
Arecales (palm)
Cyclanthales (*Carludovica*)
Pandanales (screw pine)
Arales (jack-in-pulpit)
Class: Liliidae
Liliales (onion)
Orchidales (orchids)

Kingdom: Fungi

SUBKINGDOM 1: OOFUNGI

DIVISION: Saproleginomycetes
    Class: Saproleginomycotina
        Saprolegniales (*Saprolegnia*)
        Peronosporales (*Phytophthora*)
        Lagenidiales (*Polymixa*)
    Class: Protomycetomycotina
        Protomycetales (*Protomyces*)

DIVISION: Hypochytridomycetes
    Class: Hypochytridiomycotina
        Hypochytridiales (*Rhizidiomyces*)

SUBKINGDOM 2: EUFUNGI

DIVISION: Phycomycetes
    Class: Myxomycotina
        Ceratiomyxales (*Ceratiomyxa*)
        Liceales (*Tubifera*)
        Trichiales (*Trichia*)
        Echinosteliales (*Echinostelium*)
        Stemonitales (*Stemonitis*)
        Plasmodiophorales (*Plasmidiophora*)
        Physarales (*Physarum*)
    Class: Acrasiomycotina
        Acrasiales (*Dictyostelium*)
    Class: Labyrinthulomycotina
        Labyrinthulales (*Labyrinthula*)
    Class: Chytridiomycotina
        Chytridales (*Allomyces*)
        Blastocladiales (*Blastocladella*)
        Monoblepharidales (*Monoblepharis*)
    Class: Zygomycetomycotina
        Mucorales (*Rhizopus*)
        Entomophthorales (*Conidiobolus*)
        Zoopagales (*Harpella*)
        Eccrinales (*Enterobryus*)

DIVISION: Ascomycetes
    Class: Hemiascomycotina
        Ascoideales (*Dipodascus*)
        Endomycetales (yeasts)
        Taphrinales (*Exoascus*)
        Protomycetales (*Protomyces*)
    Class: Loculoascomycotina
        Hemisphaerales (*Stigmata*)
        Histerales (*Lophium*)
        Myriangales (*Myringa*)
        Dothideales (*Dothidae*)
        Pleosporales (*Venturia*)

    Class: Plectomycotina
        Eurotiales (*Arthroderma*)
        Aspergillales (*Penicillium*)
    Class: Laboulbeniomyocorina
        Laboulbeniales (*Stigmatomyces*)
        Lecanorales (disk lichens)
    Class: Pyrenomycotina
        Sphaeriales (*Neurospora*)
        Pyrenulales (lichens)
        Hypocreales (*Claviceps*)
        Pseudosphaeriales (*Pyrenophora*)
        Erysiphales (*Erysiphe*)
        Meliolales (*Meliola*)
        Coronophorales (*Coronophora*)
    Class: Discomycotina
        Medeolariales (*Medeolaria*)
        Cyttariales (*Cyttaria*)
        Tuberales (*Tuber*)
        Pezizales (*Morchella*)
        Phacidiales (*Rhytisma*)
        Ostropaies (*Ostropa*)
        Helotiales (*Sclerocina*)
    Class: Deuteromycotina
        Fungi Imperfecti

DIVISION: Basidiomycetes
    Subdivision: Heterobasidiomycetidae
        Class: Phragmobasidiomycotina
            Tremellales (jelly fungi)
            Auriculariales (Jew's ear)
            Septobasidiales (insect parasites)
        Class: Teliomycotina
            Uredinales (rusts)
            Ustilaginellales (smuts)

Subdivision: Homobasidiomycetidae
Class: Halobasidiomycotina
    Exobasidiales (*Exobasidium*)
    Brachbasidiales (*Dicellomyces*)
    Dacrymycetales (*Ditiola*)
    Tulasnellales (*Tulesnella*)
    Aphyllophorales (*Thelephora*)

Class: Hymenomycotina
    Polyporales (conks)
    Agaricales (mushrooms)
Class: Sasteromycotina
    Lycoperdales (puffballs)
    Phallales (stinkhorns)
    Nidularales (*Nidula*)

THE GREEN WORLD

CHAPTER 1
How People See Plants

What Is a Plant?

Flower in the crannied wall
I pluck you out of the crannies
I hold you here, root and all, in my hand
Little flower—but if I could understand
What you are, root and all and all in all
I should know what God and man is.
ALFRED LORD TENNYSON

If one looks up the word *plant* in a dictionary, one is told that "any member of the vegetable group of organisms" is a plant. Other definitions include putting seeds into the ground, an industrial building, or to defraud, as by planting evidence. Common sense tells us that the botanical plant is a living organism, that it is green, and that it grows in the ground. But a tree in winter may not be green, and many plants grow in water or, like orchids and bromeliads, grow in trees or even on telephone wires. A person in a green sweater isn't a plant by any definition, but we correctly associate greenness with plants, and we know that this color is related to the process of *photosynthesis.*

Any living system has to have provided for it, or must provide for itself, several fundamental conditions for its very existence. It must reproduce itself. It requires a place to live, formally called a *habitat.* It must have appropriate mechanisms for obtaining minerals, water, and energy for maintenance and growth—it must have food. Organisms can obtain food by one of three methods. They can absorb food in liquid form from their habitats—bacteria and fungi use this method. Some organisms can passively or actively ingest particulate matter—animals use this method. Finally, some organisms can make their own food— and only green plants can do this. We can now define a plant as a living organism that can make its own food from water, inorganic minerals, and a simple source of carbon as carbon dioxide. This definition is essentially a definition of photosynthesis and needs only the additional qualifier that the energy needed to convert carbon dioxide into food is derived from light.

In order to delimit a plant more precisely, a few additional attributes must be included. Not all of the organism need carry on photosynthesis, and the organism need not carry on photosynthesis all the time. This allows for roots, for wood in the center of an oak tree, and for deciduous trees in winter. There are some structural attributes we associate with plants that serve to separate them from animals. We think of animals as moving about and plants as staying in one place. While this is certainly true of evolutionarily advanced organisms like lions and dandelions, there are algae that swim and marine invertebrates that are bound firmly to rocks. As research tools and research techniques become more sophisticated, and correspondingly more expensive, subtle differences between the cells of plants and animals show up, but these exciting developments in molecular biology are not part of our mind's-eye definition.

The gross form of plants is very different from that of animals. Animals have appendages—heads, legs and arms, tails, or wings—and plants are composed of four major organs: roots, stems, leaves, and flowers. In addition to being a structural entity, each plant organ has functional roles necessary for survival and reproduction of the individual. There is a division of labor in flowering plants, a division more profound than exists among nonflowering plants like algae. This can be seen in the water economy of plants.

THE ROOT

All living cells require water. Since plants do not possess circulatory systems to recycle water, they need many liters per day. Water keeps chemicals in cells in solution, participates directly as a reactant in biochemical processes, maintains cell pressures, and is the solvent in which minerals are taken into the plant. Plants like seaweeds are surrounded by water and can take it in directly. More complex plants— ferns, conifers, and the flowering plants—live on and in the ground and have developed an organ, the root, to take up water. There are many different kinds of root systems ranging from the fine fibrous systems of lawn grasses to trees with tap roots that can extend down for 30 m. One common-sense idea of a plant is that if a plant has roots it is restricted to one spot. We even adapt this idea to people when we say they "put down roots."

In spite of the variations seen in the pine tree or lily, the basic structural plan for root systems is surprisingly similar for most plants, especially at the root tip. Because root structures have varied physiological roles, they are divided into three zones (Figure 1.1). At the tip is an *apical meristem*, a region where cells are actively being formed by cell division—this is the *generative* or *meristematic zone*. Distal to the meristem is a *root cap* that serves as a buffer protecting the meristem as the root grows through the soil. The chemical and physical sensors that determine the direction of root growth are found in the root cap. As new cells are formed in the meristem, they begin to enlarge; this occurs in the *zone of elongation*. It is in this zone that some cells begin to show an altered form that leads to the initiation of a primary root *vascular system* that can transfer water from soil up to the stem and leaves. In the *zone of maturation*, these young vascular cells become arranged within a circle—the root vascular *stele* (Figure 1.2)—and further change their form to permit them to function as water-conducting tissues. Some roots have extensions from the outer or epidermal layer of cells called *root hairs* which may increase the amount of root surface for water absorption. More mature roots can become woody and will show further internal alterations to increase the amount of water-conducting tissue—the *xylem* (Greek for wood).

Figure 1.1
Diagram of a primary root
from a dicotyledenous plant.
The three segments are not
drawn to the same scale; the
root tip has been enlarged to
show the meristem and root
cap.

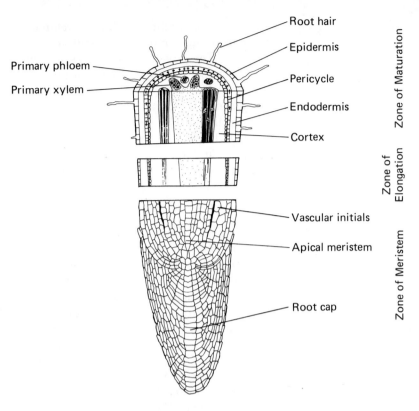

Root hair

Epidermis

Primary phloem

Primary xylem

Pericycle

Endodermis

Cortex

Vascular initials

Apical meristem

Root cap

Zone of Maturation

Zone of Elongation

Zone of Meristem

Vascular stele

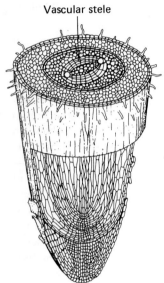

Figure 1.2
Diagram of a young conifer
root. Although basically similar to the dicot root, the conifer's vascular bundle organization is different.

The root, then, has several roles in the economy of the plant. It serves to anchor the plant firmly and permanently in the soil. It also serves as the primary organ for uptake of water and dissolved minerals from the soil, and, as anyone who has ever eaten a beet, a carrot or a radish knows, some roots evolve into storage organs, converting sugars made in the leaves to starch.

THE STEM

A moss plant, closely pressed to the moist bosom of Mother Earth, has the advantage of not having to move water any great distance, but its photosynthetic surfaces are down in the muck, susceptible to physical damage and capable of being covered and rendered inoperative. Over evolutionary time, certain tissue systems elongated and stems came into being, giving their possessors an advantage in competing with other plants for sunlight and for atmospheric carbon dioxide needed in photosynthesis. As seen in conifers and oaks, stems must have a number of structural and functional attributes. They must be strong, capable of supporting a heavy load of leaves and of withstanding strong winds. They must contain large numbers of water-conducting and food-conducting cells to supply the leaves. In many instances,

Figure 1.3
Diagram of an apical meristem or stem tip.

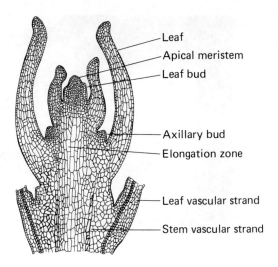

Leaf
Apical meristem
Leaf bud

Axillary bud
Elongation zone

Leaf vascular strand

Stem vascular strand

the stem serves an additional role as a food-storing organ. For continued growth, the stem has an apical meristem (Figure 1.3) comparable to that at the root tip, but it usually has the capacity to initiate leaves. A head of cabbage is a grossly enlarged stem tip (Figure 1.4). Leaves originate as small bumps on the apical meristem and develop into typical leaves. The vertical distance between successive leaf initials is the *internode.* The point where the leaf forms is a *node* or joint. As cells in the stem tip elongate, the distance between nodes increases. In the angle formed by the stem and the leaf, a bud is produced shortly after the young leaf is initiated. This *axillary bud* is itself a potential stem tip and

Figure 1.4
Section of a cabbage head showing its organization as an apical meristem. Courtesy U.S. Department of Agriculture.

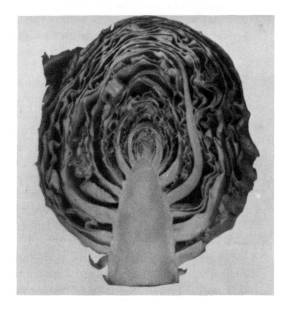

Figure 1.5
Anatomy of a tree trunk. From the outside, the tissue systems are: bark, phloem, cambium, and xylem. Courtesy St. Regis Paper Company.

its growth leads to branching. As stem tips elongate, there is a progression of internal differentiation with initiation and development of vascular stele containing the tissues that transmit water from the vascular tissues of the root. Axillary buds develop into side shoots and the leaves differentiate vascular tissues that complete a connected water-transmission system from close to the root tip up to all leaves and the tips of branches.

With increasing stem length, increase in stem diameter is brought about by the division of a special layer of cells, the *cambium* (Figure 1.5), which can form cells internally that differentiate into xylem cells and externally that develop into *phloem* (bark) cells (Figure 1.6). Xylem carries water up the stem and phloem carries the products of photosynthesis down from the leaves—a dual plumbing system is contained in the vascular stele. In a perennial woody plant such as a tree, the cambium is active during the growing season, producing larger xylem cells in the spring than it does later in the season, giving rise to the annual rings we see in cross section (Figure 1.7). The cambium also produces new phloem cells. Some of the older phloem cells become cork cambium cells which can then begin to form new cork or bark cells which increase the protective function of this tissue system. Wood in the center of a tree loses its capacity to conduct water as chemicals accumulate that turn the light yellow cell walls a dark red-brown. This forms the heart wood. However, as long as there are two or three annual rings of light-colored, unplugged sapwood, the tree can remain functional—this explains why hollow trees can continue to live and grow.

The branching patterns, the type of twigs, and the form of leaves allow one to distinguish between a spruce and a pine or between an oak and a maple as far away as they can be seen. These patterns of gross form are under genetic control as modulated by the environment. Genetically identical seedlings of a maple tree will, at maturity, look very different if one of them develops in a sugarbush surrounded by other

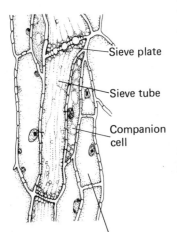

Companion cell

Face view of sieve plate

Sieve plate

Sieve tube

Companion cell

Phloem parenchyma cell

Figure 1.6
Diagram of the conducting cells of the phloem.

Figure 1.7
Diagram of part of a wood stem showing two annual rings. The large xylem vessels are formed in the early spring.

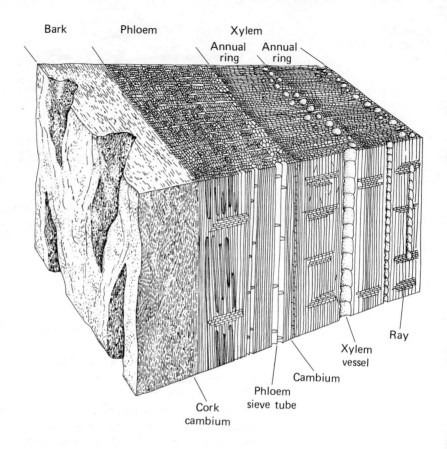

Bark Phloem Xylem
Annual ring Annual ring
Ray
Xylem vessel
Cambium
Phloem sieve tube
Cork cambium

trees and the other grows in an open field where it can fully express its genetic potential. Indeed, the solitary oak or maple silhouetted against a blue sky with white fleecy clouds gently drifting by is a powerful image of long life and resistance to the hazards of the environment. We speak of someone as "sturdy as an oak," and other images such as "graceful as a willow" or "powerful as an ash" are plant concepts applied to people.

THE LEAF

Before flowering plants evolved, plants with only stems (*Rhynia*—now found only as a fossil) carried out photosynthesis with *chloroplasts* in the cells of their stems. Although several theories have attempted to explain the evolutionary progression from a leafless stem to one with needles or flattened leaves, none is fully adequate. The "typical" leaf with its flattened blade is reflected in human speech: *leaves* of a book in English is paralleled by the German *blatt* and the French *feuillet* or their exact equivalents in many other languages. The annual dropping of the leaves of deciduous trees occurs as a death image in all cultures; O'Henry wrote a mawkish short story with this theme. The bursting of

Figure 1.8
Diagram of a leaf.

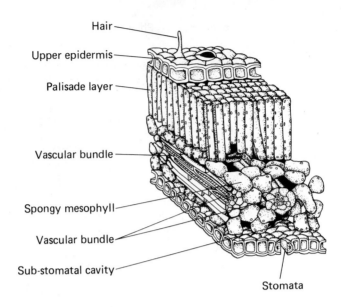

Hair
Upper epidermis
Palisade layer
Vascular bundle
Spongy mesophyll
Vascular bundle
Sub-stomatal cavity
Stomata

buds and the formation of young leaves in the spring is a resurrection symbol in all religions.

Structurally, the leaf is well adapted for its two major functions—the formation of food by photosynthesis and the elimination of water (Figure 1.8). The upper and lower leaf surfaces are bounded by sheets of epidermal cells covered by a waxy, cutinized layer that cannot be wetted and prevents the excess loss of water. Both water loss and up-take of the carbon dioxide for photosynthesis occur through *stomata*—literally "little mouths"—found most abundantly on the lower surface of most leaves. By changes in the water pressure within the guard cells on either side of the stomata, they open or close, controlling the amount of water vapor leaving the leaf and the amount of carbon dioxide entering (Figure 1.9). Since the water diffuses out as vapor, minerals taken up by the plant in soil water remain in the plant. Typically, the leaf has

Figure 1.9
Diagram showing stomata in the opened and closed positions.

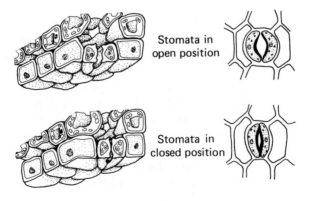

Stomata in open position

Stomata in closed position

three tissue systems sandwiched between the upper and lower epidermis. One or more rows of columnar cells, the *palisade layers*, make up the major chloroplast-containing cell system and carry out most of the photosynthesis. Less tightly packed cells, the *spongy layers*, are also responsible for the movement of water from the veins, which are leaf vascular tissues, to the stomata where water evaporates back into the atmosphere.

AN ESSAY ON WATER MOVEMENT IN PLANTS

From the time a seed is put into the ground until withered stalks and leaves are harvested for cattle food in the fall, a single corn plant moves over 250 liters of water from the soil into the air. Reduction in the availability of this water will result in a stunted, poorly developed plant with small kernels on a shriveled cob. On a hot summer's day with a brisk wind, a date palm can lose 400 liters of water and a ragweed plant 8 liters, while a large desert cactus will lose little more than a spoonful. This huge range in water requirements of different plants indicates that control of water movement is one of the adaptations plants have made to allow them to occupy the wide diversity of habitats that Earth affords.

Yet the processes by which water is taken up by a root system, moved through plants, and eliminated are basically the same. The physical problems that must be overcome are formidable: water moves through a series of microscopically small pipes, xylem vessels and xylem tracheids (Figure 1.10), encountering considerable resistance to flow, and rises to heights of 150 meters (in large redwood trees). The entrance of water occurs through the outer layers of the roots under a small pressure developed by the difference between the concentration of dissolved material within and without the root—the water potential due to solutes—and moves between root cells until it enters the stele containing the root xylem tissue. Root pressure rarely exceeds two atmospheres, about enough to move water three or four feet up the root because of the resistance to flow. Water molecules have a physical attraction to each other, allowing us to think of a column of water as a unit, but pushing this column to the heights required is beyond the capacity of any pump. The only physical way that this can be done is by pulling it up by a suction force. The source of this pulling power is the process of transpiration in leaves. Assuming that there is a continuous column of water from the soil solution up to the leaf via the xylem systems (Figure 1.11) in root, stem, petiole (leaf stalk), and leaf, evaporation of water from the surfaces of the spongy cells of the leaf out through the stomata engenders suction pressures capable of pulling the water column up to considerably greater heights than the tallest trees now on earth.

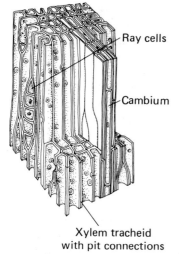

Ray cells

Cambium

Xylem tracheid
with pit connections

Figure 1.10
Diagram of the xylem cells of a conifer with tracheids and bordered pits.

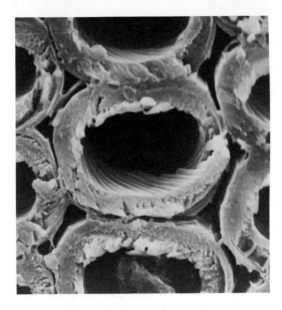

The control over the amount or volume of water moved through a plant resides primarily in the leaves. Signaled by temperature, light quality and intensity, and other environmental factors, stomata may be fully open allowing maximum water transpiration or fully closed allowing minimal water loss.

Depending on the genetic constitution of the plant and on environmental conditions during leaf development, there are variable numbers of stomata per leaf. Leaves formed when a tomato plant is growing in damp shade have more stomata than those formed when the plant is in the hot sun. Most cactus plants have dispensed entirely with leaves and lose very little water, and other desert plants like the ocotillo develop leaves during the rainy season and drop them when the water supply becomes limited. Fuzzy hair coatings on leaves slow down the loss of water, and sunken stomata below the leaf surface (as in pine needles) are among many other structural adaptations of plants for coping with environmental situations in which water is limited. Plants growing in situations where water is adequate or in excess have other leaf adaptations befitting their life styles.

THE FLOWER

The evolutionary alterations of leaves that resulted in the formation of the flower, the fourth of the plant's major organ systems, is also part of our mental construct of plantness. Flower form, visually and microscopically, keys precisely with flower function as discussed in the following sections of this chapter. The archetypical concept of the flower is also a source of verbal and cultural images that can be traced in the art of many cultures. The fruits and seeds that develop from flowers have cultural and verbal implications that transcend the botanical and biological roles of these "bearers of the next generation" (itself a cultural phrase). "Planting seeds of doubt," "sowing wild oats," "the germ of an idea" are all related to the concept of the life cycle of plants, the flow of the seasons, death, transfiguration, and renewal. Although the source of these images is obvious when one thinks about it, humans have often used plant images in very general ways and have deliberately stylized plant concepts to the point where their origin is virtually unrecognizable. Sturdy, unyielding pillars that hold up buildings are stylized tree trunks; swirling, intricate, colorful paisley patterns are leaves; gnarled hands are roots; and the flower has become as sexual a symbol as the devious mind of man could devise.

ADDITIONAL READINGS

Cronquist, A. *Introductory Botany*, 2nd ed. New York: Harper & Row, 1971.
Galston, A. W. *The Life of the Green Plant*, 2nd ed. Englewood Cliffs, N.J.: Prentice-Hall, 1964.
Heath, O. V. S. *Stomata*. Oxford Biology Readers. New York: Oxford University Press, 1975.
Huxley, A. *Plant and Planet*. New York: Viking Press, 1975.
LIFE Nature Library. *The Plants*. New York: Time-Life Corp., 1965.
Maheshwari, P. *Plants and Human History*. Hyderabad, India: Osmania University Press, 1965.
Quinn, V. *Leaves—Their Place in Life and Legend*. Philadelphia: Lippincott, 1957.
Ray, P. M. *The Living Plant*, 2nd ed. New York: Holt, Rinehart and Winston, 1972.
Rayle, D., and L. Wedberg. *Botany: A Human Concern*. Boston: Houghton Mifflin, 1975.
Rickett, H. W. *The Green Earth*. Lancaster, Pa.: Jacques Cattell Press, 1943.
Rutter, A. J. *Transpiration*. Oxford Biology Readers. New York: Oxford University Press, 1972.

Cells

The body is a State in which each cell is a citizen.
RUDOLPH VIRCHOW

Plant organs were obvious to fifteenth-century observers with their crude microscopes, and the general organization of roots, stems, leaves, and flowers could be made out, even if dimly. Robert Hooke looked at a thin slice of cork under a microscope in 1665 and reported in his *Micrographia* that the pores or cavities into which cork was partitioned looked like the rooms or cells in which monks lived (Figure 1.12). As a good protestant, this allowed him to imply that Catholics, only recently banned from the social and political life of England, were ministered to by vegetables—people without intellect. Hooke turned

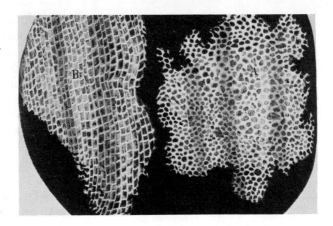

his microscope on bits of charcoal, fruits, and on vegetables and found that this same "schematisme" or organization was present. Almost in passing, he noted that fresh tissues from green plants are "fulled of nutritive juices." Almost 200 years later it was recognized that the living cell was the juice and what Hooke called a cell was the nonliving cell wall, an excretion product of the living matter in many ways comparable to fingernails and hair.

Mathias Schleiden, professor of botany at the University of Jena, announced in 1838 to a most skeptical scientific world that the cell, the fluid stuff, was the fundamental living unit of all plant structures, and that leaves, stems, and roots were made of similar units. As an additional heresy, he suggested that botany was more than classifying plants and that the examination of how these units came into being and how they were organized would yield much useful information. He pointed out that each cell was a self-contained unit and that each new cell added to the structure of organs must be derived from a preexisting cell. These assertions were previously made by several others. Within a year, Theodore Schwann, professor of anatomy at Louvain University, extended Schleiden's cell theory to animals. "Growth," he wrote, "is not the result of a force having its ground in the organism as a whole, but each of the elementary parts possesses a force of its own, a life of its own, if you will. . . . the totality of the organism can indeed be a condition, but on this view, it cannot be a cause."

As microscopes improved in the late nineteenth century, botanists turned them on all sorts of living tissues, building up a picture of what Hooke's "nutritive juices" actually contained. The plant cell could be viewed as a box. Indeed, the wall about the living cell is made up of several layers of cellulose fibers impregnated with other chemicals, the whole box being essentially wood, for wood in the modern sense is composed of the dead cell walls of specialized plant cells.

Directly inside the cell wall is a thin membrane, the *cell membrane*, which encloses the rest of the living cell (Figure 1.13). Composed of a

Figure 1.13
Diagram of the organization
of a plant cell and the fine-
structure of cell organelles.

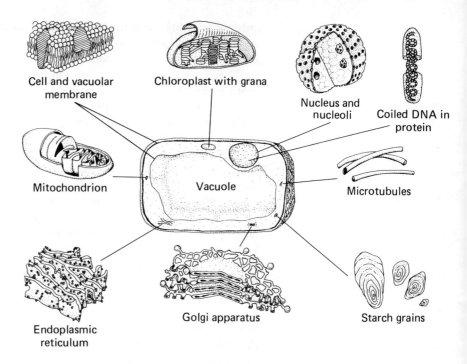

Cell and vacuolar
membrane

Chloroplast with grana

Nucleus and
nucleoli

Coiled DNA in
protein

Mitochondrion

Vacuole

Microtubules

Endoplasmic
reticulum

Golgi apparatus

Starch grains

sandwich of fat and protein, this bounding membrane is part of the living cell that serves to regulate the movement of water in and out of the cell, excluding some chemicals and facilitating the passage of others. It is extremely sensitive, losing its functional integrity with heat (e.g., the release of red juice when beets are boiled), to cold (when frozen strawberries are thawed), or to chemicals (when lettuce wilts as vinegar is added to a tossed salad). There is a second, structurally similar membrane that serves as the other boundary of the living cell—the *vacuolar membrane*—which has the identical functions of regulating the movement of dissolved material into the *vacuole.* Plant cells are different from animal cells in the possession of a vacuole, and this difference bears directly on how plant cells enlarge.

THE VACUOLE

Within any animal organ, the liver for example, there is little variation in the sizes of cells of a given type; a cell divides and its progeny reaches mature size. Plant cells of a given type, cortical cells in a stem for example, may vary by a factor of three to five in size, starting out as small units and growing in volume as water accumulates in the vacuole, much like a balloon increases in size as air is blown into it. There is, however, one important difference between the balloon and the cell. As the plant cell enlarges through water uptake, its cell wall also grows. Thus, for a plant cell to grow in length and/or diameter, not only must water accumulate in the vacuole by expedited passage through the cell and vacuolar membranes, but the cell wall must loosen and new cellulose fibrils added. Both processes involve growth-regulating chemicals produced primarily in the apical meristems of stem and root tips and in the young leaves.

Vacuoles are of variable size. At one extreme are individual cells of the watermelon in which each cell is over 95 percent water and at the other are newly formed cells of meristems which lack a vacuole, forming one only when the cell enlarges. Although as cells mature their cell walls become thicker and stronger, a plant like a coleus or geranium does not have sufficient wall strength to maintain its upright position without the water pressure developed as stem cell vacuoles accumulate water. If water is limited, individual cells lose their vacuolar water, water pressure is reduced, and the plant wilts. Cells with high vacuolar water pressures are said to be *turgid* and the opposite condition is termed *flaccid*. A plant under water stress, when water is suboptimal, will do poorly and will grow slowly, with individual cells failing to achieve their maximum potential for elongation. Water stress can occur from many causes: inadequate soil moisture, a damaged root system, derangement of water translocation, and excessive water loss through the leaf stomata. In orange fruits considerable citric acid and sugars accumulate in the vacuole as side products of cell metabolism, and the brilliant red pigments (anthocyanins) in geranium petals or the equally bright betacyanin pigments of red beets are dissolved in the vacuolar sap. Tannins used in leathermaking and a variety of crystals are also found in the vacuole. Thus, the vacuole is not merely a passive bag of water, but is a participant in the metabolic economy of the cell.

THE CYTOPLASM

The living material of a cell is bounded by the two membrane systems with the cytoplasm, consisting of an emulsion of protein dispersed in water, much like Jello without artificial colorings and flavorings. These proteins are not inert, but are *enzymes*, each functioning to synthesize or degrade one of the multitude of compounds that exist within cells. To some extent, cytoplasm can be simulated by adding appropriate enzyme proteins, fats, sugars, etc. to water, but the test tube simulation is not alive—it looks, smells, even tastes like plant cytoplasm, but it doesn't behave like a living system because the reactions that occur in the test tube are neither regulated nor integrated, and life is both. We don't know all the constituents of living stuff, how it is organized and how it is integrated. Perhaps some day . . .

THE CHLOROPLAST

Within the cytoplasmic fluid-gel are discrete bodies, each bounded by the same fat-protein membrane system that delimits the cell and the vacuole. A few of these *organelles* (little organs) are large enough to have been seen rather early; Robert Brown observed in 1831 a large, roundish organelle he called the nuclear body, although he didn't think it was important. Green bodies, somewhat football-shaped, were seen even earlier and given the name *chloroplastids*—green bodies. It was obvious that they had something to do with the self-feeding (autotrophy) of plant cells.

By 1850–1860 it was agreed that all plant and animal cells contained a nucleus and that only plant cells contained chloroplastids. This was as far as observations could go until the latter part of the nineteenth century when light microscopes became powerful enough to magnify a thousand-fold with good sharpness. With this new tool botanists were able to make out smaller bodies. Individual starch grains could be stained with tincture of iodine and visualized as small blue-black dots floating freely in the cytoplasm. Small ovoid bodies, about the size of bacteria, were named *mitochondria* (a Greek construction: *mitos* - thread; *chondrus* - cartilage), although they were frequently confused with bacteria and their function was unknown. A microscopist named Camillo Golgi saw what appeared to be roundish blobs with rough edges, and this apparently functionless organelle, the *Golgi apparatus,* was named for him. A series of threads and tubes, the *endoplasmic reticulum,* could be seen if the microscopist strained very hard and had a good imagination.

The invention of the electron microscope in Germany in the 1920s and its perfection in the late 1940s allowed magnifications of over 400,-000 times, permitting confirmation of earlier findings and detailed descriptions of the fine structure of small organelles. As high-speed centrifuges and new biophysical techniques became available, the organelles of plant cells could be removed from the cells for detailed examination under the electron microscope. With the structure of these organelles known, biochemists began to probe their functions, ushering in an era of macromolecular biochemistry that is still an active research area.

These cell organelles are found in both animal and plant cells, and although there are structural and functional differences between those in the cells of tigers and tiger lilies, they are sufficiently alike so that researchers who work at the cell level can study either plants or animals with a reasonable assurance that the information obtained is applicable to both. The chloroplast, however, is unique to plants.

In spite of tremendous diversity in life styles, in habitats, and in form among various plants, the internal structure and the function of the chloroplasts is virtually identical. Surrounded by a double protein-fat membrane, the chloroplast contains a semiliquid *gel* in which an array of enzymes active in photosynthesis are found. The chloroplast pigments are contained on and in stacks of flattened hollow disks, each disk called a *thylakoid*, each stack called a *granum*; the granum resembles a stack of coins. Other thylakoids connect stacks of grana. Thylakoid membranes are similar to other plant membranes in that they are composed of protein and fatty substances to which the pigments are attached. The chemical composition of individual pigments is known with some precision. Most abundant are two chlorophylls, one bright green (chlorophyll *a*) and the other a blueish-green (chlorophyll *b*). Several chemically related yellow xanthophyll pigments, several orange-red carotene pigments, and several bright red cytochrome pigments are also present. The pigments and nonpigmented chemicals are spatially arranged to facilitate the capturing of light energy and the transfer of chemical energy that occurs in the process of photosynthesis.

THE NUCLEUS

The most prominent organelle is the *nucleus*, a spheroid-shaped body enclosed in a double membrane with many fine pores and containing the *chromosomes* (the hereditary apparatus of the cell) and one or more smaller bodies, the *nucleoli*. Chromosomes are made up of the nucleic acid DNA, arranged as a coiled helix or spiral surrounded by a protein overcoat. The number of chromosomes per nucleus is constant for any particular plant or plant species, although the number may be as few as four or as many as several hundred. With certain exceptions, all cells in a plant have not only the same number of chromosomes, but each cell contains exactly the same genetic information, which raises exciting research questions of how and why cells become functionally and structurally specialized. Based on the arrangement of four different chemicals in the DNA subunits (two purines and two pyrimidines), the genetic information is coded in such a way that it can be transcribed onto the other type of nucleic acid—RNA—which is synthesized in the nucleoli. These information-retrieving RNAs move out into the cytoplasm through nuclear pores and direct the synthesis of protein enzymes which, in turn, control all cellular activities. Protein synthesis itself takes place on the *ribosomes*, small spheres attached to the endoplasmic reticulum.

Multiplication of cells occurs by division, a process called *mitosis*. Typically, the dual (diploid) sets of chromosomes in the nucleus each duplicate themselves giving four sets and then transferring two sets to the new cell formed by cytoplasmic division. The evolutionary logic of this process is impeccable; it insures that each new cell is not only the product of a preexisting cell, but also that the new cell has exactly the same hereditary complement as its progenitor. Stability and uniformity is maintained. Variation, with the consequent possibility of giving rise to a cell better adapted to the totality of the environment, can occur by a change in the ordering of the genetic chemicals and a mutation, or sport, may arise. The navel orange, most of our apple varieties, and many useful economic plants arose by somatic mutations and have been propagated by cuttings or grafts.

The linked series of enzymes that carry out respiration are carried on the convoluted surfaces of the membranes that create the maze (cristae) of the mitochondrion. Their numbers vary, with many per cell when the cell is metabolically very active to relatively few when the cell is quiescent. When activity is stimulated, and during mitosis, the individual mitochondria divide, with the new mitochondrial synthesis directed by nucleic acid information from the nucleus and by DNA information contained within the mitochondrion itself.

Functional roles of the other cell organelles, aside from the chloroplast, are less well understood and active research is still underway on these problems. We know that Golgi and microtubules accumulate where new cell membranes and walls are being formed, and there is considerable evidence that these organelles play roles in wall formation. Still something of a mystery is the structure and function of very minute structures that even with the resolution of the electron microscope are seen as blobs or tubules—the *spherosomes*.

AN ESSAY ON AUTUMNAL COLORATION

In the eastern provinces of Canada and adjacent northeastern states of Maine, New Hampshire, Vermont, and northern New York, October is known as "leaf season," when these areas are invaded by people looking at the colors of leaves on maples, oaks, sumacs, and other deciduous trees. Leaf season is of economic importance to the tourist industry and a good year may bring a million people and several million dollars into the state of Vermont alone. It isn't a case of "Well, if you've seen one tree, you've seen them all," for natives of Sherbrooke, Quebec or Burlington, Vermont, also drive or walk through the woods with as much pleasure as the visitor from British Columbia or Florida. The brilliance of the foliage is unique not only because of the tree species involved, but also because the fall weather in the Northeast is nearly perfect for the development of the colors.

Careful analysis of the colors that, together, provide the beauty of an autumn sugar maple leaf indicates that there are five basic colors. The maple leaf will show green, orange-red, yellow, bright red, and brown; and from these five colors can be obtained the delicate shadings and mixtures that bring dollars to the economy and pleasure to the viewer. Three of these are present in the leaf throughout the spring and summer.

The summer leaf is seen as bright green because the chlorophyll molecules preferentially absorb blue and red wavelengths of sunlight, allowing green wavelengths to be reflected from the leaf or transmitted through it. The high concentration of chlorophylls in chloroplasts mask or cover up the orange-red pigments of carotene (seen in its full hue in the carrot) and the yellow pigments of the xanthophylls (seen in yellow Delicious apples and in daffodil petals). As days get shorter and cool nights occur, chlorophyll is broken down faster than it is synthesized, and oranges and yellows become visible. Hues of brown are due to breakdown products of chlorophylls and other leaf chemicals. Just these four colors give the leaves a good deal of color; the wonderful yellow-golds of poplars shining through the deep green of conifers in the Rockies represent the amalgam of these four colors.

But a fifth pigment can bring Staghorn sumacs into flame and add that special touch to a hillside. The red pigment is anthocyanin, the pigment of geranium petals, Macintosh apple skin, and rhubarb petioles. As both day and night temperatures continue to decline, all pigments fade as leaf enzymes degrade them. As cells lose their capacity for biosynthesis, they no longer synthesize growth substances that can move down the leaf into the petioles. With reduction in the concentration of these substances, special cells at the base of the petiole become corky and die, and the leaves fall (abscise) from the trees. Winter is well on its way. Skiing, anyone?

CELL GROWTH

The diagrams of roots, stems, leaves, and flowers clearly show that cells are certainly not all the same. All cells in an individual plant start out as a product of mitotic division of a preexisting cell (we shall avoid the question of where the first cell came from, although there is considerable speculation on this point) and hence have exactly the same *genome*—the same genetic complement. As these cells enlarge, each may begin to differentiate into one of a variety of possible cell types, their final form keying with their final function in the plant. Similar differentiations lead to the formation of a tissue or tissue system which may be made of one or a few cell types, just as a mammalian organ like a kidney contains several cell types each performing certain functions. The existence of plant tissue systems was recognized in the late seventeenth century by Marcello Malpighi, John Ray, and especially Nehemiah Grew. It wasn't until the nineteenth century that the arrangement of cell types into tissue systems and thence into organs was fully appreciated, and it wasn't until the twentieth century that the functional significance of tissue systems began to be understood with any precision.

The study of the alteration—the differentiation—of a simple cell, as described above, is the study of growth and development, although some scholars call it the dismal science because as soon as we think we have an answer, someone comes along and demolishes the theory. At least theoretically, we are fairly sure that the forces that bring about cell differentiation are chemical and physical, and they operate through the genetic machinery.

An example can indicate this basic concept. A Ph.D. candidate in France took a bit of tissue from a lilac stem and, under sterile conditions, grew it on an agar jelly containing mineral salts, sugars for energy, and vitamins that stem cells cannot make. The cells continued to divide and to enlarge, forming a somewhat firm but amorphous mass of cells. Such tissue cultures can be maintained for many years without significant structural or functional change. Dr. Guy Camus then grafted the apical meristem of a lilac onto the tissue culture and found that in a line extending from the bud down through the cell mass, certain of the cells began to change, differentiating into rows of functional xylem and phloem cells; the xylem moving water and minerals up to the now-growing stem derived from the meristem and the phloem carrying sugars from the young leaves down through the tissue culture mass.

Professor Ralph Wetmore and his students at Harvard postulated that the chemical impetus for differentiation was growth-regulating chemicals, formed in the meristem, which diffused down into the tissue culture. They demonstrated this by substituting a glass tube containing these chemicals for the meristem. Subsequent research in several laboratories showed that there were several chemicals involved, one (auxin) favoring differentiation of xylem cells, and others (gibberellic acid and cytokinin) favoring differentiation of phloem. As research data accumulated, it was learned that other chemicals, produced in one tissue and moving to another, could negate or reinforce these responses; that temperature, the pressures of cells against each other, or even the concentration of oxygen and carbon dioxide could modulate the responsiveness of cells to the process of differentiation.

Light, absorbed by complex molecules called *phytochromes,* is a powerful factor in differentiation. Conversion of a vegetative apical meristem that normally continues to form new cells into a meristem that will give rise to a flower is controlled by the relative balance of light and dark and the ratio between bright red wavelengths and far red wavelengths. If every plant cell in an organism is genetically the same, the physical and chemical effectors of differentiation must, ultimately, act to suppress or to activate genes that result in the formation of specific enzymes—the proximate cause of cell differentiation. This is the real cutting edge of all biological science—the coordination of information obtained by the biochemist and the molecular geneticist with that obtained from studies of growth and development into a comprehensive field theory that will allow us to predict (and possibly to control) differentiation and development.

ADDITIONAL READINGS

Salisbury, F. B., and R. V. Parke. *Vascular Plants: Form and Function.* Belmont, Ca.: Wadsworth, 1965.

Steeves, T. A., and I. M. Sussex. *Patterns in Plant Development.* Englewood Cliffs, N.J.: Prentice-Hall, 1972.

Thompson, D. W. *On Growth and Form,* 2nd ed. New York: Cambridge University Press, 1942.

Wardlaw, C. W. *Morphogenesis in Plants: A Contemporary Study.* London: Methuen, 1968.

How Plants Get Their Names

> What's in a name? That which we call a rose
> By any other name would smell as sweet.
> SHAKESPEARE, *Romeo and Juliet*

Each year, plants become a part of the Christmas holidays. One is the poinsettia, another the Jerusalem cherry, and another, luckily, is mistletoe. Holly, Scotch pine, sage in the turkey stuffing, sweet potatoes, and many other plants have festive connotations. But where did these names come from? The Jerusalem cherry certainly isn't a cherry; it looks like a miniature tomato and, like the tomato, it originated in South America, not the Holy Land. Poinsettia is a French-sounding name; Mistletoe is vaguely Germanic; and the Scotch pine isn't from Scotland, but from Asia Minor. Culinary sage is a member of the mint family and has nothing to do with being wise, and sweet potatoes are only relatively sweet. We also know that poison oak isn't an oak, that club mosses aren't mosses, and that eggplants are not related to hen fruit.

The word Jerusalem is given to many plants that did not come from the Middle East. There are Jerusalem cherries, artichokes, oaks, and sage corns. Another Jerusalem cherry is really a hot pepper, another sage and about 50 other plants share the same common name. Although the plants themselves are usually unrelated, they do have a few things in common. Many of them come into flower or fruit around Christmas time in Europe, giving them an association with the festival and hence the connection with the city in which the festival originated. The Jerusalem oak is a weed related to the common pigweed, but it has leaves that are vaguely oaklike, and it was collected in North Africa during the Crusades by surgeons who were searching for medicinal plants. Crusaders appended the name Jerusalem to many plants they collected—after all, they had to show something to the taxpayers when they got home. As empires expanded in the fourteenth and fifteenth centuries, many plants that came from far-away places were called Jerusalem something-or-other.

NEED FOR A CLASSIFICATION SYSTEM

By the middle of the seventeenth century, Spain, Holland, France, Belgium, and Britain were well embarked on their imperialistic goals. Tremendous voyages of conquest and exploration extended the sway of these countries over lands scattered around the world, and the treasures of these subjugated countries, including the plant wealth, were pouring back into Europe. The Royal Society of London, organized in 1660, prevailed on the British Admiralty to allow biologists to accompany the fleets to collect plants and animals for universities, museums, botanical gardens, and zoological parks. In keeping with ideas of the Age of Reason, enthusiastic Fellows of the Royal Society believed that they could understand the world by cataloguing the contents of the earth—animal, vegetable, and mineral. It is easy to snicker at their innocent arrogance; today's catalogues are many times thicker, and yet we understand less each passing day. Remember, however, that they had the courage to dream a great dream, and there have been few enough great dreams.

Problems of cataloguing began to overwhelm the scientists; Europe was being engulfed with new plants. Not only were plant hunters commissioned by scientific organizations, they were also sent out by ministeries of agriculture, by navies looking for trees suitable for masts, by medical societies hoping to locate new drugs, by industry searching for new sources of raw materials and, not least of all, by wealthy patrons. These rich and powerful individuals built ostentatious homes with extensive gardens, and it became social one-upmanship to possess plants new to Europe.

New problems arose. How could one determine whether a dried plant from Tierra del Fuego was the same as a plant in a salt-encrusted specimen box, marked RUSH, which had been in transit for several years from the Malay Straits? Were there similarities between the plants from Brazil and those from the west coast of Africa? Obviously, there had to be some systematization, the development of a method which could do several things simultaneously. Linguistic differences had to be eliminated; your moss rose might be my portulaca and her purslane.

The classification language had to be the scholar's unchanging Latin. It was clear that plants were related and that green peppers, tomatoes, potatoes, and Jerusalem cherries should be put into a group (what botanists call a *taxon*), while others, like oaks, beech trees, and willows were less closely related to each other, but showed some characteristics in common. A pigeon-holing system was needed. Equally importantly, the system had to allow for rapid information retrieval so that the plant namer, the taxonomist, could determine not only into which group to put a hitherto undescribed plant, but to ensure that the same plant wasn't given a different name by a different taxonomist.

LINNAEUS

The man who finally put it all together was the son of a Lutheran pastor, educated at Upsala University, Sweden, in medicine and who eventually became professor of medicine and botany at Upsala: Carl Linnaeus, raised to the nobility as Carl von Linne. In order to arrange and classify plants, he developed a hierarchial system starting with the least complex grouping, the *species*, and then grouping species into *genera*, genera into *families*, and families into *orders*. Giving each new kind of plant a double name—a binomial—was conceived in 1623 by a Swiss, Caspar Bauhin, who, following the same principle by which we have two names, called the genus of roses *Rosa* and then delimited each species with a second name, *rugosa* or *multiflora* or *virginiana*.

Linnaeus also incorporated taxonomic ideas of a German (Joachim Jung) and an Englishman (John Ray) who had earlier suggested that in order to classify plants one must use some stable characteristic of the plant. The choice was the flower, since flowers are the organs that are the least susceptible to modifications caused by differences in growing conditions. Using the flower and its parts also eliminated a problem that existed from the time of Aristotle who had classified plants into trees, shrubs, and herbs. The difficulty with Aristotle's classification can be seen among members of the rose family. There are many plants obviously related to roses; trees (apples, peaches), shrubs (blackberry bushes), and herbs (strawberry, cinquefoil) or some that have all three growth forms like the *Spiraea* group.

Linnaeus's great books *Species Plantarum* and *Genera Plantarum* appeared between 1727 and 1753 and still form the basis for our classification of plants. Although he did not originate the idea of herbaria (storage places for pressed, properly named plants), he did make a herbarium in Upsala that is the foundation of our taxonomic system. A taxonomist who thinks that he or she has discovered a species new to science can independently determine in which taxon (genus, family, order) the plant must belong by the size and arrangement of its floral parts. He or she then compares this plant description with published descriptions and, in critical cases, compares the plant with a known, named herbarium specimen. Today, much of this information is available from computer printouts.

FINDING A NAME

Even with a taxonomic system, choosing names is, as any parent can testify, a difficult problem. By the time Linnaeus started his work, there were about 5000 plants widely known in Europe, most of them with common names—pea, wheat, willow, birch—and many of these names were the same throughout Europe. Yet, many were not alike and communication failures were common. Linnaeus classified, using the Latin binomial system, over 18,000 plants, and by 1790 there were over 50,000 plants yet to be named. The only way that this could be done was by shuffling names around, a concept that really makes good sense. For example, let's look at the junipers, whose genus name is *Juniperus*. *Juniperus chinensis* is a species from China, *Juniperus japonica* comes from Japan and *J. virginiana* (red cedar) from North America. *Prunus japonica* is a Japanese plum and *Iris japonica* is a Japanese iris. *Arabica, californicus, pennsylvanicus, columbia,* are obvious place names. Plants in common use for centuries in Europe are called *vulgaris* (Latin for common) and here we have *Phaseolus vulgaris* (beans). A spring-flowering plant might be *vernalis* and a fall-flowering plant would be *autumnalis*. *Deliciosa, edulis,*or *esculentum* refers to a plant's palatability.

Figure 1.14
Bird-of-paradise flower, *Strelitzia reginae,* a member of the banana (Musaceae) family. Brilliantly colored, the bright blue "tongue" of two fused petals encloses yellow stamens and is joined to the orange-colored third petal.

Many plant names reflect an outstanding characteristic of the plant. *Thymus* is the latinized genus of the common thyme, and *T. serphyllum* is creeping thyme (serpent and leaf). Sandwort is *Arenaria serphyllum; Houstonia serphyllum* is a creeping bluet. *Microphyllum* means small leaf—there is a *Thymus microphyllum* and also a big-leaved thyme *T. macrophyllum. Glaucum* means smooth, *hirsuite* is hairy, and *pubescent* is fuzzy. *Rotundifolium* means round leaved, *quinquifolium* is five-leaved, and so it goes. Color names are as common as Ms. Brown or Mr. White except that the names are latinized. The yellow birch is *Betula lutina,* black birch is *Betula niger,* and white birch is *B. alba. Acidus* means sour, *altus* is tall, and *barbarus* is foreign. *Denticulatus* is slightly toothed, *dentifera* is tooth-bearing, and *bidentatus* is two-toothed—not especially imaginative, but very effective. Some names roll grandly off the tongue. How about *entomophilis* (insect-loving) or *erythrocephalus* (red-headed) or *foetidissimus* (very foul-smelling)?

People's names, latinized of course, are very common. One can honor one's professor, spouse, good friend, or lover. Names have been given plants for political and economic reasons. The South African bird-of-paradise plant (Figure 1.14), *Strelitzia,* was named for Charlotte Sophia, wife of England's George III. She was from the German house of Mechlenberg-Strelitz, and the plant was named by a man who was then given a soft, well-paying job as royal gardener. The poinsettia is a French family name once removed. Joel Roberts Poinsett was an American from South Carolina who, as a diplomat, was sent to Chile in 1809 and brought the plant back to the United States. A specimen ended up in a German botanical laboratory where it was named in Poinsett's honor.

THE SAGA OF PETER KALM

The mountain laurel, *Kalmia latifolia* L., was named by Linnaeus (hence the L. after the species name) to honor one of his students, Peter Kalm, who was among the early plant hunters in North America. Peter was a Calvinist minister in Upsala who traveled with the master on collecting trips to Russia and the Ukraine in 1744. He was commissioned to go to the American colonies to bring back plants that could grow in Swedish gardens. The Swedish Academy of Sciences agreed to field test the imported plants, much as our agricultural experiment stations run tests today. With an expert gardener, Lars Jungström, he left Stockholm "in the name of the Lord and right after dinner" in early summer of 1748. Shipwrecked off the coast of Norway, he found another ship and arrived in Philadelphia in September. Presenting letters of introduction to the famous Dr. B. Franklin, he was wined and dined and shown the sights with, as he said, "a great uneasiness at the thought of learning so many new and unknown parts of natural history." Jungström, a self-proclaimed sophisticated man, refused to believe in poisonous snakes or skunks, and insisted that the three-leaved plant he collected at the edge of town was innocuous—until he came down with poison ivy. Kalm was, like about 15 percent of the European population, immune to poison ivy and joked about Jungström's discomfort until he collected poison sumac and was sick for a week.

By October 1748, they had collected throughout the relatively stable Pennsylvania countryside. The winter was taken up with pastoral duties in New Jersey where another peril faced him. He fell in love with the young widow of the pastor he had replaced because, it was said, she was an excellent cook. Leaving this attractive widow in the spring, he traveled overland to New York, hired a keelboat, and moved against the spring flood up the Hudson River to Albany. There he hired a canoe and two guides and moved into the wild region of the upper Hudson River valley, land of the unconquered Mohawks who remembered the French and Indian War with some pleasure. The river gave out near Glens Falls, and Kalm and his party traveled by compass and prayer through the unpathed forests to Fort Anne. Kalm noted dispassionately in his diary that the black flies were particularly annoying. Nevertheless, his collecting baskets continued to fill.

Fort Anne, where they were to replenish supplies, had been deserted the previous fall. Kalm, Jungström, and the two guides built with hand knives and a small hatchet a very tippy birchbark canoe to hold four men, the gear, and of course the precious plant collections. The canoe was used for only a short time for their maps were wrong and the explorers had to travel overland to Fort St. Jean at the north end of Lake Champlain. The walk to Montreal was, they wrote, very comfortable. The French governor-general of Quebec was delighted to have such distinguished Swedish visitors and grandly offered to have His Majesty, the king of France, pay all expenses of the party during its stay in New France. This was more than kind, for Kalm had been out of touch with civilization for several months, had no money, and, in fact, hadn't thought about money when he left New Jersey. While botanizing around Montreal, a Dr. Gaultier served as Kalm's guide, and as a token of appreciation, Linnaeus named the northern wintergreen *Gaultheria procumbens* in his honor.

In early fall, Kalm and companions left Quebec and retraced their steps back to Philadelphia. After conferences with Franklin and the noted American botanist, John Bartram, Peter married his waiting widow and sailed the following spring, arriving in Stockholm with collections intact. Several hundred new plants were added to our listings and a few bear Kalm's name. There is a lobelia, a grass, a plant related to St. John's wort, and, of course, the mountain laurel.

Peter Kalm was not the only plant hunter in North America. André Michaux and his son François, David Douglas for whom the Douglas fir is named, Thomas Nuttall who explored the Great Lakes area, and many others collected all over the world. One searches almost in vain for women in the seventeenth and eighteenth centuries who attempted to break the male hold on science. There was one American,

Jane Colden, born in New York of Scottish immigrant parents who grew up near Newburgh, New York. Cadwallader Colden, her father, was, like many gentlemen-physicians, a journeyman botanist who grew or gathered the herbs which along with prayer and a contaminated scalpel were his stock in trade. A Tory, he served as the provincial lieutenant-governor of New York until the Revolution. In 1773, Dr. Colden sent to Sweden for a copy of Linnaeus's *Genera Plantarum* and, fascinated by it, took to roaming the Catskills on botanizing expeditions. His daughter joined him and word spread through the colonies that the maid Jane, affectionately called Jennie, was a master of the Linnanean system. Dr. Colden initiated a correspondence with Linnaeus and a plant in the heliotrope family, *Coldenia*, was named in his honor, although apparently Jennie found it. Jane married a widower doctor at the age of 35, bore a child when she was 42, and died in childbirth. She left a portfolio of drawings, leaf prints, and detailed descriptions of plants of New York. The collection is now in England. In spite of efforts by her friends, Jane Colden has never been memorialized in a plant name.

ADDITIONAL READINGS

Anderson, B. *Wild Flower Name Tales.* Colorado Springs, Colo.: Century One Press, 1976.

Bailey, L. H. *How Plants Get Their Names.* New York: Macmillan, 1933.

Coats, A. M. *The Plant Hunters.* New York: McGraw-Hill, 1969.

Dawkins, R. M. "The Semantics of Greek Names for Plants." *Journal of Hellenic Studies* 56(1936):1–11.

Eifert, V. S. *Tall Trees and Far Horizons.* New York: Dodd, Mead, 1965.

Gardner, E. J. *History of Biology,* 2nd ed. Minneapolis, Minn.: Burgess Publishing Co., 1965.

Geiser, S. W. *Naturalists of the Frontier.* Dallas, Tex.: Southern Methodist University Press, 1948.

Hawkes, E. *Pioneers of Plant Study.* London: Sheldon Press, 1928.

Lemmon, K. *The Golden Age of Plant Hunters.* New York: A. S. Berman Co., 1969.

Marafioti, R. "The Meaning of Generic Names of Important Economic Plants." *Economic Botany* 24(1970):189–207.

Morwood, W. *Traveler in a Vanished Landscape: The Life and Times of David Douglas.* New York: C. N. Potter, 1973.

Smith, A. W. *A Gardener's Book of Plant Names.* New York: Harper & Row, 1963.

Steele, A. R. *Flowers for the King.* Durham, N.C.: Duke University Press, 1964.

Bees, Birds,
Anthropologists,
and Allergists

  The buzzin' of the bees
  In the Sycamore trees . . .
  Big Rock Candy Mountain

A flower's biological role can be performed only if pollen from the anthers reaches the stigma at the top of the female organ, the pistil, and its tube grows down to the ovary where fertilization occurs. This is a dull scientific fact that fails to give even a hint of the mind-boggling diversity of methods by which pollination takes place. The obvious way for pollination to occur is for the pollen produced in the same flower to travel the small distance between anther and stigma. The problem with this method is that it precludes introduction of new genetic characteristics into the offspring, since both male and female parts of the flower are derived from the same vegetative plant. Where self-pollination does occur, as in garden peas, the offspring are genetically identical—as alike as two peas in a pod—and while this is advantageous for canners, it means that adaptability of the plants to environmental stress is reduced. Self-pollinization in many plants is prevented by one or more processes. Pollen may be self-incompatible, incapable of causing fertilization of the egg of the same plant, but effective when it reaches the pistil of the flower on another plant. Some plants separate the male and female flowers, restricting them to different locations on the same plant (the silks and tassels of corn, for example), or having male and female flowers on different plants (holly is a good example). Pollen may mature earlier or later than the pistil or anthers may hang out and away from female organs.

ALLERGISTS

If self-pollination cannot occur, pollen must reach the flowers of another plant, that is, cross-pollination must occur. The most obvious movement of pollen through space occurs by its being passively carried by wind or water. This is a chancy proposition and, in order to secure a reasonable amount of pollination, the number of pollen grains produced is enormous. If you have ever parked an automobile under a pine tree in late spring, you have some idea of the amount of pollen produced by a wind-pollinated plant. All grasses, including cereal grains, are wind pollinated and so are most of our forest trees like oaks, maples, birches, and beeches.

Air-borne pollen can travel long distances; 30 miles is not unusual. Pollen has been captured by balloons more than ten miles above

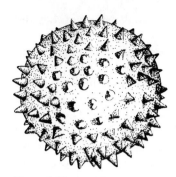

Figure 1.15
Pollen grain of giant ragweed, *Ambrosia trifida.*

the earth, and just a few high-pollen-producing plants can blanket a ten-mile radius. Wind-pollinated flowers are the cause of considerable human discomfort as any hayfever victim can testify. A hyperallergic person may react in spring to pollen from maples, willows, and oaks; in midsummer to grass pollen; in early fall to pollen from several weeds; to breathe again only when plants die back in early winter. The most common of these pollen allergies is due to the ragweeds, which by some strange quirk of botanical sadism were given the genus name *Ambrosia* (Figure 1.15)—"food of the gods." There are several species of ragweed with wide, overlapping geographical distributions, and their pollen production is prodigious. Some people are so exquisitely sensitive that merely a few pollen grains per milliliter of air will cause severe allergic reactions. In some cases, allergies to specific pollens may be inherited; allergic children frequently have allergic parents or grandparents.

The allergic reactions are caused by proteins in the outer walls of the pollen grains. These proteins move from the pollen to the sensitive lining of the nose where they react with antigens already in the victim's system. Desensitization by a series of inoculations is frequently successful in lessening the severity of the reactions. It has been suggested, so far without rigid proof, that the minute, sharp bumps and "spines" on the pollen wall are physical irritants to sensitive cells in the nose.

POLLINATION BY ANIMALS

It is, however, pollination mechanisms involving animals that attract the most attention. Evolutionary evidence suggests that the flower evolved about 150 million years ago (Jurassic period) and that flowering plants began to dominate plant life forms about a million years ago (Cretaceous period). Insects evolved about 400 million years ago, and most of the different kinds of insects were evolutionarily well advanced about 300 million years ago (Carboniferous period). Bees, wasps, flies, and butterflies—all insect pollinators—are today, obligatorily dependent upon plants for their food, and yet they were on earth 200 million years before there were flowers. The obvious question, "What did they eat before flowers evolved?" simply cannot be answered.

This question aside, we do know a good deal about the interaction of insects and plants. Pollen is a fine nutritional source, containing proteins, fats, carbohydrates, and minerals. It, and the sugar-rich nectar of flowers, constitute virtually all the food of bees and a significant part of the food of other insect pollinators. The association is obligatory for many plants, for a large number of wild plants and many of our food crops are absolutely dependent on insects for pollination. This means that over 60 million years or so, both the flower and its pollinator adapted structurally and functionally for mutual survival. If, as modern scientific thinking suggests, evolution is not directed but is a random process that occurs by pure blind chance, the statistical probability of

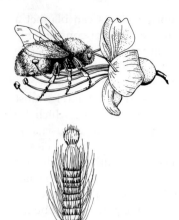

Figure 1.16
Bee visiting flower having extended stamens. Pollen collects on the bee's hairy body. Lower diagram shows the tongue of a honey bee with pollen-catching hairs.

so many specific adaptations occurring is very small—and yet they did occur.

Many of these adaptations are so precise that they boggle scientists' minds. Bees, for example, are attracted to plants for the nectar—the classical "drink of the gods"—secreted by cells at the base of the showy petals (Figure 1.16). Since bees work hard, requiring sugar for energy, nectar must be at least 18 percent sugar to make its collection efficient. Most bee-visited flowers have about 20–25 percent sugar in their nectars, and some, like the horse chestnut, have so much dissolved sugar that the nectar is syrupy. As long as the flowers of a species produce nectar, bees will remain exclusively with that species, thereby ensuring adequate pollination. Apple flowers produce nectar daily, just before bees leave the hives. Bees have color vision, seeing best in the blue and near-ultraviolet ranges, and bee flowers tend to be just those colors. Botanists used to worry about the fact that many bee flowers are red or even pale yellow or white and thus should not be seen by the insects. Only recently was it discovered that red and white flowers reflect solar ultraviolet rays and, to a bee's eye, appear as bright spots against a dull gray background of leaves and grass. In some plants, the ultraviolet-reflecting areas of petals are arranged in streaks or a series of dots which, like painted stripes on an airplane landing field, guide the bee directly into the center of the flower (Figure 1.17). The perfumes of flowers are not designed to attract humans, but are keyed to the organs of smell of the insect pollinators. The smell of rotting meat, offensive as it is in skunk cabbage, is irresistible to flies. Pollinating insects are structurally and functionally adapted for pollination. Their long tongues are usually precisely the right length to tap the nectar pool at the base of their host flowers. Hairs on their bodies collect pollen as

Figure 1.17
Flower of the orchid, *Habenaria orbiculata*, with a grossly elongated petal serving as a landing strip for pollinating insects.

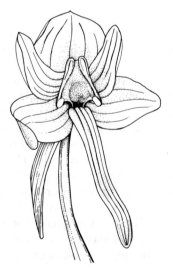

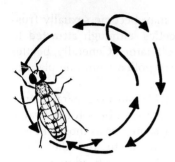

Figure 1.18
The "dance of the bees" takes two forms, a round dance and a waggle dance.

Figure 1.19
Pollination of wild geranium, *Geranium maculatum*, by a skipper moth (*Poanes spp.*). The nectar-sipping tongue of a moth is shown below.

they feed, and the sex organs of the plants are keyed in length and in structure for their task.

AN ESSAY ON BEES

The most brilliant research on interactions of bees and flowers has been done by Karl von Fritsch, professor of zoology at the University of Munich. Fritsch began his work before World War I, but the scientific community only got around to awarding him the Nobel Prize in the 1970s. Von Fritsch asked the simple question of how a bee can tell her hive sisters that she found a fine nectar source on her travels. Placing dishes of sugar water outside his laboratory, von Fritsch found that if the sugar was visited by a single worker bee, within a short time many bees from her hive would be around the sugar. He constructed a glass-walled hive to observe the method of communication. When the food source was less than 100 meters from the hive, the scout bee "danced" for the others in what von Fritsch called a round dance, and when the source was at a greater distance, she did a wagging dance (Figure 1.18). The kind of flower was communicated by the perfume odor clinging to her body.

Von Fritsch found that direction was supplied by the worker bee dancing at an angle that related to the elevation and direction of the sun. Thus, when the sun was just above the flowers, the straight portion of the dance was vertical with the bee's head up, and when food was not in a line between the hive and the sun, the angle was relative to the angle from hive to flower. So precise is the dance that bees can tell the others the straight line distance between hive and flowers even when a hill or a mountain ridge requires that the bees fly around the obstruction. Here the angle is given and the distance is the sum of the two legs of the angle. Additional information is conveyed by sound communication; buzzing can be changed in pitch and intensity. Carrying an analogy between bee and human communication farther than is probably justified, bees have "dialects," with different species of bees varying the patterns of their dance. Obviously the behavior of bees has been as exquisitely adapted to pollen and nectar gathering as the flowers have adapted to the visits of the bee.

Bees are not the only insect pollinators; moths and wasps are also important pollinators of many flowers (Figure 1.19). The flowers visited by these insects are usually different species than those visited by bees. The mouth parts of these insects are different from those of bees and are structurally adapted to the flowers they visit. Flower shape and strong scents are the main attractors for wasps. Certain orchids have flowers that so nearly mimic the shape of the body of the female wasp that the not-too-bright male wasp attempts copulation with the flowers,

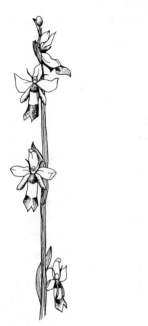

Figure 1.20
The wasp-mimicking orchid, *Ophyrs muscifera*.

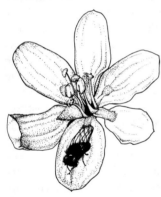

Figure 1.21
Pollination of a lily, *Zigadenus fremontii*, by a fly in the Syrphidae group.

one after another (Figure 1.20). The male may become sexually frustrated, but the orchid gets pollinated. Beetles, although attracted to flowers by their odors, are not important pollinators. Generally, beetles eat the soft parts of flowers and rarely carry pollen from one plant to another.

Flies which pollinate plants (Figure 1.21) can be separated into two groups. Those with long tongues compete with bees for the same flowers; their mouth parts are very much the same size as those of bees, and the flowers are not specifically adapted to the flies. The short-tongued flies are attracted to flowers that give off odors that we perceive as decay. The short-tongued flies are fairly crude, flying about and bumping into the anthers; one scientist described them as "unindustrious, unskilled and stupid." Some flowers show specific adaptations to take advantage of these attributes. The flowers of the Dutchman's Pipe (*Aristolochia macrophylla*) attract flies into tubular flowers, and they become imprisoned there for several days in a floral trap where the flies buzz around madly trying to escape. When the flowers release them, the flies, not having learned a thing, enter another flower of the same species.

Birds are important pollinators, too. In the tropics, birds may be more important than bees. In the Americas, the bird pollinators are the jewels of the avian world—the hummingbirds (Figure 1.22). They are replaced by the sunbirds of Africa and Asia, the honey creepers of the Pacific Islands or the honey eaters and lorikeets of Australia. Most birds have excellent vision and a poor sense of smell; flowers visited by birds are usually red or yellow, colors birds see best. Hummingbirds rarely perch on flowers. They suck nectar while hovering with their wings beating so rapidly that they appear to us as a blur. Flowers visited by hummers are usually tubular and hang out and down, facilitating the insertion of the long slender bill and even longer tongue. African sunbirds cannot hover, however, and the flowers they visit are formed with a landing platform next to the flower. Since Sunbirds (Figure 1.23) probe for nectar with pointed bills, the delicate parts of the plant are placed out of harm's way, but the anthers are where the bird cannot fail to brush its head against them. Somewhat different, but equally appropriate, floral adaptations are found in flowers visited by honey creepers (Figure 1.24) and honey eaters. Even though these bird pollinators are not closely related, their heads and bills are very much alike. Lorikeets

Figure 1.22
Female ruby-throated hummingbird.

Figure 1.23
African sunbird feeding at
heath flowers.

Figure 1.24
Hawaiian honey creeper.

and other parrotlike birds are not very selective feeders and use their
strong beaks to cut holes in the sides of the flowers to get at the nectar.
In spite of extensive damage to petals, pollination is ensured because
anther stalks are short and next to the nectar sacs so that the lorikeets
have their heads dusted with pollen as they feed.

In addition to man, one other mammal is a pollinator. Some bats
(Figure 1.25) feed on nectar and on small insects present in the pools of
nectar. Since bats are nocturnal, the flowers they visit are usually
night-bloomers and their petals are usually white. The muzzles of pollinating bats are slender, their tongues are long and extendable, and
their front teeth, which would otherwise be in the way, are either short
or, in some species, undeveloped. Bats are guided to flowers by sight
and by fruity odors. Of assistance to the bat is the fact that the flowers
are usually large and raised above the foliage. In a few cases, bat-visited
plants drop their leaves at the time of flowering making the flower more
accessible.

POLLEN ARCHEOLOGY

Because of their thick outer walls, pollen grains are resistant to
decomposition and may last for long periods of time. Since they are
also sculptured, identification of a pollen grain down to the genus level
is possible. These facts have been the basis of a scientific specialty, the
field called pollen archeology, started in Denmark by Johannes Iverson.
Iverson reasoned that since we can identify pollen of specific plants,
and since pollen can remain intact for thousands of years, we should be
able to study the ebb and flow of plant life in a particular location by
analyzing the pollen obtained in archeological sites. One such study
will serve as an example. During the last ice age, Denmark was essentially scraped down to rock by the ice. As soil formed, the land was covered by forests, mainly spruce and fir as determined by pollen resting
on rock. As forests became dense, game decreased and man, then a

Figure 1.25
A bat, *Leptonycteris*, feeding
on flowers of the organ-pipe
cactus, *Lemaireocereus thurberi*.

hunter, retreated to the coasts where he fished for a living. Sometime later, during the Neolithic period, people returned and began to clear the land and burn the logs, as evidenced by a layer of spruce and fir charcoal just above the pollen of these species. For a period of time, pollens were mainly of herbs and of barley and wheat, proof that this land was being farmed. New weeds, those from the Danish coast, were mixed with local weeds. The farmers moved on, and their fields began to return to their natural state, first to weeds and then to species of trees which typically follow forest clearings; willows, aspen, and birch. The fact that birch was found is also an indication of clearing by fire, since birch requires high levels of soil phosphorus which is found in wood ashes.

Apparently the people didn't completely abandon this land, because pollen of clover and other pasture plants was found, an indication that the land was in pasturage for goats and sheep. Later, the land was completely abandoned, and birch gave way to hazel and then to elm, linden, and finally to oak. During this transition period, there was some renewed cutting and burning (charcoal again), and the land grew small crops and blueberries, a plant that seeds in after fires. Some climatic shifts occurred, for pollen of sedges and spores of sphagnum moss show that this area became marshy. A shift in topography allowed the land to drain and forests of spruce and fir developed. Again, men moved back, cleared the forests, and planted barley and wheat, but pollen of weeds native to Russia was found mixed with grain pollen. This suggested that the new farmers weren't Danes, but were likely to have been Finns, driven west by Russian invaders of their land. This prompted a search for evidence of human artifacts which, when found, were unmistakably from eastern Europe. In addition, pollen of pasture grasses was mixed with the cereal pollen, an indication that these people practiced some kind of crop rotation or left land in pasture for several years between plantings of grain. The eastern invaders interbred with the natives and the land became permanently agricultural by the end of the Neolithic period.

ADDITIONAL READINGS

Dimbleby, G. W. *Plants and Archeology*, New York: Humanities Press, 1967.
Faegri, K., and L. van der Pijl. *Principles of Pollination Ecology*, 2nd ed. Elmsford, N.Y.: Pergamon Press, 1972.
Gilbert, L. E., and P. H. Raven. *Coevolution of Animals and Plants*. University of Texas Press, 1975.
Greenewalt, C. *Hummingbirds*. Garden City, N.Y.: Doubleday, 1961.
Heizer, R. F. *The Archeologist at Work*. New York: Harper & Row, 1963.
Iverson, J. "Forest Clearance in the Stone Age." *Scientific American* 194 (1956): 36–41.
Meeuse, B. J. D. *The Story of Pollination*. New York: Ronald Press, 1961.
Percival, M. S. *Floral Biology*. Elmsford, N.Y.: Pergamon Press, 1965.
Perlman, D. *The Magic of Honey*. Los Angeles: Nash Publishers, 1971.
Procter, M., and P. Yeo. *The Pollination of Flowers*. London: Callens, 1973.

Stanley, R. G., and H. F. Linskins. *Pollen.* New York: Springer-Verlag, 1974.

von Fritsch, K. *Bees,* Ithaca, N.Y.: Cornell University Press, 1972.

West, R. G. *Studying the Past by Pollen Analysis.* Oxford Biology Reader #10. New York: Oxford University Press, 1971.

The Sex Life of Flowers

The flowers that bloom in the spring
Tra La.
GILBERT AND SULLIVAN

Sometimes when a child asks "Where did I come from?" parents embark on a lecture on sex when the kid merely wants to know whether he or she came from Chicago or Toronto. In many instances, this discussion takes the classical birds-bees-flowers route, which is fine as far as it goes, but it is basically about how reproductive entities get together and not about sex as biologists understand it. Biologically, sex is the process by which two reproductive nuclei, each bearing one-half the complement of chromosomes, fuse to form a fertilized cell containing the full complement of hereditary material. All the rest, including the information in How-To books, is recreation, not procreation.

The correlation between intercourse and initiation of the next generation was obscure for many centuries. That there was a relationship was obvious; when men and women live together, children are born. But the relationship itself? In 458 B.C. Eumenides of Aeschylus said: "The mother of what is called her child is no parent of it, but nurse only to the young life that is sown in her. The parent is the male and she but a stranger, a friend, who if fate spares his plant, preserves it till it puts forth." That the male serves as the procreator is seen in the biblical story of Onan who masturbated and "spilled his seed on the ground." Aristotle concluded that males are warm and complete while females, being colder or more passive, are less complete. She contributes matter and a place for development, and the male contributes his seed which is the future child and its soul. This concept of the female as a mere receptacle held sway in Europe for 1500 years. What it did for the position of women can only be surmised, but it certainly didn't advance their dignity. The Koran states: "Have we not created you of a repulsive drop of seed which we placed in a sure depository until the fixed time of delivery?"

Identification of the child as the product of a seed implanted into the soil of the womb was accepted throughout most of the world, and the parallelism between copulation in humans and sowing crops was also widespread. Woman is the field, the furrow is the vulva, and the seed is the fructifying principle. So, too, did Oedipus "sow his seed in

Figure 1.26
Fifteenth-century concept of the birth of the souls of little children in flowers. Taken from a woodcut by Meydenbeck made in 1491.

the sacred furrows" and Sophocles referred to the furrows of paternity. The act of planting seeds in the earth was homologous to human copulation, and this parallelism was fervently made manifest in spring rites.

PLANTS DO NOT HAVE SEX!

Dating from about the second century A.D., man was concerned with the concept of intercourse as sin. Since the relationship between copulation and child-making was obscure, many believed that intercourse was unnecessary and a part of original sin. Because it was also pleasurable, it was even more sinful. And the intercourse of animals? . . . Were they also contaminated through Adam and Eve, and was copulation in animals necessary for procreation? This was a hard question, but since animals didn't have souls—or lesser ones—it wasn't too immediate a problem, and it was pushed under the rug of scholasticism. Plants, on the other hand, were pure entities by definition and were clearly innocent of any taints (Figure 1.26). Their reproduction could not involve sex, for if even the lilies of the field reproduced by sexual means, every creature on earth was drowned in sin, and this couldn't be rationalized with the idea of a benign Creator. It didn't make sense that plants were sexual and hence sinful, and if this were promulgated by theologians, maybe these doctors of the Church were wrong about original sin and, perhaps, wrong about other things the priests thundered from their pulpits. It was incumbent upon learned theologians to insist that plants were without sex. And insist they did!

Theophrastus, pupil of Aristotle, described in detail the ancient North African method of initiating fruit formation in the date palm by dusting the female plant with a frond of the male flower. St. Augustine had read Theophrastus and lived in North Africa, and yet he explicitly denied that plants had any trace of a sex life. St. Thomas wriggled out of the problem during the thirteenth century by insisting that crude cross-breeding of plants was unnatural and thus contrary to moral law. Asians didn't have this hangup; the Chinese *Classic of the Golden Emperor* stated that the fine dust of plants was the male entity, and Hindus of the Vedic period of 1000 B.C. were aware that "the golden powder" was needed to ensure fruit development.

Denial of sex in plants was still dogma by the end of the sixteenth century. Andreae Cesalpino, an influential professor of botany and medicine, stated in 1590 (following Galen who followed Aristotle) that plants could not have sexual phase. The fruit, he insisted, was formed from buds produced by the pith and inner bark. In 1583 Dodoens wrote, "The flower we call the joy of trees and plants. It is the hope of fruit to come. . . ." William Harvey, newly famous for his discovery of the circulation of the blood, stated that the male principle conferred upon the passive human egg some vital force that caused it to develop into an embryo. The male principle in humans and other animals was discovered in 1667 by Antony van Leeuwenhoek and was identified as such, but his report to the Royal Society of London was either ignored or it was assumed that the little "beasties" were parasites. Another Hollander, Jan Swammerdam, agreed with his friend that the things Leeuwenhoek saw were sperm and reported that Aristotle was right after all, because one could see a little man, a *homunculus*, coiled tightly within the head of the sperm. Yet another Dutchman, Reinier de Graf, found the human egg cell—actually he saw the ovarian follicle—and asked whether, perhaps, and with all due respect to Galen and Aristotle, this female cell might not play a more important role in embryo development than had previously been believed.

WELL, MAYBE THEY DO!

Marcello Malpighi, born in Bologna in 1628, was one of the first of the enlightened breed of scientists to focus a microscope on plants. He published detailed, accurate drawings of flower structure, but concluded that the flower itself served only to purify the juices of plants, this allowing seeds to develop. "The pollen dust," he said, "is a mere secretion and may be compared with the menstrual discharge of women." At the same time, the English plant anatomist Nehemiah Grew advanced the most appalling idea that the pistil of a flower is the female part and that stamens and their pollen were male. He spoke of male and female "juices" but copped out on the question of how seeds were formed except to suggest that plants might reproduce "like snails." This was essentially hermaphroditism, a neat idea since Grew shrewdly noted that stamens and pistils were frequently found in the same flower. Grew's ideas were, however, demolished by the pompous Joseph de Tournefort, professor of medicine and director of the Jardin des Plantes in Paris, who insisted that Aristotle could not be denied, that the scholastics were correct, and that this heretic protestant, this country parson, this Grew, was dealing in superstition. John Ray came to his fellow Englishman's defense, stating that plants must have sex since all other living things on earth did, but "for the life of himself," he didn't understand the mechanism.

Yet, the stability of the form of flowers and the precision of their timing in spite of variations in the vigor and growth habit of the plants within a species suggested that there was something fundamental about flowers. Fifty years before Linnaeus, Joachim Jung used the flower as the main criterion for classifying plants, although he refused to comment on its function. In 1694 Rudolph Jacob Camerarius wrote *A Letter on the Sex of Plants* in which he discussed the ideas of the ancients and concluded that they really didn't know what they were talking about. He reported experimental studies done in the gardens of the University of Tübingen where he removed the female parts of complete flowers and obtained no fruit. He also grew plants in which the pollen-containing flowers were separate from pistil-containing flowers and found that fruits developed only when they had been dusted with appropriate pollen. Being aware of what happened to heretics even at the end of the seventeenth century, Camerarius simply left off at this point without reaching any conclusions on whether what he had reported represented sex in plants.

In 1730, at age 23, Linnaeus published his *Introduction to Floral Nuptials.* In spite of the deliberately provocative title, the volume was entirely on plant anatomy and created a stir only in the limited botanical world. Linnaeus read Camerarius's writings and decided that plants could best be classified by comparing and contrasting the numbers and arrangements of their sexual parts. His views on sexuality in plants was most literal. Thus, he described the condition of monandism as "one husband in a marriage," and used the term polyandry when a plant showed "twenty or more males in the same bed as the female." For the composite family—sunflowers and others with sterile ray and fertile disk flowers—he said that "the beds of the married occupy the disc and those of the concubines, the circumference." In 1760 he suggested a "new employment for botanists to attempt the production of new species of vegetables by scattering the pollen of various plants over various widowed females." Such shocking words made life difficult for botanists of the day. Many botanists were medical men whose interest in plants had been sparked by medicinal plants or they were dedicated amateurs—gentle souls all—with a high proportion of ministers in their ranks. Linnaeus's taxonomic system had many sexual implications that were abhorrent, but the system worked—and worked beautifully. There ensued a series of intellectual juggling acts in which the implications were thundered down while the method continued to be used. The Rev. Samuel Goodenough, Bishop of Carlisle, took indignant pen in hand to write a letter to a friend "to tell you that nothing could equal the gross prurience of Linnaeus' mind," and the celebrated Goethe worried publicly about the potential embarrassment of botany texts to chaste young minds.

YES, THEY DO!

The topic was, however, ripe for clarification. Father Spallanzini made "trousers of waxed fabric" for male frogs and prevented fertilization; the collected semen was used for artificial insemination of frogs. Thomas Fairchild cross-pollinated a carnation with a pink in 1717 and got a hybrid. Showing mixed characters, "Fairchild's mule" was famous throughout Europe. Gardeners on royal estates were reporting to their lords of the hybridizations of fruit trees, flowers, and medicinal plants. Buffon in France, von Haller in Switzerland, von Baer in Germany, and others had overthrown the male seed idea, isolated animal eggs, and worked out the fundamentals of fertilization and early embryonic development by the middle of the eighteenth century. Two botanists gave the death knell to Aristotelianism.

Joseph Koelreuter, curator of the botanical gardens in Karlsruhe, speculated in 1761 on sex in plants as a result of his extensive cross-breeding of medicinal herbs. His experiments involved pollen transfer, selective removal of floral parts, and an attempt to keep records on his crosses. His thoughts on plant sex are, upon modern reading, quite accurate, but they were so contrary to Church doctrine, and thus so abhorrent, that his contemporaries ignored him and his ideas.

In 1793, Christian Sprengel published *Nature's Secret Revealed: The Structure and Fertilization of Flowers.* Sprengel was the son of a Brandenburg Calvinist minister and was convinced that since nature is preconceived by the wisdom of God down unto the smallest thing, there was a purpose for everything on earth, even flowers. Sprengel had a lot of time on his hands. Although he had studied theology, he was a mere schoolteacher, a recluse, and a most irascible citizen without a social life. So, as he spent his days looking at the interactions between bees and flowers, he discovered that plants with flowers having both male and female parts could have the parts develop at different times to mandate cross-fertilization, and he saw that some plants were specifically designed for wind pollination. We can now be a bit condescending about Sprengel's failure to distinguish between pollination—the transfer of pollen to the pistil—and fertilization—the fusion of male and female nuclei—but he specifically used the word fertilization with the clear implication that plants did have sex. The philosophers of the day, churchmen all, had little patience with the concept of detailed research and the laborious reporting thereof. They couldn't understand Sprengel's book and, therefore, could safely ignore or denounce it. They did both, to the point where poor Sprengel became more of a recluse and even more irascible and eventually faded away.

By the early years of the nineteenth century, the last bastions of scholastic belief were falling. If the common people could have read the learned discussions denying sex in plants, they would not have believed that educated men could be so blind. Of course plants had sex. Why were they dusting pollen on flowers to get better crops or being so concerned that the bees worked in blooming apple orchards? The governor of Pennsylvania, James Logan, published experiments on pollinating Indian corn. Gregor Mendel was fully aware that pollen was necessary for the development of peas. But Victorian England and its American counterpart stood firm. Alfred Lord Tennyson could write of "the wild flowers of a blameless life," while others, like Croisat, noted that the sex organs of plants were "blatantly displayed and act like husbands and wives in unconcerned freedom." Jamie Alston, professor of botany at Edinburgh, attacked the whole idea of sex in plants, although he used Linnaeus's system to reorganize the university herbarium. A Mrs. Lincoln of Albany, New York, wrote a famous botany text, *Familiar Lectures in Botany,* in 1829 for young ladies enrolled in female seminaries. It went through ten editions until 1849 and sold 275,000 copies. Botanical language contained several unmentionable words, and so she substituted "germ" for "ovary," and the word "placenta" was never defined, although it had to be used. Mrs. Lincoln, in fact, never discussed reproduction in plants at all.

The Victorian lady who diligently cultivated orchids in her conservatory was either unaware of or repressed the knowledge that the name is derived from the greek, *Orchis,* meaning testicles. In rural England, however, native orchids were called dog's stones or bull's bags. A careful look at the Unicorn Tapestry in the Metropolitan Museum of Art clearly shows the orchid as a sexual symbol. In his book, *On the Various Contrivances by which British and Foreign Orchids are Fertilized by Insects and On the Good Effects of Crossing,* Charles Darwin demonstrated that insect pollination was a prelude to a sexual fertilization in every way comparable to that of animals. This really struck home because the orchids, of all nature's creations, were held up to the youth of England as a perfect example of botanical art for art's sake—pure, innocent, chaste. That orchids have traps, nectars, perfumes, and structures specifically to attract insects that ensure pollination and hence fertilization was as distasteful to a Victorian curate as the thought that the Queen's children were conceived by the same procedure as were those of the least of his congregation. Since Victorian young ladies received instruction in botany as a suitable avocation, and the curate was frequently the local botanist, how was he to explain to this girl the function of the pollen grain, the discharge of the sperm nuclei through the micropyle, and the union of sperm and egg nuclei? No, this he could not do. By 1870 botany was no longer part of social instruction. Painting, music, poetry, and needlework were more appropriate; our legacy is some charming petitpoint, some horrible watercolors, and a tradition of genteel poets. It wasn't until after World War I that many women entered botany departments.

❧ AN ESSAY ON SEXUAL REPRODUCTION IN PLANTS

From structural and developmental viewpoints, the flower is a wondrous thing. Prior to the evolution of the flowering plants, sexual reproduction in algae, fungi, mosses, and liverworts and ferns required that fertilization be dependent on the presence of free, liquid water in which sperm swim to the egg. As plants developed water-transporting systems and grew taller, evolutionary changes occurred that replaced the naked sperm cell with a specialized entity, the pollen grain. This could be transferred to the female structures without the need for free water. Thus originated the cone and the flower.

Both female and male flower parts evolved from modified leaves organized on a structurally specialized shoot, the receptacle (Figure 1.27). Typically, flowers can be separated into four structures. Two of these, *sepals* and *petals,* are leaflike with veins and sometimes stomata or

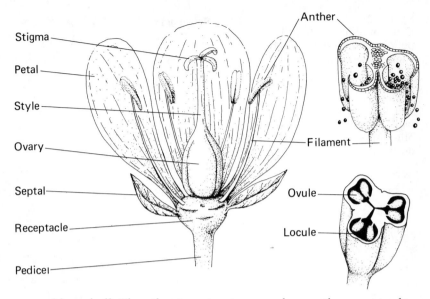

Figure 1.27
Diagram of a flower and its parts.

Stigma

Petal

Style

Ovary

Septal

Receptacle

Pedicel

Anther

Filament

Ovule

Locule

even chlorophyll. The other two structures are less easily recognized as modified leaves. *Stamens* are male organs connected to the rest of the flower by a filament which is all that remains of a leaf midrib. In the enlarged *anther*, certain cells undergo a reductive division in which the chromosome number is reduced to one-half that of cells that comprise the vegetative plant body. From this pollen mother cell a group of four pollen grains is formed which develop walls that prevent desiccation of the delicate cells within them. The female structure, the *pistil*, is divided into three parts: the sticky *stigma* to which pollen grains attach, the *style* down which the pollen tube grows, and the *ovary* in which the eggs develop. As with the formation of pollen, cells within the ovary undergo reduction division resulting in the formation of a large cell with eight nuclei, each nucleus containing half the complement of chromosomes of vegetative cells. One of these nuclei, usually the center one in a group of three, is the true egg nucleus. The two nuclei at the center of the egg cell are called *endosperm nuclei.*

When a pollen grain lands on the stigma, sugars and other chemicals stimulate it to germinate with the formation of a pollen tube that grows down the style, directed by substances diffusing from the ovary (Figure 1.28). During growth, the reproductive pollen nucleus divides again resulting in the formation of two sperm nuclei. When the pollen tube reaches the ovule, it bursts at the tip releasing the two generative nuclei, one of which fuses with the polar or generative nucleus to reinitiate the normal, vegetative, diploid number of chromosomes. This is all there is to sex! The other sperm nucleus combines with the two endosperm nuclei to initiate a special tissue, the endosperm, which has one and a half times the diploid number of chromosomes per cell and serves to nourish the developing embryo derived from the sexual fusion of sperm and egg nuclei.

Figure 1.28
Diagram of fertilization. The pollen grain, landing on the stigma, grows down through the style to the ovary where it bursts, releasing the two haploid generative nuclei. One generative nucleus fuses with the egg nucleus and the other fuses with the two endosperm nuclei.

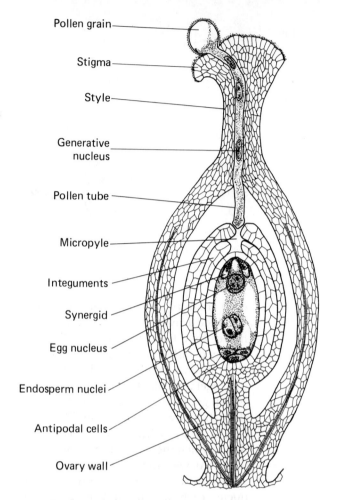

Pollen grain

Stigma

Style

Generative nucleus

Pollen tube

Micropyle

Integuments

Synergid

Egg nucleus

Endosperm nuclei

Antipodal cells

Ovary wall

Each ovule contains one egg nucleus and develops into one embryo contained in one seed. There are exceptions to this rule, but these unusual situations can be ignored here. The fertilized egg undergoes a beautifully integrated series of divisions that results in the formation of the embryo. This consists of a primary root (the radicle), a stemlike structure (the hypocotyl) that connects the root with the leaves, and a small tip that is the true stem. The embryo also has one or more *cotyledons* that, in plants like beans, constitutes the bulk of the seed with the rest of the embryo tucked inside for nourishment and protection. A seed has aptly been termed "a young plant in a box with its lunch." The box is the multilayered seed coat formed from the walls of the ovule. In pines or other cone-bearing plants, seeds are borne on cone scales exposed to the elements; these plants are termed *Gymnosperms* or "naked-seed plants." In the flowering plants, the *Angiosperms*, the seeds receive additional protection by being enclosed within the enlarged ovary walls to form a fruit, defined as the ripened ovary plus contained seeds.

Cronquist, A. *Introductory Botany*, 2nd ed. New York: Harper & Row, 1971.

Jaeger, P. *The Wonderful Life of Flowers*. New York: Dutton, 1961.

Raven, P. H., R. F. Evert, and H. Curtis, *Biology of Plants*. 2nd ed. New York: Worth, 1976.

Salisbury, F. B. *The Flowering Process*. Elmsford, N.Y.: Pergamon Press, 1963.

Meat-Eating Plants

The most wonderful plants in the world.
CHARLES DARWIN

About every five years, newspapers and magazines dig out old stories about man-eating plants. If one traces these stories back in time, one finds that they originated about 1869–1880 in England. During the Victorian age, upper-class men and women were intrepid travelers, visiting out of the way parts of the Empire. No one knows why they had this urge to travel; one author suggested that British food was so bad that any excuse for avoiding it was taken. Many of these gentlemen and gentlewomen wrote books about their journeys, with titles like *The Travels of a Gentlemen in Central Africa* or *Lady Whatsis's Journey to the Spice Islands*. They are charming, fancifully illustrated books written in impeccable, grammatically correct British English with frequent smug homilies on the virtue of being English and hence superior to other mortals.

Several parodies of these travelogues appeared in which strange customs of the natives and even stranger plants were described. The authors got away with them because only a handful of Europeans had been tourists in many of these countries. One such parody, privately printed in the gay nineties and reprinted in 1924 in America, was authored by a C. S. Osborn (a pen name) and was entitled *Madagascar: Land of the Man-Eating Tree*. Captain V. de la Motte, Fellow of the Royal Geographical Society, mounted an expedition to find the plant based on the assertion of Dr. Karl Liche who, in 1878, claimed personally to have seen the tree and who wrote about natives sacrificing young girls to the monster. A Mr. Dunston reported that his dog was eaten by a similar shrub in Nicaragua, and Mariano da Silva reported that he saw a tree in Guiana which lured monkeys into its crown, devoured them, and dropped their bones to the ground. L. Frank Baum used the idea in 1913 in his children's book *Patchwork Girl of Oz*.

Man-eating plants were promoted by circus side shows along with the wild man of Borneo and the half-man–half-woman, and reporters wrote tongue-in-cheek articles to fill space in newspapers. Illustrators had a field day with the art work. The plant was drawn about ten feet tall, sometimes with a gaping mouth (fortunately without

Figure 1.29
An artist's conception of a man(woman)-eating plant.

teeth), and possessing a number of vinelike tentacles. One or two of these sinuous arms was wrapped about the waist of a woman, usually blond and with a goodly portion, but never all of her blouse torn away—decorum in the Victorian era could be strained, but never ripped asunder. The maiden, for maiden it had to be, had a look on her face variously interpreted as terror, or, perhaps, some perverse sexual ecstasy (Figure 1.29).

Regretfully, the whole thing is a myth.

There are plants which do ingest animals, but only one carnivorous plant can handle anything bigger than a beetle. The plants have excited study far beyond their numbers or their ecological importance. At least part of this developed in the late eighteenth century when they were discovered by plant hunters. The early interest centered about their unusualness—the "ain't nature grand" syndrome. There are things to be said in favor of a wide-eyed, almost childlike appreciation of the wonderful diversity of living things as exemplified by the Victorians. Equally good things can be said in favor of the wide-eyed excitement of modern scientists about the capacities of living systems.

The problem that occupied Victorians was the same one that people have long wondered about; just what is a plant? Plants are organisms that make their own food, and animals are organisms that eat their food. Where, then, does one place carnivorous plants? They are green, they carry on photosynthesis, and they have roots, stems, flowers, and fruit. But they use animals as part of their nutrition, and they grow poorly without animal victims. Are they, as Charles Darwin's grandfather Erasmus asked, living beings partaking of both the animal and plant kingdoms? No, answered Francis Ernest Lloyd, F.R.S., F.L.S., D.Sc., they are truly plants and the structures for *insektenfressung* are modified leaves.

"Are not these plants examples of special creation?" asked Victorian anti-Darwinians who desperately needed arguments to counter the godlessness of scientists who were saying that diversity of form and function was a consequence of evolution. This was a difficult question. By 1880 two German botanists, Engler and Prantl, had developed a comprehensive scheme of plant classification which organized all known plants into genera, families, orders, and classes and showed how one classification group (a taxon) was related to other taxa in an evolutionary or phylogenetic way. It was apparent that all insectivorous or carnivorous plants were not closely related to one another. The 500 or so plants could be, indeed had to be, placed in taxa with other plants whose flower organization—the basis for classification—matched that of various types of non-insect-eaters. The taxonomic distribution did not follow any pattern, and if one believed in evolution, insectivorous plants evolved not just once but several times. Furthermore, not all of these plants captured and ate insects in the same way, suggesting that each new and independent evolution resulted in plants whose structural adaptations were different.

The physiology of these plants was studied by Charles Darwin, whose book on the topic is a masterly exposition of careful experimentation. Darwin attempted to find out whether insects were really necessary for their nutrition and, if so, what part of the insect did the plant need. He found that they would not dissolve cotton, linen, or hempen fibers, so they did not utilize cellulose. Hard-boiled egg yolk or pieces of suet were not dissolved, indicating that fats could not be digested. But, if Darwin added hard-boiled egg whites or bits of raw meat, juices from the plants quickly dissolved them. Withholding proteins, whether insect bodies or meats, resulted in decline and eventual death. Darwin concluded that the plants need protein, and recent research in Germany has shown that it isn't the protein that is required, but the amino acids that make up proteins. The juices secreted contain protein-dissolving enzymes which separate animal proteins into their constituent amino acids. These are absorbed by the plant and reconstituted into plant proteins. Darwin asked why these plants, which have roots and could take up inorganic nitrogen fertilizers, could not make their own amino acids, a question that is still not adequately answered. He noted that insectivorous plants live in swamps and marshy areas where available nitrogen supplies are low, but he also observed that many other plants live in the same habitats and survive with the inorganic nitrogen available to them.

ℰ AN ESSAY ON EVOLUTION

Given the diversity of structure and physiology of insectivorous plants and their taxonomic and phylogenetic distribution, the capacity to use insects as a primary source of nitrogen supplies cannot fully be explained by modern evolutionary theory. There is a low but real statistical probability that any cell or any organism can, through genetic mutation, lose the ability to synthesize any of the thousands of compounds which a cell or organism requires. This usually results in the death of that cell or that individual. But cells or organisms cannot say, "Well, since I am losing the ability to synthesize amino acids, I had better alter my structure and my biochemistry so that I can obtain my amino acids from insects." We believe that alterations in the genetic constitution leading to structural and physiological changes involves not just a single genetic alteration, but many mutations. Equally important, we believe that these mutations are undirected and occur by chance. Therefore, the "need" for mutation cannot activate specific mutation or series of mutations; environment cannot instruct or imprint genetic information upon the hereditary apparatus, for genetic information is self-generating. For example, some bacteria have the genetic capability of synthesizing enzymes that can destroy life-threatening antibiotics or other chemicals which never existed until man—who came

Figure 1.30
Northern pitcher plant, *Sarra-cenia pupurea*.

along eons after the bacteria—created them in test tubes. This concept of mutations occurring independently of the needs of the organism is fundamental to modern biological science. If we are correct, then the various and wonderous alterations resulting in the independent origin of the very different taxa of insectivorous plants was pure happenstance, a calculable series of probabilities. The oak tree, the orchid, and the human look, act, and function as they do because of the operation of pure blind chance.

PASSIVE TRAPS

There are four generalized types of insectivorous plants: passive traps, semiactive traps, active traps, and, strangest of all, fungus traps. One writer has characterized the passive trap plants (Figure 1.30), such as the various genera of pitcher plants, in dramatic terms: "They lie in wait—jewels of nectar just out of reach beyond slick funnels of death. Their victims try to reach the nectar treasure, but fall prey to the ingenious traps." After this exercise in the macabre, Randall Schwartz goes on in less impassioned terms to describe the pitcher plants as having modified leaves forming a vase-shaped funnel containing at the base a sweet, perfumed liquid that attracts insects. Once the unwary bug goes down to feed in the liquid, it can't get back up again because stiff hairs, pointing downward, line the walls of the pitcher. Depending on the species (Figure 1.31), the pitchers may be small and delicate or, in one tropical species, large enough to drown a mouse. Once the animal is dead, proteolytic enzymes secreted by cells lining the trap digest the animal, and its amino acids are absorbed. There are five genera of pitcher plants. Several are in the genus *Nepenthes* found in tropical areas

Figure 1.31
Southern pitcher plant, *Sarra-cenia flava*. Courtesy U.S. Department of Agriculture.

of the Pacific; the cobra plants (*Darlingtonia spp.*) are found in marshes of Oregon up to British Columbia, and a third genus, *Sarracenia*, has species found in marshes and swamps in North America.

SEMIACTIVE TRAPS

Four different genera are categorized as semiactive traps, the best known being the sundews (*Drosera*). These are found in western North America from California north to Alaska, on the great plains north to Saskatchewan, and in the coastal areas of the mid-Atlantic states and Florida, frequently growing in the same swamps or tundras as pitcher plants. Leaf blades of sundews are edged with long hairs having round, sticky tips (Figure 1.32). When an insect lands on the leaf, its feet get gummed up with secretions on the leaf surface, and as it struggles to wrench itself free, more sticky material is secreted. At the same time, by signals which we cannot decipher, the hairs begin to fold inward, holding the bug firmly to the leaf. The knobs at the ends of these hairs secrete enzymes which digest the insect into a smelly "blob," and proteins are broken down and absorbed. The whole process takes just a few hours.

SNAP!

The best known active trap is the Venus flytrap (*Dionaea muscipula*) found in swamps and marshes of the Carolinas (Figure 1.33) where, because of pollution and industrial development, its habitat is being destroyed. The leaf looks like a bear trap, hinged in the middle and bearing a row of stiff spines along each edge. The surface of each leaf

Figure 1.32
Sundew, *Drosera spp.*, with detail of a single leaf.

Figure 1.33
Venus flytrap, *Dionaea muscipula*, with detail of a single leaf bearing trigger hairs.

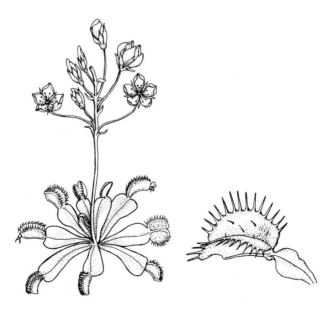

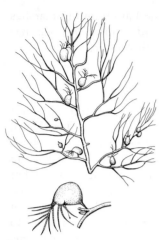

Figure 1.34
Bladderwort, *Utricularia spp.*,
with detail of the bladder
and sensitive hairs.

segment bears three hairs, and when an insect blunders into the hairs, the two halves of the leaf snap shut in a matter of a second, capturing the insect in a cage. The signal mechanism in the flytrap is complex; an insect must strike one hair twice or two hairs once each. The closing of the trap involves the rapid uptake of water from cells along the hinge which raises the osmotic pressure of the cells. Nothing is known about the method whereby the signal from the sensitive hair is transmitted to these cells; no plant has a nervous system, and no plant has a brain. Yet, this signal must, somehow, be processed and transmitted. Once caged, enzymes are secreted by leaf cells, the insect is digested, and, when the trap opens again, the dried-up exoskeleton simply blows away.

THE AQUEOUS MOUSETRAP

There are two other insect-trapping systems. The first is another flowering plant (Figure 1.34), the bladderwort (*Utricularia*), found in fresh water, occasionally on very wet soils, or, as with one species, only in water at the base of overlapping leaves of bromeliads—members of the pineapple family. The insect-catching part of the bladderwort looks like a small balloon (about 5 mm in diameter) with a series of hairs sticking out of one end. When a swimming insect brushes against these hairs, a trapdoor opens and the bug is literally sucked into the bladder. After the insect is digested, the trapdoor opens, expells the remains and water to create a partial vacuum that shuts the door again. Professor Lloyd was fascinated by these plants and, remembering the old saying about the path that would be beaten to the door of him who invented a better mousetrap, decided to see if he could build one on the *Utricularia* model. He did, and it worked very well, but the cost of production was excessive.

FUNGAL TRAPS

The final group of insect-eaters is rarely seen, although they are present in numbers greater than all the other insectivores put together. They are microscopic fungi discovered in 1888 by a German mycologist, Friedrich Zopf, who described them as mycological rabbit snares. Their food is a rarely seen invertebrate worm, the nemotode or eelworm, which lives in soil and causes root disease in plants (Figure 1.35) and other diseases in animals. Professor Zopf found that one of these fungi (*Arthrobotrys oligospora*) has hyphal threads that form loops covered with an adhesive. When an eelworm, attracted by the secretion, attempts to go through a loop, it constricts like a cowboy's lariat. The loop tightens, and the struggles of the worm increases the pressure. This snare is triggered within a tenth of a second and few eelworms escape. Research in several countries is now directed towards seeing whether the fungus snares can control eelworms in fields of potatoes, pineapples, and tomatoes, where the worms cause severe malforma-

Figure 1.35
Rootknot of tomato caused
by infection with nematodes.
Courtesy U.S. Department of
Agriculture.

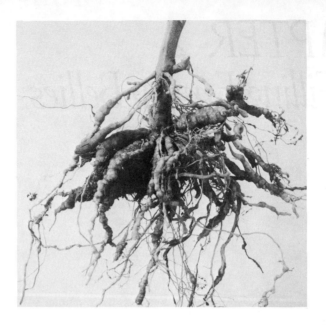

tions of the roots that can so restrict water movement that the plant
dies. Mycologists in France are working with animal pathologists to
find out whether related fungi can capture the larvae of worms which
cause serious lung diseases in sheep. In 1976 another group of fungi
was found whose motile spores are attracted to the eelworm, swim into
the worm's body, and grow into a fungal mycelium that utilizes the
worm as food.

ADDITIONAL READINGS

Barron, G. L. *The Nematode-Destroying Fungi.* Guelph, Ontario: Canadian Biologi-
cal Publications, 1977.
Erikson, R. *Plants of Prey.* Australia: Lamb Publications, 1968.
Lloyd, F. E. *Insectivorous Plants.* Waltham, Mass.: Chronica Botanica, 1942.
Maio, J. J. "Predatory Fungi." *Scientific American* 199(1958):67–72.
Monod, J. *Chance and Necessity. An Essay on the Natural Philosophy of Modern Biol-
ogy.* New York: Knopf, 1971.
Poole, L., and G. Poole. *Insect-eating Plants.* New York: Crowell, 1970.
Pryor, S. "Carnivorous Plants and the 'Man-eating Tree.'" *Botany Leaflet* 23.
Chicago: Field Museum of Natural History, 1939.
Schnell, D. *Carnivorous Plants of the United States and Canada.* Winston-Salem,
N.C.: J. F. Blair, 1976.
Schwartz, R. *Carnivorous Plants.* New York: Praeger, 1970.

CHAPTER 2
Filling Empty Bellies

Nutrition and Malnutrition

⸘ A man doubtful of his dinner or trembling at a
creditor is not disposed to abstracted meditation or
remote inquiries.
DR. SAMUEL JOHNSON

Although we can only speculate on the origin of agriculture, it is
reasonable to assume that conversion of hunting humans to gathering
humans to farming humans was intimately tied to wheat and its near
relatives in the grass family (Gramineae). When human beings progressed from chewing the cereal grains found in their wanderings to
tending a deliberately planted crop, the need to traverse large areas in
search of food disappeared. Wheat and the other cereals have long storage lives and, if kept dry and not eaten by the rodents that accompany
humans wherever they go, palatability and nutritional value is retained
for more than 20 years. A stable food supply made possible by farming
ushered in the agricultural revolution that transformed human lives
and led to what we still call civilization.

Settled man has been known to exist for about 12,000 years. Excavations in Jarmō, a hilly area in modern Iraqi Kurdistan, and in other
sites in the Fertile Crescent have yielded primitive sickles, grain storage
bins, carbonized wheat kernels, and barley and millet heads. By 8000
years ago, the stable agriculture of the ancient Middle East permitted
the establishment of urban centers whose inhabitants provided goods
and services for the farmers. These urbanized people had the leisure
time, the money, and especially the assurance of a stable and adequate
food supply to release human energy for nonagricultural pursuits.
These included the technique of smelting metals, weaving, and the initiation of the fine arts. Music, sculpture, painting, philosophy, systems
of finance, arts of governance, trade, organized religion, and other
amenities of life are grounded in the small, hard, palatable, and nourishing cereal grains. This foundation is still shoring up the world. Our
cities would be depopulated and our vaunted cultural life would disappear without the cereal grasses.

At a minimum, adult humans require close to 2200 Calories per
day in a diet that will allow a life reasonably free from disease. People
without this level of nutrition die at early ages, see most of their children die before their first birthday—if they are born at all—and can
expect a painful history of slow but inexorable decline in health and
vigor. Of these 2200 Calories, protein should be 10 percent of the total,
about 60 grams of complete protein per day. In some societies, ours
being the outstanding example, a much larger proportion of the caloric
intake is protein, to the point where we are in a positive nitrogen balance, excreting excess nitrogen. For most of mankind, however, protein
is scarce and expensive, and the limited supply must somehow be
bulked out to fill empty bellies. Generally, carbohydrate is used as the

extender of fish, fowl, and mammalian protein. Indeed, most of the classic cuisines of the world were developed as protein extenders. At a mundane level, a hamburger illustrates this bulking out of protein. But cast your eyes on a casserole dish like the French *cassoulet* or a German potato and ham dish. The first uses high-carbohydrate beans, the second uses the potato. Or, if you prefer *haute cuisine*, what about Bouef Wellington and King Cole's pie containing blackbirds? The southern Italian, strongly influenced by the wonderous reports of Marco Polo, adopted the noodle made from millet and wheat brought back from north China, and the Italian cooks had the superb taste to add the tomato that the Spanish had brought back from Peru to create a great cuisine. The northern Italian, the central American, and the American Indian uses *polenta*, cornmeal, to supplement protein, while the Scots put oats into sheep's stomachs together with meat—and other things— to feast on *haggis*.

The legumes—lentils, peas, the soybean of China, the beans of the New World Indians—have also become part of our diets. The diets of many tropical peoples are based on cassava, the *manihot* we use only in tapioca pudding. Taro, from which the *poi* of the Polynesians is made, and bananas are other tropical carbohydrate staples, feeding millions of people. These, plus cereal grains, potatoes, and a few others are so important for survival that the success or failure of a crop is literally a matter of life or death. Is it any wonder that the anxiety of planting time and the joy of a successful harvest are foci of religious activity, a basis for ritual and superstition, and a major concern of government?

In today's world, large numbers of people know intimately the Four Horsemen of the Apocalypse. The one named Famine thunders nightly through their dreams. In its coldest terms, famine results when large numbers of people take in less carbohydrate, less protein, less fat, and less of the essential vitamins and minerals than they require for their daily activities. An active adult can lose up to a kilogram of body weight per week on a 900-Calorie diet. If the low calorie intake is accompanied by inadequate protein, fat, and minerals, weight loss is more rapid. A month of 900 Calories per day for people of normal weight will reduce weight and resistance to disease to dangerous levels. Two months will result in the death of 50 percent of the people, with children and the aged dying first.

Even without overt starvation, malnutrition takes a frightful toll. Severely malnourished children are dull, stunted, apathetic, and have a reduction in brain size from 5 to 10 percent. In malnourished women, menarche is delayed, menstruation is irregular, and the rate of spontaneous abortion increases. In young children, brain damage is caused, not by low carbohydrate intake, but by inadequate protein. The human requires several amino acids that cannot be synthesized in our bodies but must be taken in as part of our food. Since most plant proteins are low in several of these amino acids, an exclusively plant diet during the first two years of life results in poor muscle development and inade-

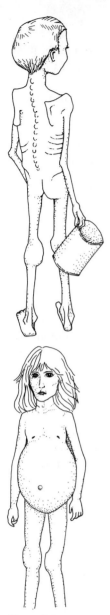

quate formation of brain cells. *Kwashiorkor*, grossly manifested by swollen bellies and a reddish cast to skin and hair (Figure 2.1), is an African word which is translated as "taken too early from the breast" when the supply of nutritionally complete milk protein is replaced by an inadequate plant diet. If supplied with daily proteins or with meat, the child can recover physically, but will never recover mentally. The skin-and-bones type of semistarvation called *marasmus* is also accompanied by poor brain development. We need not go to other countries to find nutritionally related brain damage. There are about 10,000 children in North America who will never reach the genetic potential for intelligence that they received from their parents. A proportion of the difficulties experienced by children in school and in their subsequent careers can be ascribed to undernutrition in utero and during the first years of life.

The litany of mass starvation—famine—is long and unpleasant. An average of 50,000 people starve to death each year, and at least twice that number die from diseases they can no longer resist. More die from the ravages of another of the Horsemen, Pestilence, whose hooves stir up rats with plague-bearing ticks and fleas. Cholera, typhus, plus a host of other diseases weaken people, and general disorder visit the starving. In addition to the irreplacable loss of humanity and the unfulfilled potential of individuals, mass starvation is a concern of all of us.

ADDITIONAL READINGS

Flanagan, D. "Food and Agriculture." *Scientific American* September 1976.
New York Times. *Give Us This Day.* . . . New York: Arno Press, 1976.
Prentice, P. *Hunger and History.* Caldwell, Idaho: Caxton Printers, 1951.
Revelle, R. "Food and Population." *Scientific American* September 1974.
Zinsser, H. *Rats, Lice and History.* Boston: Little, Brown, 1935.

Figure 2.1
The two major types of starvation in children. The boy shows the skin-and-bones syndrome of *marasmus*, and the girl has the swollen belly and staring eyes of *kwashiorkor*. Unless protein feeding is started within a week, such children would die in less than a month.

Wheat and Rice— The Staffs of Life

And he gave it for his opinion that whoever could make two ears of corn or two blades of grass to grow upon a spot of ground where only one grew before, would deserve better of mankind and do more essential service than the whole race of politicians put together.
JONATHAN SWIFT, *Gulliver's Travels*

Two cereal grains, wheat and rice, can and do hold the horsemen in abeyance. They are the staffs of life for well over half the peoples of the world. The fervent plea in the Lord's Prayer, "Give us this day our daily bread," means exactly what it says.

WHEAT

Wheat consists of a group of 14 species in the grass family (Gramineae). Archeological evidence for the cultivation of wheat dates back at least 8000 years in Iraq, 7000 years in the Nile basin, 5000 years in the Indus Valley, and 4500 years in China. Wheat has been cultivated in Europe for close to 4000 years. It was introduced into the New World by the Spanish in 1520, to the United States in 1602, and to Canada in 1630, although Norsemen may have brought it to North America in the eleventh century. The first cultivated wheat (Figure 2.2) was probably the one called einkorn with one grain per spikelet. Genetic crosses, probably natural ones, with wild grasses gave rise to the emmer and durum wheats. Additional natural crosses occurred, since wheat is a wind-pollinated plant, and one or more of these gave rise to the bread wheats. Based on husk color, kernel hardness, and the time of planting, there are four major types of bread wheat. Hard red winter and hard red spring wheats have up to 14 percent protein including the gluten proteins that all bread needs for leavening. Soft red winter wheats have less gluten and are used to make cake and cracker flours. White wheat is the softest of the group and provides very silky flours used primarily for fine cakes. Durum wheat has low protein and low gluten and doesn't rise well, but it is the best wheat for making spaghetti, noodles, and other pasta.

In North America, winter wheats are planted in the fall, they germinate, and, because of the ability of the seedling to resist low temperatures, they overwinter and start growing again when frost is out of the ground. They are ready for harvest in late spring. This late fall planting is obligatory; the young seedlings require a cold period for early development and maturation of the grain. The cold requirement

Figure 2.2
Four important species of wheat. From the left: *Triticum monococcum* (einkorn wheat), *T. dicoccum* (emmer wheat), *T. durum* (durum wheat), and *T. aestivum* (bread wheat). Einkorn has 7 chromosomes, emmer and durum have 14 chromosomes, and bread wheat has 21 chromosomes.

Soft Red Winter

Hard Red Spring

Hard Red Winter

Durum

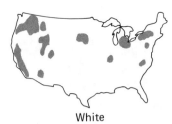

White

Figure 2.3
Distribution of growing areas of the major types of wheat in the United States.

is called *vernalization* which means to "make springlike" and is a translation of a Russian word, *Jarovitzation*, coined by the Russian scientist who first studied it.

Spring wheats are planted in spring and mature in fall. They are *photoperiodic*, requiring that the daylengths become shorter before the flowers will form. Temperatures in the Dakotas and on the Canadian prairies are too low even for the more cold-resistant winter wheats, and these lands are planted to hard red spring wheats (Figure 2.3). Winter wheats are grown on the lands south of Nebraska. White wheat is planted on the west coast and some soft winter wheat is planted in Ohio, Pennsylvania, and Maryland. Durum wheat is grown in the Dakotas, Saskatchewan, and Manitoba. Durum accounts for about 5 percent of the total North American wheat crop, white for 10 percent, soft red winter for 15 percent, and the hard red winter and hard red spring wheats, the best for bread, account for 60 percent of the harvest.

Because of differences in planting and maturation times, wheat harvest starts in Texas in April and continues north until the final harvest in Canada in late September and early October. Large companies rent combines and expert crews for harvesting, and a crew may live in Texas until spring harvest starts and then never stop working until the men reach Edmonton, Saskatoon, or Winnepeg in October. During this period they will harvest over 100 million metric tons of wheat. To put this figure into perspective:

15,000 grains of wheat/pound
60 pounds of wheat/bushel
1 bushel equals 2150 cubic inches
37 bushels equals 1 metric ton (2205 pounds)
1 bushel of wheat equals 70 one-pound loaves of bread

The bushel measure is an old Anglo-Saxon volume, based on a cylinder now in Winchester, England, equal to four pecks which, in turn, are equal to eight quarts. As North America joins the rest of the world in metric measurements, we will become familiar with the metric quintal, equivalent to 100 kg (220.5 pounds). Another measurement to remember is the hectare, an area of land equal to 2.5 acres.

World-wide, about 800 million metric tons of wheat are harvested yearly, with North America producing 20 percent of the total, the Soviet Union another fifth, and China about 10 percent. To produce this amount, the world utilizes 25 percent (250 million hectares) of its cultivated land. Close to three-quarters of the wheat produced is consumed where it is grown. As animal food, it takes two pounds of grain to add a pound to a chicken, four pounds for a pound of pork, eight pounds for a pound of beef, and eight to nine pounds for a pound of human weight. A pound of beef can add less than three ounces to the weight of a human.

Only a few countries produce enough wheat to feed their own people and still have exportable surpluses. Australia exports 6–7 million metric tons, Canada produces 12 million metric tons and exports close to 10, and the United States can export close to half its production, sending abroad 20 million metric tons per year. Although the Soviet Union is still the world's foremost grain producer, increased populations, a series of poor harvests, and a desire of her citizens for more meat produced by feeding cereal grain has made the USSR a major importing country.

RICE

Rice (Figure 2.4), the other staff of life, has been cultivated for at least 6000 years in China, India, and Japan. Rice entered southern Europe by 700 B.C. and North America in 1685 where it was grown in the Carolinas, Georgia, and Louisiana. These states plus California are primary areas of rice cultivation. There are 25 botanical species in the genus *Oryza*, but over 99 percent of cultivated rice is *O. sativa*. Subspecies *indica* is long-grained and becomes dry and fluffy when cooked; it is grown in China, and India, and in North and South America. *Indica* is not photoperiodic, but it is temperature-sensitive and cannot be grown where warm growing seasons are short. Most *indica* rice is derived from a type called 'Champa,' discovered in China about 1000 A.D. 'Champa' matures in about 130 days after planting instead of the more usual 160 days, permitting rice cultivation in central China instead of its being restricted to the southern part of the country. Subspecies *japonica* is short-grained and tends to be glutinous and sticky when cooked. It is the primary rice of Japan, Korea, and Indonesia. *Japonica* rices are not temperature sensitive, but they are photoperiodic, requiring decreasing day-lengths in late summer for flowering.

Rice is consumed in small amounts in North America, averaging less than 4 kg per person per year. It is the mainstay of close to a third of the world's population, who eat an average of 176 kg per person per year. There are over 120 million hectares devoted to rice cultivation

Figure 2.4
Rice, *Oryza sativa*. Courtesy U.S. Department of Agriculture.

(Figure 2.5). Of the 300 million metric tons of rice produced each year, 100 million are grown in China, 40 million in India, and smaller amounts are produced in southeast Asia, the Philippines, and other tropical or subtropical countries. Thailand, Burma, and the Philippines are primary exporting countries with smaller exports from the United States and Brazil. China and India can scarcely meet internal needs.

Under optimum conditions, a third of a metric ton can be produced per hectare, but the world's average is a variable figure less than this amount. Rice cropping is controlled more by water and fertilizers than by other factors. A minimum of 20 inches of rain is required. The ten inches of rain received in India during the late and weak monsoons of 1971–1972 reduced the crop by half. In spite of massive foreign aid in the form of wheat and corn, five million people in India, Pakistan, and Bangladesh starved to death. For maximum yields, 100 kg of nitrogen are needed per acre, with 60 kg coming from nitrogen in the air fixed into inorganic compounds by bacteria and blue-green algae growing in rice paddies. For close to a third of the rice-growing lands, irrigation is necessary, and for another third some method of removing excess water must be used. Traditional rice, in subtropical land, requires 160 days, 'Champa' rice requires 130 days, and, prior to 1965, no rice produced a crop in less than 120 days, meaning that most lands were restricted to one crop per year.

Nutritionally, even the oriental rice does not come close to wheat as a food, being much lower in protein, vitamins, and minerals. Fully polished rice is over 90 percent carbohydrate, and the diets of rice-eating peoples must be supplemented with vegetables, oil, and protein from meat, bean curd, or nuts. Weight-for-weight or nutritionally, rice is a more expensive crop than wheat, requiring considerably more hand labor. Nevertheless, the culture of over a billion people would be damaged by attempts to alter their food habits, and the wet tropical lands used for rice cultivation are not suitable for wheat or barley.

FOOD PRODUCTION

Rice has rarely been stockpiled in large amounts, since it has never been produced in great excess over current yearly needs. But, from the middle of the nineteenth century when the North American plains and prairies were broken by the plow until close to the middle of the twentieth century, wheat production in North America greatly exceeded the demand from the countries of North America and Europe. By 1976, however, the granary was virtually empty, with a three-week's supply in storage compared with a six-month's supply on hand in 1950. The reason for this decline in cereal grain stockpiles is obvious; the population of the world is on the ascent portion of an exponential or geometric growth curve and shows no sign of leveling off (Figure 2.6).

Figure 2.6
Exponential growth of the world's population. The period from 8000 B.C. to 1750 A.D. has been compressed. Data from the United Nations.

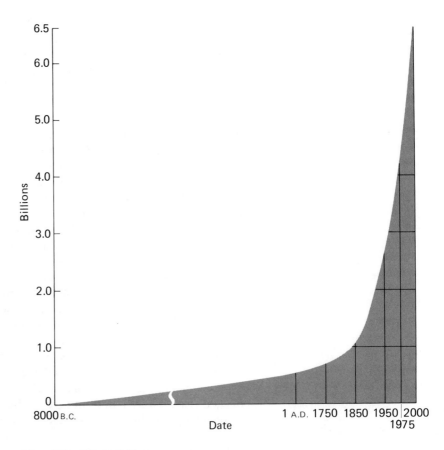

In addition, the period between 1950 and 1976 has been marked by shortages of pesticides and fertilizers and by a long period of bad weather on a world-wide basis.

FOOD AND PEOPLE

The relation between food and population was formulated by an English curate, Thomas Robert Malthus (1766–1834), whose book, *The Principle of Population*, went through several revisions by the author and even more extensive alterations in concept by those who followed him. Basically, Malthus said that since natural procreation in the human species is stronger than the possibility of maintaining life, there is competition in human affairs that may result in misery. Man is potentially capable of increasing in numbers at a geometric rate, while food supply could not, he believed, increase at better than an arithmetic rate. At least part of the controversy about this formula was due to Malthus's seventeenth-century assumptions. The countries with which he was familiar, mostly those of western Europe, were each trying to be self-sufficient in food, trading only small quantities each year and relying on trade of goods and variable quantities of speciality plants (tea, coffee, spices) to balance payments. Malthus was also unaware of the potential for population growth in what we now call underdeveloped countries and the disincentive for population increase in developed countries. Even when we plug in these additional variables, we come up with essentially the same conclusion. World-wide population growth is increasing exponentially and food production, although exceeding the arithmetic rate assumed by Malthus, is not increasing in quantity and quality at the same rate.

In 1950, world wheat production stood at 5 billion bushels per year, and there were 2.5 billion souls on Earth. In 1975, wheat production rose to 14 billion bushels, and there were 5 billion people. This suggests that Malthus was wrong; the rate of population increase for this 25-year period was close to 1.9 percent, and food grain production increased at the rate of 2.5 percent. There are, however, other factors to be considered. There is no reason to assume that the rate of population increase will stabilize and every reason to believe that this rate will increase to 2.8 percent by the end of the century. To feed the yearly net increment of close to 80 million people per year will require an additional 20 million tons of grain each year, an amount equal to the total Canadian crop or an increase of 40 percent in the amount of wheat now exported yearly by the United States. India's population is increasing at one million per year with about a third of these new human being rice-eaters. An additional five million metric tons of rice must be grown each year—every year. This means that the world's rice crop must increase by nine percent per year. Put another way, Indian rice production must be doubled by the year 2000. China's population is increasing by ten million per year, and her rice crop must triple by the year 2000.

These increases will not, however, raise the nutritional level of undernourished people. Total caloric intake is not a complete measure of nutritional status. Cereal grains are fine carbohydrate sources, but processed flour and polished rice are inadequate sources of fats, minerals, and vitamins, and their protein is "incomplete" with suboptimal amounts of several required amino acids. Only when we look at world-wide production figures can we say that we are producing food at a greater rate than we are producing ourselves. Even here, Africa as a whole has a population increase that exceeded food production by one percent in the 1950–1975 period, and equally large gaps are appearing in the food-people ratio of southeast Asia and South America. In addition, the affluent countries of North America, Europe, and the Soviet Union are diverting a portion of available grain to feed domestic animals in order to satisfy the ever-increasing demand for meat. The United States and Canada use about four percent of their wheat for this purpose, and other countries are following this example. With a backlog of a few month's supply of wheat in storage—and this only in North America—drought, floods, plagues of insects, or an outbreak of plant disease can send those countries teetering on starvation's edge over the brink in one growing season. That the shortfall in food predicted by the year 2000 may arrive sooner is evident from current reports of famine; undernutrition is rarely newsworthy.

Even if North American grain farmers were to increase production by an almost impossible 5 percent per year, the shortfall would still exist in the tropics where 90 percent of the population growth is taking place. Today, the problem is one of distribution and this problem seems insurmountable. There isn't enough money in the world to permit North America to play the philanthropist permanently. Someone has to cover the cost of seed, chemicals, machinery, and labor necessary to plant and harvest a crop, to prepare it for food use, and to ship and distribute it. In 1975, the total available food supply could have furnished every person on earth with 2600 Calories per day, including 68 grams of protein—a complete and adequate diet. But the undernourished peoples of Africa, Asia, and South America do not have the foreign exchange to meet even a third of the costs of this diet. The economies of the food-exporting countries could not be maintained if they were to give the food away.

The solution, if there is to be one, must involve a two-pronged attack, one directed towards population control (a very long-term solution) and the other directed towards increasing the food supply in the countries where it is needed. People aren't having more children than they used to; in the developed countries they are having fewer. Between 1925 and 1975, infant mortality decreased world-wide from 300 per 1000 live births (the U.S. mortality rate was 10 per 1000) to 50 per 1000 (the U.S. rate was less than 10 per 1000). The average life span in 1925 was well under 50 years and has increased to over 60 years for men and 65 for women. Both increased life expectancy and decreased infant

Figure 2.7
The Chicago Board of Trade, the largest grain market in the world. Courtesy U.S. Department of Agriculture.

mortality are due to advances in medicine and public health. In addition, all of us want the best possible lives for our children and all of us have rising expectations as a result of what we see is available. Since human breeding is beyond the control of government, means for facilitating local production of food is, given our present moral codes, the basic factor in a future that can be contemplated with any equanimity.

The economics and geopolitics of the import-export trade in food grains on an international level is beyond the understanding of virtually all of the world's people, but some indication of its importance must be considered. For the United States, income from wheat exports, which amount to 65 percent of the international wheat market, was $23 billion in 1975, just about balancing the cost of imported petroleum. Canada, the world's other wheat granary, showed a surplus in its balance of payments. Since, in both countries, grain exports are handled by a free enterprise system, the fewer than a dozen multinational corporations that control the bulk of the grain trade can strongly influence geopolitical decisions and policies of both importing and exporting countries. Commodity markets, dealing in futures of grain and other foodstuffs, allow for speculative financial activities (Figure 2.7) which can drive the price of grain up or down with little regard to the needs of hungry people. The food-grain cereals are obviously an economic weapon, with exporting countries using them to obtain concessions from countries desperate to feed their people. All of the exporting countries have used this weapon and continue to do so with unknown consequences.

GREEN REVOLUTION

With this gloomy and perhaps over-pessimistic view, is it any wonder that the Green Revolution was greeted with so much joy and hope? The possibility of doubling and even redoubling the production of wheat and rice galvanized the scientific, political, and human communities. When Dr. Norman Borlaug was awarded the Nobel Peace Prize in 1970, the whole world nodded its agreement, for nothing is more important for world peace than a decent standard of nutrition and the assurance that an adequate food supply will be available for one's present and future family.

For thousands of years, wheats planted in different parts of the world were unstandardized local strains, moderately well adapted to the climate and growing conditions of the area. After wheat was introduced into North America, there was a period of selection, as immigrants selected seed each year from their crops. Most of these selected wheats were fairly good producers in good years, but few were highly resistant to drought and even fewer had the genetic potential to resist plant diseases. Among the introductions which did have desirable characteristics was the famous 'Turkey Red' wheat. We can date its introduction with some precision. German-speaking people, Mennonites, immigrated from Holland to the Russian Crimea in 1568 to escape religious persecution. Catherine the Great annexed the Crimea in 1783, expelled the Turks, and, to promote development of the Crimea as the breadbasket of Russia, gave these thrifty Mennonite farmers tax relief and exemption from military service. In 1870, Czar Alexander II refused to renew the concessions and the Mennonites immigrated in 1873–1875 to Manitoba and the United States, taking with them the hard red wheat which they had named 'Turkey Red.' They were attracted to Kansas because land could be obtained cheaply through the Homestead Act of 1862. The Atchinson, Topeka and Santa Fe Railroad offered passage to Topeka for $11 in order to get good farmers to grow wheat that the railroads could ship east. 'Turkey Red' was well adapted to the semiarid prairie and became the standard for hard wheat. 'Red Fife' was obtained by David Fife of Ontario in 1830, 'Hard Red Calcutta' was crossed with 'Fife' by Charles Saunders of Canada to give 'Marquis' and these, plus 'Blue Stem,' were standard wheats of North America by 1900. Another parent of our modern wheats was 'Kharkov,' a drought-resistant strain brought back from Russia by Mark Carleton, a U.S. Department of Agriculture agronomist. Carleton also introduced a hard durum wheat, 'Kubanka,' used for pasta. These, in various crosses, formed the solid foundation for virtually all North American wheats until the end of World War II.

All of these wheats had some recognizable drawbacks, but these couldn't be overcome. All of them were tall wheats, between 120–140 cm in height. This meant that a considerable proportion of the water and nutrients taken from the soil was used to develop the straw and not the grain. Because they were tall, there was a great tendency to lodge, that is, to be knocked over by high winds and hail. For efficiency in grain production, a shorter stemmed plant is desirable, and this character was first found in Japan. There is a children's toy called a rolly-polly in the West that has a heavy, rounded bottom that causes the toy to rock when hit by a child's fist, but it can't be knocked over. In Japan the toy, and hence the wheat, is called 'Daruma.' Crossed with taller wheats possessing drought-resistance and some resistance to wheat rust fungi, the progeny was named 'Norin.' When 'Norin' was crossed with the descendants of 'Marquis,' 'Turkey Red,' and 'Fife,' a plant that was only 90–120 cm tall resulted. This semidwarf wheat was named 'Gaines' and was the first of the miracle grains.

These crosses were made to grow wheat adapted to the temperate zone. They were photoperiodic or thermoperiodic (having a requirement for specific temperatures for development) and could not grow in the tropical countries where the need for home-grown wheat was most acute. Recognizing this problem, the Rockefeller and Ford Foundations set up an experiment station in Mexico to see if wheats could be bred for tropical conditions. Headed by Dr. Norman Borlaug, teams of geneticists, agronomists, soil specialists, and botanists drew heavily on genetic studies of H. Kihari and H. Fukasawa in Japan who had developed methods to allow controlled breeding research. What was needed were strains that were short-stemmed, high-yielding, resistant to tropical plant disease organisms, and adapted to both drought and flood conditions. The plants also had to be independent of both temperature and photoperiodic signals for flowering and fruiting, because the tropics have even temperatures and very small seasonal changes in daylength. The first successful strain was developed by 1944 as a hybrid between Mexican strains and several North American varieties.

Rice production also came under scientific scrutiny, supported by grants from the Rockefeller and Ford Foundations. The International Rice Research Institute in the Philippines and the Aduthruai Experiment Station in Madras, India, developed short-stalked strains, but they were exquisitely sensitive to applications of fertilizer. In 1966, the first miracle rice, IR-8, was distributed, and by 1968–1969 close to a fifth of the world's rice-growing areas were planted with IR-8 (Figure 2.8). Reports of increases in yields of up to 300 percent began to appear.

Figure 2.8
Experimental plots of dwarf rice (IR–8) compared with a standard tall cultivar (CS–M3). Courtesy Dr. D. Marlin Brandon, Cooperative Extension Service, University of California, Davis.

Since these rices had no photoperiodic or temperature requirements, it was possible to grow two or sometimes three crops a year on the same land, because the new rices could mature in 100 days instead of 130–180 days. Dr. Borlaug, ever the cautious professional, stated that while the new grains could not solve the impending crisis in food, they could buy the world time to work on the other prong of the problem, that of population control. And, so it seemed, until the early 1970s, when yields began to decline.

Some of the difficulties are inherent in the new plants themselves. In order to obtain high yields, growing conditions must be just right. If a plant is to produce more grain per head, nitrogen and phosphorus fertilization must be very high; the new wheats and rices require two to three times more fertilizer and three times as much pesticide than do traditional strains. Adequate water at exactly the right time is also necessary; a drought for even a short time will reduce the crop. Lacking foreign exchange to purchase fertilizer, costs which became particularly acute as fossil fuel prices skyrocketed in the early 1970s, tropical countries stood by helplessly as yields decreased. The costs of irrigation, fertilizers, pesticides, and advanced machinery became the limiting factors for a successful Green Revolution.

There were other problems, unforeseen by the scientists and virtually beyond the control of political leaders. These problems resulted from the failure to consider cultural, social, religious, and economic factors. In Thailand, for example, it is believed that rice grows from the womb of the Rice Mother, *Mae Phosop*, whose body is impregnated by the seed inserted by the planter. Since it is "well known" that machinery and even man can frighten a pregnant woman, the use of mechanical cultivators necessary for maximum yields is contraindicated, and even weed-pulling must be done by women (how convenient for men!). Planting and harvesting throughout the world is inextricably bound to religious belief. The natural flow of the seasons is marked and highlighted by planting and harvesting festivals and ceremonies. The shorter growing seasons for the new cereal grains and the possibility of double-cropping does not coincide with these ceremonies. It is as if westerners were forced to have Easter parades in August. Techniques for planting, cultivating, and harvesting the miracle grains differ from those used for centuries, and tradition is not lightly discarded by people of any culture. The early strains of miracle rice looked and tasted different from traditional strains, to the point that people wouldn't eat them, but tried to sell them on a glutted market or fed them to their pigs and chickens, essentially negating the advantage gained by increased yields. This last problem has been solved by newer strains, particularly IR-22, whose consumer acceptance is much greater.

Equally fundamental problems face full acceptance of the new strains, and some of these difficulties are likely to become more acute. Among the most severe is the matter of available land to support wheat or rice. In developed countries, virtually all potentially arable land is under cultivation, but in underdeveloped countries, this is not the case. Indonesia has 120 million hectares that are potentially arable, but only 10 percent are under crops. Indonesians lack the money to purchase machinery to irrigate, drain, and open up the land. Spiraling fertilizer and pesticide costs also must be included in the balance sheet. Even assuming that these matters can be solved by massive inputs of funds from more developed nations, and that additional grain is produced, the probability exists that in the short term, there may be a glut of grain coming to market. This can disrupt international trade, as the Philippines found in 1973 when they started to export rice. Were Canada and the United States to lose a proportion of their overseas grain customers, effects on balances of payments would severely disrupt their economies and could activate a depression in those sectors of the economy that depend on agriculture. Thailand and Burma earn their foreign exchange by selling rice. Funds received from these sales are used to purchase the goods of other countries. With cultivation of miracle rice by traditional importers, the market can dry up. They, like the United States and Canada, cannot afford to maintain large stockpiles of grain because it would interfere with the free market by stabilizing prices (but not costs) and would inhibit profits. Solving this problem requires that both importers and exporters must ask what new products and for what new markets they must plan.

Land tenure is a staggering difficulty, made more acute by the Green Revolution. Until the end of World War II, the bulk of the world's arable land was owned by a minute fraction of the world's population. Landlords would either hire labor for their fields or set up a tenant farmer system in which the landless peasant lived and worked the land on terms that were rarely favorable and usually kept peasants in serfdom. With the Green Revolution, land values in developing countries almost quadrupled, resulting in an increased reluctance of landowners to support governments that wish to redistribute land. Land ownership allows the owner to impose his or her will on the landless, and it is in the owners' best interest to dispossess tenants in order to increase the pool of cheap labor. As anyone knows who has ever tried to obtain a bank loan, bankers want collateral. Peasants, totally lacking in resources, cannot provide bank security, while the landowner can easily obtain credit to pay for machinery, pesticides, and fertilizer, etc. Unfortunately, the Green Revolution has widened the gap between the rich and the poor.

Genetic engineering has itself created problems. The new grains are close to being *isogenic*, that is, they are genetically uniform. This is not true of traditional grain; since both wheat and rice are wind-pollinated, their progeny is genetically variable. When disease strikes, or in times of drought, at least some of the genetically mixed plants will survive to yield some grain. The highly inbred miracle grains do not possess genetic plasticity, and an abnormal occurrence can wipe out the entire crop. This is what nearly happened in the United States in 1973 when a new strain of fungus causing corn blight reduced the genetically pure crop by almost 90 percent in severely affected areas of the corn belt.

The major fungus scourge of wheat is called stem rust. In early spring, spores that have overwintered in the soil can infect just a few young seedlings, but as the fungus develops in these plants, it produces another type of spore, red in color, that makes the leaves and stems look like they have been streaked with rusty iron. Uncountable billions of these spores are released into the air where they can travel hundreds of miles before infecting other plants (Figure 2.9). These, in turn, produce more billions of spores to infect more plants. The disease was known to the ancients. On April 25, Robigus—a grain deity—was placated by a ceremonial procession to his temple. Red wine was poured on the altar and a red dog sacrificed. Romans believed that the barberry plant (*Berberis spp.*), because of its reddish leaves and bright red berries, should be kept away from the wheat and no red clothing should be worn during the growing season. Proof of the fungus nature of wheat rust was provided by Pier Michele in France in 1729, and the infective spores were seen by the Tuslane brothers in 1845. But these discoveries gave no clue as to why barberry plants seemed to be related to the disease. The eradication of barberry, first mandated in France in 1660, seemed to be mere superstition. The Canadian mycologist, J. H. Craigie, demonstrated in 1927 that the life cycle of the wheat rust fungus was very complex, involving five different kinds of spores and both the wheat and the barberry host plants. Just as unsettling was Jacob Eriksson's report from Sweden in 1894 that there were different races of wheat rust which infected different strains of wheat. In 1914, E. C. Stakeman extended Eriksson's work and found that there were several hundred genetic races of the fungus and no cultivated wheat could be

Figure 2.9
Effects of stem rust on wheat yield. Grains from healthy plants on left; shriveled grains from diseased plant on right. Courtesy U.S. Department of Agriculture.

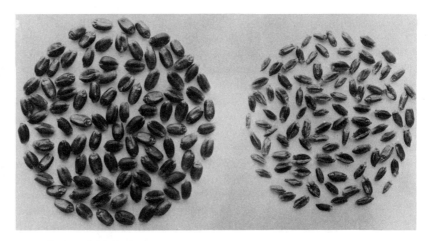

immune to all of these. University of Minnesota plant pathologists found that it takes three to four years for a significant number of mutants of the fungus to become a serious disease threat, but it takes an additional three to five years to develop a genetic strain of wheat which is resistant to the mutated fungus. Thus, the developers of the miracle wheats face a never-ending task of developing disease-resistant strains and still retain the desirable characteristics of the wheat. Like housework, the tasks are never finished.

The genetic variation that develops in cross-pollinated plants also must affect the course of the Green Revolution. Throughout the world, farmers had always saved some of the grain for the next year's seed. But, since miracle grains have been randomly pollinated and fertilized by wind-borne pollen, there can be no assurance that the crop will have the desired characteristics. This means that peasant farmers cannot save grain for seed but must buy new seed each year from producers of pure-line plants; the corn farmers of North America do this. Of course, this means establishing lines of credit with banks or borrowing from landlords or moneylenders at often ruinous rates of interest. Planning and honesty in governments are required to get new seed, fertilizer, and pesticides to peasants at the right time in countries not noted for either efficiency or rectitude.

Continuous cropping has, as we have seen, doubled the yield on a given piece of land, and the miracle grains were engineered with this in mind. Since most of the tropics have seasonal changes of wet and dry periods, this means that increased irrigation is needed in the dry season planting and some soil draining for the monsoon season's crop. Modifications in fertilizer and pesticide schedules are needed and must be explained to people whose formal education is minimal. Peasants are understandably reluctant to gamble with the very survival of their families, and those without money could not introduce new techniques even if they wished to. In underdeveloped countries, large landowners, absentee agribusiness corporations, and cooperatives could, and did, introduce the new methods and could, and did, undersell the peasant. Small plots of land were bought up for very little money and consolidated into large farms that efficiently used the new technology. These landlords could hire the now-bereft peasant for very little money. Again, the rich got richer.

All this must not be taken as an argument against the concept or the necessity for the Green Revolution, nor as a denigration of the scientists who brought the revolution into being. Feeding people is the most important task the world now has, and the techniques, the scope, and the ideas that motivated the revolution are correct and necessary. What was not well thought out was the political, economic, and social parts of the revolution. These are more difficult to solve than were the problems of genetic engineering. That they must be solved is obvious. But when and how?

ADDITIONAL READINGS

Wheat and Rice

Associates of Japanese Agricultural Scientific Societies. *Rice in Asia.* Beaverton, Oregon: ISBS, 1974.

Buller, A. H. R. *The Story of Marquis Wheat.* Toronto: Macmillan, 1919.

DeKruif, P. *Hunger Fighters.* New York: Harcourt Brace Jovanovich, 1928.

Edgar, W. C. *The Story of a Grain of Wheat.* New York: Appleton & Co., 1903.

George Washington University. *Bibliography of World Literature: Rice.* Metuchen, N.J.: Scarecrow Press, 1963.

Grist, D. H. *Rice,* 4th ed. New York: Longman, 1953.

Hanks, L. M. *Rice and Man.* Chicago: Aldine, 1972.

Leonard, W. H., and J. H. Martin. *Cereal Crops.* New York: Macmillan, 1963.

Mangelsdorf, P. C. "Wheat." *Scientific American* July 1953.

Moore, A. C. *The Grasses. Earth's Green Wealth.* New York: Macmillan, 1960.

Peterson, R. F. *Wheat.* New York: Wiley, 1965.

Trager, J. *Amber Waves of Grain.* New York: Arthur Fields Books, 1973.

The Green Revolution

Barnett, H. J. *Population Problems—Myths and Realities.* Washington, D.C.: George Washington University Press, 1973.

Borlaug, N. "Genetic Improvement of Food Crops. *Nutrition Today* 7(1972):20–25.

Brown, L. *The Green Revolution and Development in the 1970's.* New York: Praeger, 1970.

Griffin, D. *The Political Economy of Agrarian Change: An Essay on the Green Revolution.* Cambridge, Mass.: Harvard University Press, 1974.

Hardin, G. "Living on a Lifeboat." *BioScience* 24(1974):561–568.

Katz, G. *Giant in the Earth.* Briarcliff Manor, N.Y.: Stein & Day, 1973.

The Soybean

How wonderful was Heaven to give man a plant
that provides both milk and meat.
CHINESE PROVERB

In 213 B.C., Shih Huang Ti, Emperor of China, decreed that all classical manuscripts were to be utterly destroyed and memory of their content expunged. Why he so ordered is unknown, but the edict was carried out under pain of decapitation. Among the lost works was a book of poems, the *Shih Ching,* a classic said to have been edited by Confucius and quoted frequently in his *Analects.* Not until the early Han dynasty, in the reign of Wu-ti about 100 B.C., were scattered fragments of the 300 poems reassembled.

Although the *Shih Ching* poems are on the nature of man, on his place in the scheme of things, and on his role in society, the book is important to botanists because plants are frequently mentioned. At the time of Confucius, about 550 B.C. during the Ch'u dynasty, China had little to do with the rest of Asia, and it is assumed that plants mentioned

in the *Shih Ching* could be regarded as native to the country. This is not strictly true. The provenance of the poems is a bit questionable, and, while governments may not have much to do with other governments, people interact with people and exchange at the local level goes on. In the *Shih Ching,* references are made to a plant named *shu,* cultivated for the high protein and oil content of its seed. The Chinese also noticed that when millet or cabbages were grown in fields in which shu had grown the previous year, the crop was darker green and healthier than it would otherwise have been, and yields were unusually high. Shu is a legume, a bean, whose Latin binomial is *Glycine max.* In the West we know it as the soybean (Figure 2.10).

The Leguminosae, or pea family, includes many plants of importance to human beings. It is one of the larger families of flowering plants, including trees, shrubs, vines, and herbaceous genera that live in many different habitats. There are close to 500 genera and several thousand distinct species. The family is characterized by the homology of the fruit, the familiar pod. Legumes form a natural assemblage, and while one may not know the name of a leguminous plant, identification to the family level is easy and sure. More than 50 genera in the family contain plants of economic significance as food, medicine, industrial products, and as garden plants. The soybean genus, *Glycine* (Greek for sweet) has about ten species with *max* being the only one cultivated for food. A leading authority on the botany of cultivated plants defines the genus as follows:

> Erect, brown-hairy, bushy annual plants; 2–6 ft., some shoots vine-like. Leaflets 3, ovate to narrow ovate, 3–6 inches long, entire. Flowers papi-

lionaceous, inconspicuous, small, white to purple. Pods hanging on short stalks, 2–3 inches long, ca. ½ inch broad, brown, hairy. Seeds 2–4, globose, green, brown, yellow or black.

A picture is more useful (Figure 2.11).

CULTIVATION

The belief of the ancient Chinese that soybeans improved the soil was, as are most observations of perceptive plant growers, correct. The roots of soybean plants, like those of beans, clover, and alfalfa can be invaded by bacteria of the genus *Rhizobium*. This organism induces the formation of small nodular overgrowth in the roots. Within these nodules (Figure 2.12), the *Rhizobium* can convert molecular nitrogen gas from air into ammonium compounds which supply all of the nitrogenous fertilizer needed for the growth and maturation of the plants. After the bean crop is harvested, the plant can be plowed back into the ground where its minerals, including the nitrogen compounds, are released by the action of soil bacteria and become available for plants grown the following year.

Soybeans presently cultivated are photoperiodic plants. They will form flowers only when the day-length is shorter than some internally regulated minimum; soybean is a short-day plant. Most people are aware of photoperiodism. Dandelions flower in the early spring when the days are just beginning to lengthen from the very short, sometimes gloomy days of winter. Chrysanthemums flower in late summer and early fall when the day-lengths are again decreasing from the peak of midsummer. Both of these plants are short-day plants. Many grasses and garden flowers like *Petunia* and *Rudbeckia* come into flower only

Figure 2.12
Root nodules on roots of soy.
Courtesy U.S. Department of
Agriculture.

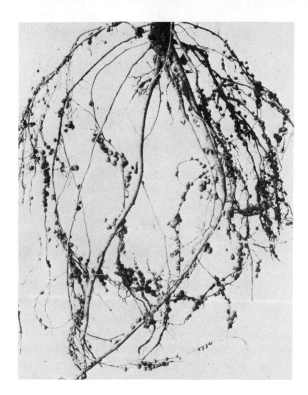

when the day-lengths are long—these are the midsummer, long-day plants. Still others—tomato, corn, cucumbers, and strawberries—are termed day-neutral, flowering when they are large enough regardless of day-length. Photoperiodism was discovered in 1920 by two U.S. Department of Agriculture plant physiologists working with tobacco and soybeans. This discovery should have been recognized by a Nobel Prize because it revolutionized agriculture. Drs. Garner and Allard found that the soybean plant would only form flowers when the day-length was less than 11 hours, a condition that occurs in the temperate zones in late summer and early autumn. The requisite short days occur in Illinois about the first of September allowing an additional four to six weeks of frost-free growing season for maturation of the seeds. Winnipeg doesn't get short enough days until October, which doesn't allow enough time before frost to mature the crop. The tropics around the equator are too hot for the plant and the subtropics never have sufficiently short days for flower induction. This restricts soybean cultivation to a relatively narrow band of 35–45 degrees north or south latitude.

The corn belt of the central states—Ohio, Indiana, Illinois, southern Michigan and Wisconsin, and Iowa—provide soil, rainfall, and temperature conditions which are close to ideal for soybean. In Illinois, where top soils can be six feet deep with excellent drainage, the plant grows best. This Chinese plant has found a happy home in the Land of Lincoln.

SOYBEANS IN NORTH AMERICA

The use of the soybean in the Western world is a recent development. Engelbert Kaempfer brought Japanese seeds to Germany in 1692. It was named by Linnaeus in *Species Plantarum* in 1753 as *Phaseolus max* (*Phaseolus* is the genus containing the common kidney bean), later to become *Soja max*, and finally *Glycine max*. As with all introduced plants, seeds were distributed to most European botanical gardens and gardens on the estates of the wealthy by the beginning of the nineteenth century. It was at first a mere curiosity, "one of these strange Japanese plants. My dear, people actually eat it." In 1804 seeds were brought to the United States by a Dr. Mease who wrote to a friend, "The soyabean bears the climate of Pennsylvania very well. The bean ought therefore to be cultivated." There was, however, no interest in the plant. Commodore Matthew C. Perry obtained additional seeds in 1854 when he "opened the door to Japan" and ushered in the Meiji Restoration that transformed Japan into a modern world power. Perry's journey sparked interest in the plants used by the Japanese. But outside the Orient, soybeans were grown only as a green manure, planted in the spring and plowed under in the summer to enrich the soil for other crops sown the following spring.

Soybeans were established in the United States only after World War I. Three million bushels (60 pounds per bushel) were produced in 1920, 70 million bushels by 1940, 500 million bushels by 1960, and 1000 million bushels were grown in 1970. Current U.S. production is 1.5 billion bushels, representing half of the total world production. Of the U.S. crop, 50 percent comes from Illinois, Iowa, and Indiana, and the value to these states now exceeds that of corn. The United States exports close to half the crop, mostly to the Orient. China, the other large producer, does not export much, since it needs all it can grow to feed its population.

SOYBEANS AS FOOD

In North America, soybeans are often recognized only when someone makes a visit to a Chinese restaurant where he or she has a bowl of soup containing a small piece of delicious bean curd. But for oriental peoples, the soybean is basic food. Consider some facts. With seed coats removed, the dry bean is 40 percent protein, 21 percent oil, 34 percent carbohydrate, and 5 percent mineral. There are high concentrations of calcium, phosphorus, and iron, and both water-soluble and oil-soluble vitamins. The carbohydrate is mostly starch plus a small amount of immediately assimilable sugar. The protein, mostly globulins (as are the easily digestable proteins of meat), has a complete

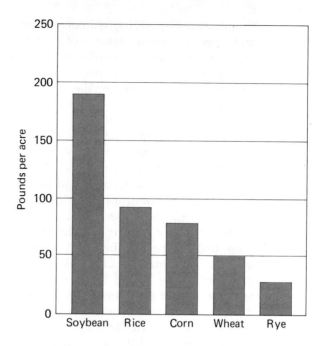

Figure 2.13
Field crops yielding large quantities of protein.

spectrum of amino acids (Figure 2.13). It is low in the essential sulfur-containing amino acid, methionine, which cannot be synthesized by man, but the amino acid balance is such that soy protein can be readily metabolized into animal protein. When supplemented with methionine, soy protein is nutritionally complete; one can live on it as the sole protein source. In cultures where expensive animal protein is in short supply, soybean, supplemented with vegetables, other belly-filling carbohydrates, and occasional bits of fish, eggs, fowl, or meat to supply nutrients not found in plants, will meet all food needs from about the age of two onward.

Soybeans are used in both West and East in many ways. Actually, North Americans consume large amounts of soy without being aware of it. Soybean meal left after extraction of the oil is virtually all protein and carbohydrate and is animal food par excellence. Millions of metric tons of soymeal are used to feed cattle and hogs, to put weight on chickens, increase egg production, and to feed cats, dogs, goldfish, and even your pet canary. Thus, soy is part of the chain that links plants to the Thanksgiving turkey or the cheeseburger. The use of soymeal is not restricted to animal food. Some of it is incorporated into sausages and hot dogs when the small print indicates "cereal additive." It is now part of breakfast foods, added to supply the protein that these products are woefully low in. Defatted soy flour is added to wheat and rye flour to increase the protein content of breads, pies, and cakes. Soy oil is a major source of the fats we use. Except for the peanut, its oil content is the highest in the plant kingdom. The vegetable or salad oils we consume are from cottonseed and soybeans. Vegetable cooking fats are manufactured by adding hydrogen atoms to unsaturated vegetable oils under pressure in the presence of a catalyst—usually nickel. When hydrogenation occurs, liquid oils become fats solid at room temperature. Margarine is a mixture of plant oils, partly hydrogenated and colored with vegetable dyes. During the extraction of soybean oil, a fraction of the oil is separated as compounds which serve as emulsifiers, foam stabilizers, wetting agents, and for other purposes. Our mayonnaise, ice cream, chocolate milk, and many other prepared foods include not only soy oils, but also their lecithin stabilizers. Large quantities of soy oil, stabilizers, and other constituents of the bean find their way into industrial products including paints and lacquers, plastics, and a long list of products (soaps, lubricating oils, glycerin, explosives, etc.).

Although these uses for soy are also found in the Orient, soybeans there are a primary protein and oil source for human nutrition. Close to 12 percent of the total protein intake of the Japanese, and at least that percentage in China, is derived from soybeans. In 1973, the president of the United States embargoed shipment of soybeans to Japan because of a presumed crisis in high-protein animal food. One of the world's important sources of animal protein food is fish meal produced from anchovies taken off the coast of Peru. Shifts in the cold ocean currents in 1973 resulted in poor catches and a large reduction in available fish meal. Farmers, poultry egg raisers, and beef cattle producers saw their costs rising and shifted to soybean meal, quickly depleting the available supply. Raising loud shouts to Washington, they succeeded in forcing the embargo. International relations between Japan and the United States were strained, and the embargo was quickly lifted. The Japanese realized that U.S. internal needs would increase with consequent increases in bean prices and that they had to have an assured supply of soybeans. Japan helped finance farmers in Brazil to start a soybean agribusiness, and the Brazilian farmer may well become real competition for the Illinois farmer. The growing conditions in that part of Brazil lying between 40 and 45 degrees south latitude is a mirror image of the American corn and soy belt.

In the Orient, food uses of the bean are varied. Whole beans are eaten after boiling or baking. Although the mung bean is the usual starting material for bean sprouts, soybeans are also used. An equally important way of preparing soybean is the bean curd. This *tofu* (Chinese and Japanese pronounce our "t" as "d") is made from selected cultivars of soybeans. Dry beans are ground into flour and the protein extracted with hot water to make soy milk. Calcium salts are added to precipitate the protein as a curd which is then pressed into cakes. Tofu is used in soups and cooked with meats, fish, and vegetables. Soy milk is fed to young children and, like cow's milk, can be made into a high protein cheese (*sufu*).

Soy in fermented products is familiar to us as soy sauce (*shoyu*). Soybeans are soaked in water, boiled until soft, and mixed with roasted and ground wheat. Other cereals like millet, rice, or barley can be added or substituted for wheat; the proportions of grain and soy paste can also vary. The critical step is the composition of the fermenting agent, called *kojii* in Japan. To prepare kojii, steamed rice is mixed with a starter culture of a fungus, *Aspergillus oryzae,* and allowed to grow for several days until the rice is covered with the mold. The kojii and salt are added to the soy-wheat paste, and the mixture is allowed to ferment slowly for 8–12 months. The product is a very dark brown, somewhat salty liquid with an aroma and taste very different from virtually anything else—you love it or loath it. Chinese and Japanese soy sauces are different, the Chinese types being sweeter. Synthetic soy sauces, the kinds sold in North America, are made by mixing strong acids with soy-wheat paste, heating to hydrolyze the starch and protein, neutralizing with strong alkali, and, after a quick aging, are bottled and sold to an unsuspecting public. Caveat emptor!

INCREASING YIELDS

As populations increase, demands for adequate protein also increase. The cost of raising cattle, hogs, fowl, and mutton increases at a rapid rate because of the relative inefficiency of conversion of plant protein to animal protein. Consequently, agricultural scientists are working out details of cultural practices designed to increase yields of soy and to the breeding of cultivars showing desirable characteristics. Through their efforts, yields have doubled from less than 20 bushels per acre in 1940 to 41 bushels per acre as of 1970. Part of this increase is due to more careful control of growing conditions, including the use of weed-killers, insecticides, and fungicides tailored specifically for soybean cultivation. Without these, production would decrease to a level where millions of the people in the world would soon become protein- and oil-starved. Most soybean cultivars are bushy plants, rarely exceeding a meter in height. The agricultural technician has been wrestling with the trade-off between plant and row spacing. If planted too far apart, valuable space is lost, and the fields quickly become choked with weeds. If planted too close together, light penetration into the plant canopy may be reduced below the level for efficient photosynthesis, and adequate soil-wetting and aeration can be impeded. Cross-breeding has provided taller plants with fewer branches allowing close spacing; and, for other growing conditions, shorter plants have been developed. To widen the latitudinal spread, the growing seasons of plants to maturity can be varied from 90 to 150 days.

In spite of numerous attempts, the one genetic characteristic that has resisted modification is the short-day control mechanism. The development of day-neutral soybeans, capable of flowering regardless of the length of day or night would permit soybeans to be grown in more northern regions or closer to the equator than has hitherto been possible. The impact of such a discovery would affect the lives of close to a billion people.

ADDITIONAL READINGS

Horvath, A. A. *The Soybean Industry.* New York: Chemical Publishing Company, 1938.

Markley, K. S. *Soybeans and Soybean Products.* New York: Wiley, 1951.

Norman, A. G. (ed.). *The Soybean.* New York: Academic Press, 1963.

Piper, C. V., and W. J. Morse. *The Soybean.* New York: McGraw-Hill, 1923.

Wolf, W. J., and J. C. Cowan, *Soybeans as a Food Source.* Cleveland, Ohio: CRC Press, 1971.

Wood, B. J. B., and Y. F. Min. "Oriental Food Fermentations," chapter 13 in J.E. Smith and D. R. Berry (eds.). *The Filamentous Fungi,* Volume I, *Industrial Mycology.* New York: Wiley, 1975.

Figure 2.14
Ceres, goddess of grain. Taken from a Pompeiian wall painting.

Figure 2.15
An early illustration of Indian maize. Taken from a woodcut in *Herbarz* by Mattioli published in 1562.

Indian Corn

The corn is as high
As a elephant's eye. . . .
RODGERS AND HAMMERSTEIN, *Oklahoma!*

All too often a name is applied which causes confusion, but becomes so deeply rooted that it just can't be changed. "Corn" is just such a word. The word "corn" is applied to the leading cereal crop grown in a particular area. Wheat, barley, oats, or millet are corn to the European, and rice is the corn of the Orient. Even buckwheat, which isn't a cereal grass, is called corn in the Soviet Union. Ruth gleaned corn in Boas's barley fields, and Joseph organized a granary system in Egypt to store the (wheat) corn for the seven lean years. Ceres, originally an Eutruscan and then Roman goddess of the corn (Figure 2.14), gave wheat to humankind. It was, therefore, quite natural for Europeans who first saw the maize plant to call it corn and, as Indian corn, it entered English (Figure 2.15). It would be more accurate for us to use the Indian name *maize,* but any attempt to change our habits would probably be futile.

The ear of corn, botanically a female flower cluster enclosed in leaflike husks, bears several hundred fruitlets, botanically kernels, on a rigid cob. The silks are the stigmas and styles through which the pollen tube grows to fertilize the individual ovules on the cob-receptacle. The male flowers, the tassel, are borne at the top of the plant. Corn is both self-fertile and cross-fertile, accepting pollen from itself or other corn plants. The pollen is wind borne, but is usually heavy enough simply to fall from the anthers down to the silks, and selfing occurs almost as frequently as does cross-pollination. The plant is botanically unique; no equivalent of the ear of corn exists elsewhere in the plant world.

Tremendous amounts of corn are grown throughout the world. Over a hundred million hectares are planted yearly with close to half this amount planted in North America. The Soviet Union, China, and South America account for most of the remainder of the crop. Worldwide, over a hundred 56-pound bushels are produced per hectare, al-

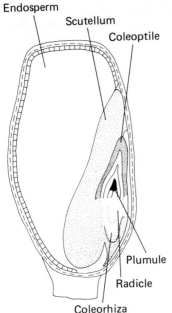

Endosperm
Scutellum
Coleoptile
Plumule
Radicle
Coleorhiza

Figure 2.16
Diagram of the maize kernel.
The endosperm consists of
starch-filled cells, the scutellum cells contain both starch,
protein, and some oil. The
coleoptile is a sheath protecting the first leaf, and the radicle is the embryonic root.

though in the corn-producing states, 250 bushels per hectare are not unusual.

Five major types of corn are grown. Two of these, flint and dent, are field types, grown primarily for animal food, for cornmeal, grits, and industrial use. In dent corn, the endosperm—the starch storing portion of the kernel—has a core of soft starch which shrinks on drying to give a dented or wrinkled appearance to the kernel (Figure 2.16). Flint corn endosperm lacks this soft starch, and the dry kernels are smooth and rounded. Sweet corn is bred for high moisture and sugar content. Flour corn has a soft endosperm starch which, upon drying, tends to be floury and is grown in Central and South America for tortillas and tamales. Popcorn is now a specialty crop whose use increased with the development of drive-in movies and TV.

The United States exports over a billion bushels of corn per year, only 20 percent of its production. That retained for domestic use has become so much a part of our everyday lives that we are usually unaware we are using a corn product. Most of our pork and beef are from corn-fed animals. We use corn oil directly in cooking and on salads, but consume even more as hydrogenated fats and margarines. Corn starch thickens gravies and even more is used as adhesives, sizings for cloth and paper, and as bulkers for other powders. A minor industry is centered in Washington, Missouri, where a specially bred strain of corn is fabricated into corncob pipes. Most beer (Chapter 7) is brewed with some corn and a considerable quantity is used to make bourbon and Canadian whiskey (Chapter 7). Stalks and leaves are converted to silage for winter feeding of cattle and are used industrially as a source of cellulose for rayon and other synthetic fibers. Corn extracted with water is used as the nutrient for the production of antibiotics and to grow those fungi and bacteria which produce the world's supply of citric, butyric, and other acids. Children in the Midwest grew up on pancakes and waffles slathered with corn syrup sometimes adulterated with synthetic maple flavoring or a trace of the "real thing." Corn sugar syrups are having a large negative impact on the cane and beet sugar industries. (Chapter 6).

ORIGIN OF CORN

Of the more than 300,000 named species of flowering plants, man has adapted for his use fewer than 400. With the sole exception of corn, the wild plant(s) from which our cultivars were derived is known. But corn, a basic food plant, is incapable of reproducing itself except under cultivation. If an ear falls to the ground, the kernels will germinate, but the seedlings will twist about themselves so inextricably that most will die. Failure to identify the wild parents of corn is a dilemma to botanists, and a considerable effort has been made to solve the riddle of its botanical origin. This aura of mystery has been a spur to continued research and is a source of considerable debate, for it is a rule of thumb in

Figure 2.17
Figure 2.17
Mature fruiting head of *Trip-sacum mexicana*, one of the presumed ancestors of corn.

Figure 2.18
Mature fruiting head of teosinte, one of the presumed ancestors of corn.

Figure 2.19
Mature fruiting head of guarany podcorn, one of the presumed ancestors of corn.

research that the less that is known about a topic, the more contentious are the debaters.

Research interest has centered on several wild plants. One, a wild grass called *tripsacum* (Figure 2.17), looks nothing like corn but under laboratory conditions can be mated with corn to yield a hybrid; thus, it has genes and chromosomes in common with corn. Another plant, teosinte (*Zea mexicana*), looks vegetatively like corn and has been placed in the same genus (Figure 2.18). It will hybridize with corn, but doesn't form the characteristic ear. Teosinte and tripsacum do not cross readily, and it is unlikely that they could have been the only parents. Another wild plant seems to be involved, possibly pod corn (Figure 2.19). When pod corn is crossed with popcorn, a plant vaguely like modern corn is produced. Geneticists are now trying to mate the pod X pop hybrid with tripsacum and teosinte to see if a true corn can be produced. This is not merely a sterile academic exercise. To improve yields, disease resistance, and adaptability to various cultural conditions, genes present in wild plants can provide hereditary potential humans can exploit.

INDIANS AND CORN

Whatever the parentage of maize, we know that it has been cultivated for a very long time (Figure 2.20). Pollen of corn has been discovered in archeological sites dated to 4000 B.C. Cobs the size of pencil erasers bearing a few kernels were recovered in Tehuacán, Mexico, and dated to 5000 B.C. Modern corn, virtually indistinguishable from that grown today, has been dated with certainty to 2000 B.C. Investigators

Figure 2.20
Peruvian maize gods. Taken from a vase dated from 500 A.D.

agree that modern corn originated in the highlands of central Mexico (Figure 2.21) and had spread south by 1500–1000 B.C. through Colombia, Peru, and Chile, as well as north into Arizona and New Mexico. By 100 B.C. corn was a staple crop of all agricultural peoples of the New World. Hunting tribes of the Great Plains traded buffalo skins for parched corn to sustain them in the winter.

Corn was so important for survival that it became a central focus of religions of Indians of the Americas—just as wheat is still part of the religions of the white man. The Aztec corn god, Centoetl, was anthropomorphized as a spirit with green and gold feathers. The plumed serpent, god of lightning and rain, was said by the Mayans to have mated with Mother Earth to give Ghanan, the corn plant, who sustained the people. Pueblo peoples believe that corn was dropped to their ancestors by a giant bird sent by a Great Spirit. Even the name maize has religious connotations. When, on 5 November 1492, sailors of Columbus's fleet went into the interior of Haiti, they saw fields of corn. They returned to the Santa Maria with "a sort of grain they [the natives] call mahiz which tasted well whether baked, dried and made into a flour." Mahiz means "our mother," and the crop was planted, tended, and harvested with reverence.

The cosmology and attitudes of the Hopi and other Pueblo peoples is, like that of the Aztecs, Mayans, Incas, Zapotecs, and other civilizations of the southwest and central Americas, replete with legends and ceremonies of the corn (Figure 2.22). Hopi children are told, "When

Figure 2.21
Glyph of a Mayan corn god sowing seed. Taken from a wall painting dated from 500 A.D. and copied from the Madrid Codex.

Figure 2.22
Hopi Indian in ceremonial dress for a corn festival. Taken from a nineteenth-century painting.

Figure 2.23
Pounding corn into meal, adopted by early New England colonists from the Indians.

you put the seeds in the ground, you give them to the Earth. She is their mother. She nourishes them so they grow strong and bring forth seeds of their own. Let us make an offering to the Earth Mother so She will feel pleased. Let us feed Her where we plant our corn so She will be strong and bring forth strong plants." Ceremonies accompany all stages of cultivation; prayers and dances for rain, dances to ward off disease, and paeans of thanksgiving are part of the rhythm of life of these corn-planting cultures. Women, closer to Earth Mother than mere males, guard the growing plants, urging Earth Mother to give birth to healthy plants. Intercourse was contraindicated during the time of pollination. A newborn child is irrevocably united with corn. The umbilical cord of Mayan babies was cut over an ear of corn, and the blood-soaked grains were planted to sustain the child until she or he was old enough to plant a crop. Thus, the child would, for the rest of his or her life, eat his or her blood-brother, the corn.

The Indians used corn in many ways. Young green corn was boiled and eaten as we eat sweet corn or was mixed with boiled beans to produce succotash. Mature corn, scraped from the cobs, was the staple for the winter season. Whole grains were boiled in water containing wood ashes that yielded lye to make hominy. Dried corn was ground in stone mortars or pounded in hollow trees (Figure 2.23) to produce a meal from which a variety of boiled, baked, or fried products could be made. The Incas first bred popcorn and it, too, spread throughout the Americas as a staple of the diet.

WHITE MAN'S CORN

The history of corn's exploitation by the white man can be followed as an exercise in the history of the United States, although here it is convenient to abbreviate and simplify it. When the Spanish conquered the Indians of Central and South America, they realized that corn was good food for their slaves and encouraged its cultivation. But they were greedy for the wealth of their colonies and allowed the elaborate irrigation systems of the Indians to fall into disrepair. At the time of Cortez, there were 25 million people in Mexico; by 1605 there were fewer than 1 million, and many died from starvation. English and French explorers of North America found corn cultivated by the settled tribes from Georgia to Nova Scotia. One ship's scribe wrote that "the grain is about the bigness of our ordinary English Pease . . . but is of divers colours, some white, some red and some blew. All of these yield a very sweet flavoure and, being used according to its kind, it maketh a very good bread." French voyageurs took avidly to cornmeal mush and cornbread, eventually using it to the exclusion of wheat flour which rises but tends to harden when wetted on portages.

Captain John Smith of the Virginia Colony realized that the self-sufficiency of the colony depended on raising a staple crop; he required all members of the company to grow corn. Since bread made with cornmeal does not rise, the colonists preferred to grow tobacco as a cash crop, selling it to England for wheat flour. A few years later and 700 miles north, the ill-equipped and mostly city-bred pilgrims landed at Plymouth on December 21, 1620. Their stores depleted, they were sustained only by gifts of corn and unearthed caches of corn. Even with this, half of them died, and by spring the survivors were desperate for food. Squanto, an Indian who spoke some English, taught them to plant several grains of corn in a hole together with a dead alewife fish to supply fertilizer. In the autumn of 1621, to praise their God for sustaining them and to thank the Indians, a feast was held which became the U.S. Thanksgiving holiday. A few years later, before 1650, several Salem goodwives were executed as witches after being accused of "hexing" the corn crop.

By 1660, the colonies were tied to a corn economy. Thomas Ash, Clerk of His Majesty's ship Richmond, wrote in 1682, "Their provision . . . is chiefly indian corn . . . of which they make a wholesome Bread and good biskit which gives a strong, sound and nourishing Diet. At Carolina, they have lately invented a way of making with it a good sound Beer; but it is strong and heady."

CORN AND AMERICAN HISTORY

Increased population, rising prices for corn as a result of European wars, and increased demand for meat, lard, and cheese, all made it apparent that additional agricultural land must be secured. The British colonies were hemmed in between the ocean and the first range of mountains to the west, and, except for the wilderness areas of northern New England (too cold) and the southern swamps (too wet), additional land just didn't exist. The reports of hunters and trappers about the wonderful land just over the mountains whetted appetites. This poorly explored territory was not English, but was owned by France. The land between the mountains and the "giant river of the west" had been explored by Joliet, Champlain, Cartier, Brulé, Nicolet, and others in the name of French kings.

Settlement to the west was also limited by the presence of only a few difficult trails from the east. The Mohawk Valley route to the Great Lakes teemed with hostile Indians. The Delaware Gap route through Pennsylvania and the Cumberland Gap trail from Maryland to the Ohio River were safer, but well patrolled by the French. Some Virginians were beginning to organize armed parties to invade the Kaintuck country and the Virginia House of Burgesses commissioned Colonel George Washington in 1753 to warn the French not to attempt to bring French-Canadians into the Ohio Valley. The French commander told Washington that this was French territory and that they would fight, if necessary, to retain the Ohio River and would incite their Indian allies to kill any English settlers who attempted to cross the mountain passes. Provoked by this—and other mutual insults—the French and Indian War was on. British victory allowed England to take possession not only of the Ohio Valley, but of all lands east of the Mississippi except for New Orleans. With supply lines stretched, and for other reasons, France ceded Louisiana to Spain.

The end of the Indian troubles unleashed a flood of immigrants. Indentured servants, having almost given up hope of finding land to secure their promised advantages, were among the first to leave the settled east. Kentucky and Tennessee were settled in 1773–1775. Virginia claimed this land and offered each settler 400 acres at $0.25 per acre on the condition that a house be built and a corn crop planted within a year. By 1789, there were 20,000 people in western Virginia and corn covered the land. It was here that bourbon whiskey was created.

Land north of the Ohio River was opened up by the military exploits of George Rogers Clark and his rangers who crossed the Ohio River and attacked the French fort at Kaskaskia, now in Illinois, on July 4, 1778. The following year, Clark took Vincennes, now in Indiana, and the Ohio Valley was open to settlement. People poured in from Virginia, Maryland, and southern New England. The Continental Congress offered land at $6 per acre—payable in corn—and by 1800 there were 45,000 settlers in Ohio and fewer, but growing numbers, in Illinois and Indiana. The topsoil of this land was amazingly fertile. It was a deep rich brown, extending down five or more feet, soil such as the settlers had never seen. Few trees had to be felled, drainage was excellent, and the climate was ideal for raising corn: 30, 40, even 50 bushels per acre was common. Pigs, cattle, horses, and chickens fattened at rates that amazed the farmers. Settlers wrote that they had discovered the biblical land of milk and honey.

Heaven, was, however, flawed. The abundant harvest could not readily be moved east to market. Mountain passes were inadequate for any conveyance bigger than a horse carrying kegs of whiskey or sacks of cornmeal. One of the first acts of the United States government was to sign an agreement with Spain to open the port of New Orleans to corn products and preserved meats shipped down the Ohio and Illinois rivers and on to New Orleans via the Mississippi. The treaty provided that the shipper could ". . . export them hence without paying any other duty than a fair price for the hire of the stores." When France again took possession of Louisiana and proved less amiable than did the Spanish, it was obvious that this port had to be secured. Jefferson's Louisiana Purchase of 1803 was a notable act of intelligence and was, as contemporary records show, occasioned largely by demands of corn farmers.

Corn was moved to New Orleans by flatboats floating down the river. Flatboatmen were a breed apart, and legends of their brawling, drinking, and wenching caused Ohio farm boys hoeing corn to dream of escape as rivermen. Since it was impossible to move the rafts back up to Ohio, they were sold for scrap lumber and the proceeds spent on "a good time" before the boatmen started the 1500-mile walk back home. The era of the flatboatmen was short lived; there was sufficient goods shipped to New Orleans to warrant introduction of Fulton's steamboat to the Mississippi by 1811. This ushered in a new era of American romanticism exemplified in the writings of Mark Twain. New Orleans was still sin city, its customers being the paddleboat crews who were more affluent and hence more sinful than flatboatmen. Bawdyhouse music became, by 1840, jazz, and it moved up the river to Kansas City.

Even the Mississippi wasn't fast enough to handle the freight from the corn fields. As early as 1802, Ohio insisted that as part of the negotiations leading to statehood, Congress had to agree to use money collected from the sale of public lands to build roads from the Ohio Valley to eastern markets. It was, the farmers asserted, unfair and expensive to have to ship to New Orleans and then reship all the way to the east coast. Since road construction through the mountains was expensive, a system of canals was started. The Hudson-Mohawk canal was built in 1817, the Erie canal by 1825, and the Ohio canal by 1830. Roads were built soon thereafter and so were the railroads. The Albany-Buffalo lines, the Philadelphia-Cincinatti lines, and the Baltimore and Ohio lines completed a road-water-rail system that permanently linked the Midwest to the East. People moved in, land prices soared, and the good, rich soil supported the growth of more and more corn.

Chicago, the "place of the wild onion," was chartered as a town in 1833. Precariously situated in the mud on the southwestern shore of Lake Michigan, it served as a trading post and fort. It was, however, a natural center for railroad builders whose lines of steel radiated out to the south and moved west. With generous subsidies and grants of land from the federal government, the rapid building program was dictated by Chicago's location. It was almost a fringe benefit that these lines traversed the best corn-growing country in Ohio, Indiana, Illinois, and Iowa and also extended to Kansas City, Dodge City, and other towns where cattle drives terminated. A few enterprising families—Swift,

Figure 2.24
Cattle feedlot in Texas with a capacity of 70,000 head. Courtesy U.S. Department of Agriculture.

Wilson, Armour, Cudahy—built cattle pens to accept range cattle and farm hogs, fed them on corn, butchered them, and shipped the meat east to the established centers of population. By 1850, the slaughter houses advertized themselves whenever the wind blew, and Chicago isn't called the "windy city" without good reason. Destroyed in the big fire of 1871, the Chicago Stockyard was rebuilt on the south side of the city where, in Carl Sandburg's words, it was "hog butcher to the world," and in the boast of one of the meatpackers, it "used everything but the squeal." Leather, glue, hog bristles for shaving brushes, fertilizer, soap, and myriad other products spawned by the corn-cattle-hog-interaction were produced in the city. As the natural center for the corn trade, a corn-products industry followed the stockyards to make Chicago wealthy albeit smelly. The Chicago Board of Trade recognized the city's debt to corn by placing a gilded statute of Ceres, goddess of the corn, on the top of its building where for a time she dominated the city. Cattle feed lots were built in many states (Figure 2.24).

PELLEGRA

Although Europeans recognized the superior virtues of corn for animal feeding, they displayed relatively little interest in trying to grow it themselves. By 1520–1525, Spanish explorers brought corn back home where it was grown in limited amounts in Andalusia. People didn't like its smell and, since Christ's bread was wheat, corn was fed to the animals. Venice bought corn from Spain for sale to North Africa and established it as a crop on Crete. Corn was being grown as cattle feed in the Danube Valley by 1630, and Portuguese Jesuits introduced it to China by 1650. Portuguese traders brought it to west Africa by 1680 as food for slaves. In northern Italy growing conditions were adequate for corn and the use of *polenta*—cornmeal—as a dietary staple developed around Milano. Only after World War II did corn become important in Europe as a major cattle and hog food. With the development of strains well adapted to European conditions, corn is now an important crop in France and the Soviet Union; China now grows over a billion bushels per year. In northern Italy and in a few isolated parts of Spain, cornmeal became the exclusive cereal food of the desperately poor. In 1730, Dr. Casel noticed that those on a corn diet became weakened after a time, and Italian physicians named the condition *pellegra*—the disease of the rough skin. Although the cause was unknown, it was obvious that corn was involved, and advising people without other food not to eat corn was useless. Pellegra was characterized not only by a roughened skin. There were disturbances in digestion, diarrhea, sore mouths, nervousness bordering on paranoia, and a characteristic butterfly-shaped rash over the nose and cheeks.

Corn cultivation almost died out in Europe by 1800, but its use as the primary food among the black slaves of the American South was on the increase. After the American Civil War, the numbers of poor, both black and white, increased in the South, and the diet of the poor was fatback, cornbread, and molasses. In 1914, alarmed (finally!) by the debilitation caused by pellegra and the increasing number of deaths, the U.S. Public Health Service placed Dr. Joseph Goldberger in charge of pellegra research. Goldberger found that pellegra was caused by the absence of something from corn that could be replaced by meat, milk, and fresh vegetables. Dried yeast also contained the antipellegra factor and was distributed to the poor. Only during the 1930s when vitamins were discovered was it determined that the antipellegra factor was one of the water-soluble B vitamins, the one called niacin.

HYBRID CORN

From the time of the Aztecs, ears of corn were saved from the previous harvest and the kernels carefully put aside for the next year's crop. Over the centuries, kernels were selected for plumpness, desirable taste, degree of starchiness, and other characteristics. By 1850, farmers vied with one another at state fairs to exhibit large, full ears with even rows of grains and slim cobs. By 1870 there was, throughout the corn country, general consensus as to what constituted good corn. Jacob Leaming's 'Lancaster Sure-crop,' James Reid's 'Illinois Yellow Dent,' and Krug's 'Famous' were preferred. Yet farmers could never be sure that all their corn would be true to type, and while yields were satisfactory, they had plateaued by 1890–1910 in spite of the introduction of better fertilizers and machinery. Agricultural experts agreed that little or no improvement in corn was possible. Yet as early as 1716, Cotton Mather of the Plymouth Colony published observations on natural crosses of corn. James Logan, governor of Pennsylvania, experimentally crossed corn types in 1735, and a few curious gentlemen in North America and Europe were defying their ministers and obscenely dusting pollen on plants to see what kind of offspring would result (Chapter 1). Charles Darwin, one of the most inquisitive of men, studied the crossing of many plants including corn and concluded that the offspring showed increased vigor, due, he reported, not to the mere act of crossing, but to interaction of whatever might be the physical basis of heredity.

William Beal, a student of the famous Harvard botanist Asa Gray, moved to Michigan's corn country and decided to study corn genetics. In 1877 he planted alternate rows of flint and dent corn, detasseled the dent so it could not self-pollinate, and found that its progeny formed some large kernels. He reported that yields of his "mule corn" were greater than those of either parent. His experiments were confirmed, but none could explain the results. In 1900 the genetic work of Gregor Mendel was rediscovered after 30 years of burial in an obscure Austrian periodical, and the impact of this new science, named genetics, permitted reorientation of plant breeding. Mendel's work provided the basis for understanding the increased vigor associated with crosses. The name mule corn took on new meaning; the progeny of crosses between corn types was, like the mule, more vigorous because it had—using the new terminology—genetic potential of two parents that complemented and reinforced each other. This acquisition of complementary characters was termed heterosis, now called hybrid vigor.

Armed with these new ideas, George Harrison Shull continued Beal's experiments in 1908. He, too, was able to increase corn yields by 50 percent. He found that herosis was most noticeable when the parent strains of corn were pure or inbred, having been deliberately self-fertilized for several generations. Donald F. Jones and Edward East found in 1915 that when they crossed the hybrid progeny of pure-line parents, the so-called double cross, their progeny was even more vigorous. When research started again after World War I, it was found that inbred, pure lines of the very different northern flint and southern dent types were, when crossed, capable of yielding corn that was 100 percent more bountiful than either parent (Figure 2.25). These hybrids did not look like the ideal corn of state fairs, and the idea of growing mule corn didn't appeal to farmers. Henry A. Wallace, publisher of *Wallace's Magazine* in Iowa, secretary of commerce in 1924 and later secretary of agriculture and vice president, recognized the potential of this new-fangled corn and publicized it in his magazine and in lectures to farm groups.

The first hybrids came onto the market in 1924, and the struggling companies that bred them almost went bankrupt. Ears were slender and kernels were very hard compared with the ideal. Yet, yields were bountiful and steers, hogs, and chickens ate it avidly. The young hybrid corn companies decided to promote their products actively. For several years they gave seed corn to farmers to plant next to the ones they had

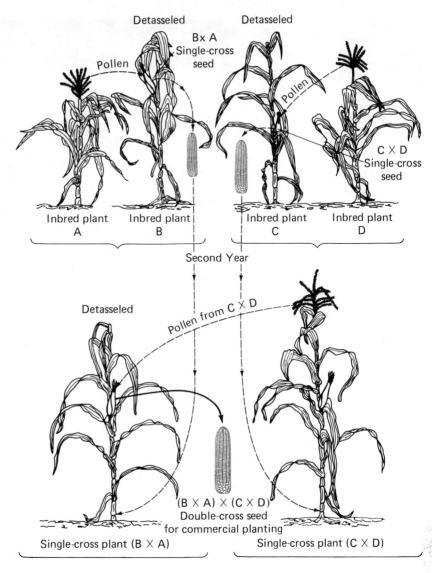

Figure 2.25
Cross-breeding for production of hybrid corn. Courtesy U.S. Department of Agriculture.

Detasseled

B × A
Single-cross
seed

Pollen

Detasseled

Pollen

C × D
Single-cross
seed

Inbred plant
A

Inbred plant
B

Inbred plant
C

Inbred plant
D

Second Year

Detasseled

Pollen from C × D

(B × A) × (C × D)
Double-cross seed
for commercial planting

Single-cross plant (B × A)

Single-cross plant (C × D)

grown for years. Free signs were given to the farmers to indicate which hybrid they were using, and on Sunday after church, the farmers would take their families for a spin in the tin lizzy to see how each corn crop was doing (Figure 2.26). At harvest time, seed companies had comparison viewings complete with coffee and doughnuts. In 1939–1945 the world went insane, and the U.S. Department of Agriculture coined the slogan "Food will win the war and write the peace." Corn farmers took it very seriously. Even with their sons away and farm labor impossible to hire, yields of hybrid corn were so outstanding that corn production tripled and, indeed, corn did feed a devastated world.

Figure 2.26
Early roadside billboard advertising hybrid corn. Courtesy DeKalb Hybrid Corn Co.

HIGH-LYSINE CORN

With the assistance of state agricultural experiment stations and the botanists and geneticists educated at the "aggie schools," hybrid corn seed companies entered the postwar market with hybrids tailored for different uses. Yields of field corn increased from the prewar average of 60 bushels per acre to over 200 bushels per acre. As yields climbed, it became increasingly difficult to use all that could be grown. Acreage restrictions had only temporary effects as farmers used better hybrids, more machinery, and increased fertilizer to grow even more corn on fewer acres. Although corn was now used for beef production, changes in America threatened to ruin the hog market. Pork never attained the popularily of beef. One major product of the hog was lard, used for shortenings, soaps, and glycerin. The more easily obtained vegetable oils from cottonseed, soybeans, peanuts, and corn could be converted to hydrogenated fats, soaps, and other products that cost less to produce, stored better, and lacked the objectionable odors of lard. By the late 1940s, American corn grain elevators were bursting at their seams, and the bottom dropped out of the corn market; it sometimes cost more to grow it than it brought on the open market. If the economy of the Midwest was to remain viable, additional uses had to be found for corn and some method developed to change the hog from a fat pig to an animal with more ham, bacon, and pork chops.

Professor Edwin T. Mertz of Purdue University received a research grant to study this problem. Mertz, a biochemist, decided to look at corn proteins. He confirmed early reports that the concentrations of several of the amino acid building blocks for protein synthesis were low. These amino acids, particularly lysine and tryptophan, were essential amino acids, ones that could not be synthesized in the bodies of hogs. Mertz also knew that protein synthesis is an outstanding example of the law of the limiting factor, propounded in the nineteenth century by Justus von Liebig, one of the pioneers in agricultural chemistry. This fundamental law in all of biology can best be understood by analogy. Suppose that one wishes to bake cakes—chocolate for example:

1. Melt 4 oz dark chocolate with ½ cup milk and ¼ cup sugar in a double boiler over hot water until thick and smooth. Cool and reserve.
2. Sift 2 cups flour with 1¼ teaspoons soda and 1 teaspoon salt.
3. Cream 4 tablespoons butter and 4 tablespoons vegetable shortening with 1¼ cups sugar. Blend in 3 eggs, one at a time, Beat for 1 minute. Add the cooled chocolate mixture.
4. Measure out 1 cup of milk and add small volumes of the milk alternating with small volumes of the dry flour mixture to the creamed shortening-egg mixture, beginning and ending with the flour.
5. Blend thoroughly after each addition using the low speed on an electric beater.
6. Pour into two well-greased and lightly floured 9-inch round layer pans and bake in a 350 degrees F. oven for 30–35 minutes.
7. Cool, remove from pans. Spread raspberry jam between the layers. Frost with dark chocolate frosting.

Now, if one has all the flour, sugar, milk, chocolate, butter, and shortening needed for five cakes and has only three eggs, only one cake can be baked. So it is for protein synthesis. If one amino acid is in low dietary concentration, it is the limiting factor for protein synthesis. Mertz concluded that corn protein could not be the sole supply of amino acids to make a hog meatier unless the lysine and tryptophan contents were raised. If this could be done, the accumulated corn surpluses would be used up, and the economy of the farms in the Midwest could be enhanced.

Together with Drs. Ricardo Bressani and Oliver Nelson, Mertz embarked on the long and difficult task of genetic engineering. They started cross-breeding experiments with corn types called "floury" and "opaque." given these names because of endosperm characteristics. When genes of opaque were introduced into standard varieties, lysine content of corn protein increased by 70 percent. They tested the new corn on rats since, many years before, Thomas Osborne had found that rats fed only corn were malnourished and that symptoms of inadequate protein formation were corrected by additions of lysine and tryptophan. The new corns allowed rats to develop normally. The complete research report was published in 1964, and small amounts of high-ly-

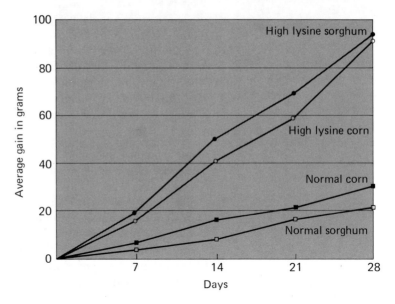

Figure 2.27
Growth rates of pigs fed normal or high-lysine corn or sorghum. Taken from data obtained by the College of Agriculture, Purdue University.

sine seed corn began to be grown for pigs (Figure 2.27). It was a few years before a medical researcher got the idea that if the new corn was good for hogs, it might be good for people. The Indians of Colombia were outstanding examples of people whose nutrition was corn based and whose children showed symptoms of the protein-deficiency disease kwashiorkor. In 1967–1968, Dr. Alberto G. Pradilla found that 6-year-old children whose development was that of 2-year-olds could recover rapidly on a high lysine corn diet; stunted bodies grew, and dull minds sharpened.

People are very much creatures of habit, and few of us willingly change our eating habits. The first types of the new corn were soft, almost floury. They didn't mill well to make the meal used by the Colombians. The ears were unduly susceptible to rots, and because each kernel was small, yields per acre were reduced. A corn geneticist on the staff of the Rockefeller Foundation's agriculture unit in Colombia transferred the high-lysine genes into cultivars acceptable to tropical peoples. Disease resistance, resistance to the stresses imposed upon plants growing in the tropics, and a host of other factors are being considered in the continuing genetic research work designed to improve corn for people and domestic animals.

ADDITIONAL READINGS

Bauman, L. F., et al. (eds.). *High Quality Protein Maize.* Stroudsburg, Pa.: Dowden, Hutchinson & Ross, 1975.

Billard, J. B. (ed.). *The World of the North American Indian.* Washington, D.C.: National Geographic Society, 1974.

Flannery, D. V. "The Origins of Agriculture." *Annual Reviews of Anthropology* 2(1973):271–310.

Galinet, W. C. "The Evolution of Corn and Culture in North America." *Economic Botany* 19(1965):350–357.

Giles, D. *Singing Valleys. The Story of Corn.* New York: Random House, 1940.

Jugenheimer, R. W. *Corn: Improvement, Seed Production and Uses.* New York: Wiley, 1976.

Mangelsdorf, P. C. *Corn: Its Evolution and Improvement.* Cambridge, Mass.: Harvard University Press, 1974

Mertz, E. T., and O. E. Nelson (eds.). *Proceedings of the High Lysine Corn Conference.* Washington, D.C.: Corn Refiners Association, 1966.

Nelson, O.E. "Genetic Modification of Protein Quality in Plants." *Advances in Agronomy* 21(1969):171–194.

Schneour, E. A. *The Malnourished Mind.* Garden City, N.Y.: Doubleday (Anchor Books), 1975.

Scully, V. *A Treasury of American Indian Herbs, Foods, Drugs and Medicine.* New York: Crown, 1970.

Wallace, H. A., and W. L. Brown. *Corn and Its Early Fathers.* Michigan State University Press, 1956.

Waters, F. *Book of the Hopi.* New York: Viking Press, 1963.

Weatherwax, P. *Indian Corn in North America.* New York: Macmillan, 1974.

Bread

For the Lord, your God, is bringing you
into a good land, a land of brooks of water,
of fountains and springs flowing forth in
valleys and hills, a land of wheat and
barley . . . a land in which you will eat
bread without scarcity.
DEUT. 8:7–9

The background that allows you to pick up a loaf of bread for your morning toast or luncheon sandwich is complex. The production of sheaves of golden wheat and the winning of the brown kernels from the heads is part of the story of agriculture. Conversion of grain into flour is part of the story of agribusiness, and the conversion of flour into bread is essentially the story of civilization.

Wheat, barley, and millet have been cultivated for our bread for thousands of years, but except for dynastic Egypt, these cereal grasses were grown by the people who ate it; grain was not a cash crop. The rulers of Egypt made grain a state monopoly with all grain becoming the property of the pharaohs who then doled it out to the people. The story in Genesis 37–45 of Joseph and the lean and fat years details a situation unique in the ancient world. In modern terms, Joseph was commissioner of the royal grain board. Several thousands of years later Rome instituted a similar system. Rome's wealth was based on wheat trade, and her military might was deployed to obtain wheat and to secure wheat trade routes. The conquest of Egypt was initiated to obtain the bountiful wheat of the Nile Valley and North Africa. Carthage was

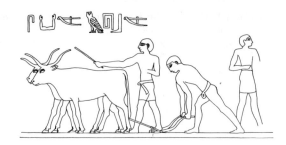

Figure 2.28
The scratch plow used throughout the Middle East and in North Africa. Taken from an Egyptian tomb painting.

invaded in the second century B.C. to obtain the wheat of the delta, and Roman legions divided Gaul and subdued the barbaric Britons for their grain. The Middle East was of less importance—these people grew millet and barley. Palestine was secured to serve as a barrier to the movement of the Phoenicians and Persians whose armies and naval forces threatened Roman hegemony in the Mediterranean. The Roman fleet needed protection while carrying grain from North Africa to Ostia, the seaport for Rome. By the first century, Rome was importing 14 million bushels per year from North Africa alone, where she held all land north of the Sahara. And every grain was needed to feed the burgeoning population. Julius Caesar instituted the practice of giving free wheat to the people of the city. The "bread and circuses" reported by Juvenal were the spoils of war. In the time of Augustus (ca. 27 B.C.), roughly a half a kilogram of grain per person per day was being given to 320,000 people, one-third of Rome's population. A few years later, bread itself was given away, because the populace complained about having to mill and bake its own bread.

TILLING THE SOIL

The agriculture methods used to grow grain were remarkably primitive. Land was plowed by the Sumarian scratch plow (Figure 2.28), pulled by a man—or his wife—and this plow remained virtually unchanged for 3000 years, except that oxen were substituted for human muscle (Figure 2.29). Grain was sown by being broadcast to the bare fields where birds ate a fair share of it and where dry weather or a sudden rainstorm would cause massive losses. Matured stalks were cut with stone, bronze, or iron sickles, were bound into sheaves, cleaned, and the seed collected. A family could scarcely find the time and energy to grow a crop sufficiently large to feed itself, much less produce for others, unless they were assisted by slaves or were forced to surrender what was literally the food from their own mouths.

The scratch plow was inefficient enough in the dry soils of the Middle East and North Africa, but it was virtually useless in the wet, clay soils of much of Europe. Sometime in the fifth century A.D., people in eastern Europe invented the simple plow which had a wooden point that cut into the ground and a share that cut the clods horizontally (Fig-

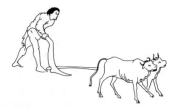

Figure 2.29
The scratch plow used in Europe throughout the dark ages was essentially the same as the Egyptian plow.

Figure 2.30
The scratch plow was modified during the dark ages by the addition of a wooden moldboard.

Figure 2.31
During the late Middle Ages, the plow was modified by the iron plowshare attached to a wooden or sheet metal moldboard. Taken from a fourteenth-century woodcut.

ure 2.30). A strong brace of oxen was needed to pull this plow through heavy soil. The moldboard, a curved side that turned over the earth, was developed allowing the soil to be cut and turned before hand sowing (Figure 2.31). Yields could easily be doubled by this new method. Oxen gave way to the horse when the crusaders found that the Arabian horseshoe provided traction for the animal pulling a plow through muddy fields. The Chinese harness system that put the strain on the animal's chest and legs instead of its neck didn't reach Europe until the thirteen or fourteenth century. Finally, the steel plow was invented in the nineteenth century (Figure 2.32), to be replaced in the twentieth century by complex machinery (Figure 2.33).

Figure 2.32
John Deere's plow of 1838 had an integrated share and moldboard of steel and was capable of breaking the tough sod of the North American prairie. Courtesy Deere & Co.

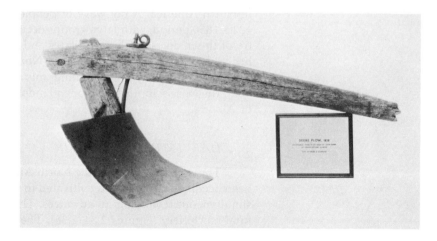

Figure 2.33
Modern disk harrow. Courtesy International Harvester Co.

From the twelfth to the eighteenth century, there were essentially no improvements in agriculture.

The feudal system in Europe did nothing for peasants except to brutalize them, their families and those that ate the wheat and barley they wrested from the soil. One can get an inkling of this in Edwin Markham's poem, "The Man with a Hoe." The Renaissance was profitable to the nobility and to the developing merchant class; peasants still toiled for other people's benefit. The invention of the scythe in the middle of the eighteenth century was the first improvement in agricultural methodology, but all this did for the peasant was to require him or her to work faster and, since the scythe cut much more grain per unit time, many peasants were dismissed. Some wandered the roads of Europe with their families. Except for the black and Indian slaves on the plantations of the southern colonies, the situation of the people in North America was much better; the concept of a peasant class never existed in North America. The freemen of most of the United States and Canada took this name proudly, for their toil redounded to their own advantage. Word of these lands of opportunity spread through Europe, and by the beginning of the eighteenth century, the first of many waves of immigrants from northern and western Europe began to arrive in North America. A big wave of people broke on these shores in the 1820–1850 period, with large groups occupying the corn-growing country of the midwestern United States and the agricultural lands of Quebec and Ontario. The hegemony of North American wheat, however, was based primarily on the immigrants who occupied the prairies beyond the Mississippi and the open lands of Manitoba to the mountains of Alberta.

HARVESTING THE GRAIN

The immense areas of the North American prairie could not have been occupied, and its land cultivated to wheat, had it not been for two simultaneous technological advances. The first of these was in agricultural machinery (Figures 2.34–2.36). The scythe was improved in 1810

Figure 2.34
Harvesting grain in ancient China. A sickle blade was attached to a stick, easier to swing than the short-handled sickle. Taken from an impression on a clay roof tile made during the Han dynasty, ca. 200 B.C.

Figure 2.35
Harvesting grain in dynastic Egypt. Grain cut with the hand sickle was threshed by having oxen tread on the sheaves. Taken from an Egyptian tomb painting.

Figure 2.36
As late as the fourteenth century, grain was harvested with the sickle. Taken from a woodcut made in Germany in 1340.

by English Quakers who added the cradle which laid cut stalks neatly in a pile, easy to gather and bind. In 1824, Patrick Bell of Scotland invented a mower-reaper which was promptly smashed by itinerate reapers who feared for their livelihoods. Nevertheless, the idea of a mechanical reaper was born, and in 1830 Obed Hussey invented a reaper with a cutting blade consisting of teeth that moved back and forth and with mechanical fingers that guided the stalks to the cutter bar. In 1834, Cyrus McCormick invented his reaper (Figure 2.37), and

Figure 2.37
A nineteenth century reaper. Courtesy International Harvester Co.

Figure 2.38
Wheat harvested with a combine. Courtesy U.S. Department of Agriculture.

almost immediately sued Hussey for infringement of patent rights. McCormick won the suit against Hussey's lawyers—Abraham Lincoln and his eventual Secretary of War, Edwin M. Stanton. McCormick's reaper equalled 60 men with scythes and, just as importantly, it was economical to use on the large fields that existed in the west. Massey-Ketcham reapers were on the Canadian market by 1845.

Farm machinery accumulated at an ever-increasing rate. Charles Newbold of New Jersey forged a cast-iron plow and moldboard, and John Lord of Chicago invented a steel plow share. John Deere, originally from Vermont, but later in Moline, Illinois, put these two inventions together in 1831 to make the steel moldboard plow, the first capable of breaking the thick, resistant prairie sod (Figure 2.32). March's harvester came along in 1858, a threshing machine in 1876, a binder in 1878, and a combined reaper-harvester and binder in 1881. There were efficient seed drills by 1885, and the first complete combines incorporating reaper, binder, thresher, and bagger of the grain was available in 1890. Tractors capable of pulling these combines came into use in 1910 (Figure 2.38). The technological portion of the wheat revolution was complete.

OPENING THE PRAIRIE

A second contributing factor to wheat cultivation was the availability of land. In Canada, John A. MacDonald dreamed a fine dream of linking the Dominion from Atlantic to Pacific with a railroad—the Canadian Pacific. In order to do so, he had to force the Hudson Bay Company to relinquish its grip on the prairies and to have the land surveyed, which the French-Indians opposed since it could only lead to the taking of their lands. In 1869, these actions, and other accumulated grievances, led to the Real Rebellion. In the United States, the securing of the prairies was an act of "manifest destiny," and rights of American Indians were taken away with scarcely a backward glance. Deliberate slaughter of the buffalo and murder or deportation of the hunting tribes themselves cleared the land. All that was needed was the iron horse. In both countries, government and robber barons connived to subsidize the lines with grants of land and money. Immigrant Irish refugees from the potato famines of 1846–1848 (Chapter 6) were hired as railroad building crews to build from east to west. Imported Chinese "coolies" built from west to east. In 1840, there were only 2500 miles of track in the United States. This increased to 31,000 miles in 1851 and to 160,000 miles in 1871. The Canadian Pacific reached Winnipeg by 1878 and British Columbia a short time later. Spurs from the main lines connected to the grain elevators by 1880, and villages arose at their bases to provide services for the farmers. Villages became towns, and some towns became cities: St. Paul, Edmonton, Winnipeg, and Grand Forks.

As grain supplies came east, more was available than could be eaten in North America, and both Canada and the United States entered the export market, competing with Hungary, the Ukraine, and Scandinavia. Bankrupt Swedish, German, and Ukranian farmers came to the prairies by the thousands and were joined by other nationalities. Horace Greeley of New York was advising, "Go west, young man, go west!" and thousands heeded him. Canadian land salesmen were advertising: "Buy land in the Canadian West. You can leave home after Easter, sow your seed and take in a harvest and come home with your pockets full of money in time for Thanksgiving." By 1900, the international economies of both the United States and Canada were tied irrevocably to wheat. This wheat was produced by the collaboratory of peoples who desired self-esteem so badly that they didn't know when they were licked, the inventiveness of blacksmiths and wheelwrights, the greed of financiers, and one helluva lot of sweat and luck.

MILLING

The grain or kernel of wheat is, botanically, a one-seeded fruit enclosed in husks (the chaff) with the seed coats (the bran) firmly bound to the seed itself (Figure 2.39). The seed consists of the embryo (the germ) which is the young wheat plant, and the starchy endosperm which is the food supply for the growth of the wheat seedling. White flour is the ground-up endosperm. The bran is indigestible by man, and the germ with its vitamins, oil, and minerals makes flour heavy and reduces its ability to rise. To make fine white flour, both bran and germ must be removed and the endosperm cells pulverized. This conversion

Figure 2.39
Anatomy of a grain of wheat.

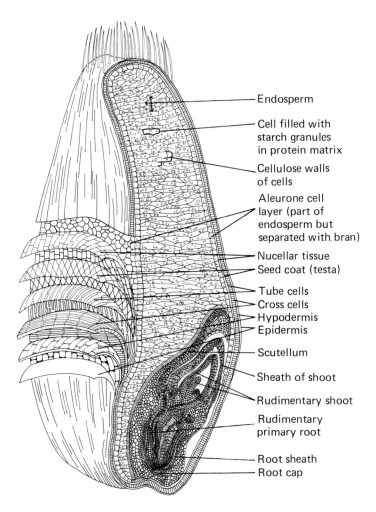

- Endosperm
- Cell filled with starch granules in protein matrix
- Cellulose walls of cells
- Aleurone cell layer (part of endosperm but separated with bran)
- Nucellar tissue
- Seed coat (testa)
- Tube cells
- Cross cells
- Hypodermis
- Epidermis
- Scutellum
- Sheath of shoot
- Rudimentary shoot
- Rudimentary primary root
- Root sheath
- Root cap

of wheat into flour has occupied man's attention for well over 4000 years.

Man learned that the kernels could be freed of their husk by passing a head of wheat quickly through a flame, the process of parching. For centuries the whole grain was eaten, sometimes raw and sometimes softened by boiling or merely soaking for several days. If grain was pounded with a rock, a coarse mixture of whole and broken kernels was obtained that would cook more quickly and was more digestible. Several thousand years ago, man learned that if whole grains were rubbed on a stone, at least some of the indigestible bran was removed, giving a better product. This is the *bulgar* of the middle east or the groats of Europe. When roasted and then mixed with water, oil, and sometimes honey, a nutritious and tasty gruel or porridge could be

Figure 2.40
The saddle stone used to grind grain into coarse flour. Taken from an Egyptian tomb painting.

Figure 2.41
The mortar and pestle are still used in primitive societies.

Figure 2.42
The handmill, a small, double-stoned grinder.

made. This was the famous Roman *pulz*. It could be eaten hot or cold, or placed on hot stones or the ashes of a fire to produce a product that had increased storage life. The first flat bread must have been hard to chew and even harder to digest. Flatbread has been excavated in homes of the Swiss Lake Dwellers of 6000 B.C. Alternatively, groats plus dried fruit could be soaked in water or milk for several days until the mixture fermented and softened forming the *frumenty* that was eaten in European farm homes until the nineteenth century.

Coarse milling of grain was, apparently, independently invented by all of the cultures that ate any of the edible grasses. The Mayans and Aztecs of the New World who ground maize into meal, the oriental who ground millet for noodles, and the peoples of the Middle East and the Mediterranean basin who grew wheat and barley, each developed a saddle-shaped grinding stone (Figure 2.40), on which grain was placed to be broken and powdered by a round stone. The resulting flour was coarse and full of stone chips. It contained all of the bran and most of the germ, but it was finer than groats and could be mixed with water to a paste that could be baked into a flat cake. The mortar was, again independently, used by several cultures (Figure 2.41) around 800 B.C., when it was discovered that flour would be finer and the tedious grinding reduced if one placed the grain between flat stones and turned the upper stone (Figure 2.42). Small millstones were so vital a part of each household that Mosaic Law stated that "No man shall take the nether or the upper millstone to pledge, for he taketh a man's life to pledge." A small rocker mill (Figure 2.43) was an obvious improvement that allowed one person to grind grain. The Greeks invented milling machines in which the upper stone, instead of being flat, had carved grooves which prevented clogging. Larger mills were built to be

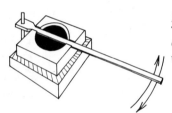

Figure 2.43
The rocker mill, invented in Greece, was used in Europe until the late Middle Ages.

Figure 2.44
A large Roman mill turned
by hand or with horsepower.

operated by slaves (Figure 2.44), and a bit later, by domestic animals.
About 200 B.C., those marvelous Roman engineers discovered a way to
harness the power of falling water to turn their mills and, with subsequent modification, the water mill was used for over 2000 years (Figure
2.45). Countries with limited water power relied on the ox or the horse
until the twelfth century when someone in Holland realized that the
windmills used to raise water could also turn millstones (Figure 2.46).

Still, the flour was dark with bran flakes. People knew that bran
was not nutritious, was a laxative, and spoiled flour. Medieval Britons
said, "Bread having moche branne fylleth the bealye with winde and
excrement," and everyone wanted bran removed from their bread.

Figure 2.45
American grist and flour mill
powered by water. Such mills
were common until the end
of the nineteenth century
when they were replaced by
steam-powered mills and
eventually by roller mills
powered by electricity.

Figure 2.46
The windmill as used in
Holland.

Coarsely woven cloth was used to bolt or sieve the flour. This removed most of the germ and the larger particles of bran. As cloth technology developed, bolting fabrics became finer, and the bread became whiter and, correspondingly, more expensive; the wealthy ate bread that was, at best, a light brown in color. The poor ate coarse bread similar to our whole wheat—and they resented it.

Even the mere existence of a mill was a source of trouble. The miller was a man who could compel the soul of the free stream to be his slave. As late as 1671, Estonian peasants burned down a mill because it offended the brook who told the river who complained to the sea—and the sea refused to release water causing a drought. Windmills weren't much safer. In Germany, many mills were outlawed by the nobility who feared that the wind, angered by being used, would wreak its revenge by sending hail to flatten the crops, or would become so enraged that it would become a hurricane. The Teutons who worshipped Odin as wind god, and the people who prayed to Jupiter or Zeus as god of the elements, asserted that they could hear their god protesting in the thunder of the turning stones. The occasional dust explosions in flour mills were, obviously, signs of this anger. Superstition was reinforced by economic considerations. Roman law, adopted throughout Europe, provided that *"Cuius terra, Eius molina"*—who owns the land, owns the mill. The feudal lords and the clergy owned both the land and the mill and could compel the peasants to have their wheat ground into flour at their mills. The miller was a servant of the noble and was paid in flour. To survive, millers routinely short-weighted the peasants. In Chaucer's "Reeves Tale," it was said of the miller, "A theef he was for sothe of corn and mele," and it was often asserted that "the miller keeps a pile of sand behind his door." The peasant who broke a tooth on a chip of stone was kept from beating the miller because of a fear that the miller had a compact with the devil. When, in central and eastern Europe, rye flour was made and ergotism spread throughout the community (Chapter 5), it was the miller who brought this curse upon the people.

The desire for white bread was almost overwhelming except in rye-eating lands of northern Europe and slavic countries; and even here, when wheat bread was wanted, it had to be as white as possible. Incised, grooved millstones were replaced in the fifteenth century by the French *buhr* stones, and the introduction of silk bolting cloth from China permitted the exclusion of much of the bran. Malisset in Paris invented the graduated grinding process in 1760. Here, the germ was removed by the first passage through the stones set three mm apart, the bran was removed in a two mm pass, and fine flour was ground in a one mm pass. Flour was now almost as white as it is today. It was, however, expensive, and the French *sans-culottes* greatly resented the fact that the Sun King and his court were eating white bread while they, who made the flour possible, were still chewing a high-bran product. One of the early acts of the postrevolutionary French government was to require that *pane egalité*, the bread of equality, was to be white for all citizens. In

England, the Tory governments of the nineteenth century were indignant to hear that the inmates of the poorhouses were demanding white bread.

The triple grinding process of Malisset was slow, and stone chips from the mills tore the delicate bolting cloths. In 1830, Müller and Sulsberger in Switzerland fitted porcelain and steel rollers together, set them in pairs at Malisset's graduated distances, and produced flour quicker. The government of Hungary subsidized construction of roller mills and, for close to 40 years, had a monopoly on high-grade flour. This monopoly was broken in 1870 when the governor of Minnesota invited Hungarian technicians to build the first roller mill in Minneapolis. Minneapolis's success was short lived. Charles Gaskill soon realized that the immense power of Niagara Falls could be harnessed to roller mills, so Buffalo, New York, became the center of milling. Steam-powered mills were developed by 1870, first in Italy to grind the very hard durum wheats into seminola flour for pasta, and later to handle the hard red wheats grown on the North American prairies. Electricity entered the milling process soon after the turn of the century, and the water-turbine generators of Niagara Falls assured Buffalo's prominence as a world center of flour milling.

The very white flours desired by most people required that the product not have the light yellow hue caused by natural pigments in the endosperm cells. Over the years, many agents were employed to bleach out these xanthophylls. Alum, ammonium carbonate, and even chalk were added to the point where the whitening agent actually became an adulterant. Tobias Smollet, an English author, had his character Matthew Bramble say,

> The bread I eat in London is a deleterious paste, mixed up with chalk, alum and bone ashes, insipid to the taste and destructive to the constitution. The good people are not ignorant of the adulteration, but they prefer it to wholesome bread because it is whiter . . . and the miller or the baker is obliged to poison them.

Chemical bleaching agents were introduced in 1903. There were many reactions against finely milled, bleached flour. In the middle of the nineteenth century natural food advocates were recommending a return to Christian morality and whole wheat flour. The most vocal of these food fundamentalists was Sylvester Graham who went so far as to assert that the depauperate white bread of the 1840s was an evil that led to drunkenness and sin (type not specified). Graham correctly observed that removing the germ resulted in flour that lacked "nutriemental value." A bowel fanatic, he also recognized that white bread did not provide the "bulk that promotes good intestinal cleanouts." He invented Graham Crackers as a healthy and tasty substitute for white bread, but since one can't make a sandwich with them, their popularity as a substitute for a slice of bread was short lived.

After World War I, when North America assumed the responsibility of feeding a world in which other dietary sources of minerals and oils were in short supply, it was realized that Graham had a point. It wasn't until the 1930s when vitamins were discovered that it was recognized that white bread had nutritional drawbacks. The virtue of bran is questionable; a moderately well-rounded diet will provide all the bulk cellulose that the body requires, but people demanded white bread without the vitamin-rich wheat germ. In a series of conferences involving nutritionists, biochemists, botanists, government, and the milling and baking industries, it was decided that vitamins and minerals removed with wheat germ should be replaced. Enriched breads were developed when synthetic B vitamins and minerals were added and milk solids and high-vitamin yeasts were used. New milling techniques allow 70 percent of the natural vitamins to be retained in the flour. Today, virtually all bread flours are enriched during milling or baking, and wheat germ is sold separately as a food supplement. This apparently topsy-turvy way of handling the problem is really quite logical, for, when wheat germ is retained, the flour cannot be stored for long periods of time because wheat germ oils will turn rancid.

THE MIRACULOUS LEAVENING

In very general terms, raised or leavened bread is produced when a dough of flour and water is acted upon by carbon dioxide gas which forms gas bubbles in the dough. Unless these bubbles are stabilized, raised dough would collapse like a poorly made soufflé. The dough constituent that confers elasticity and stability to the bubbles is a protein called gluten, which constitutes about 9 percent of wheat flour. To produce bread with small, even holes, the gas is punched out of the dough and a second rising occurs before baking. There are a number of leavening agents. Air itself is a leavening agent for beaten products such as angelfood cakes, in which the batter is stabilized by the proteins of egg white. Baking powder is a combination of baking soda (sodium bicarbonate) and an organic acid which reacts with water to form carbon dioxide:

$$NaHCO_3 + \text{tartaric acid} \rightarrow \text{Sodium tartrate} + H_2O + CO_8$$

If baking soda is used alone, the same chemical reaction occurs except that the acid is supplied in the dough mixture as buttermilk, fruit juices, or molasses.

For most raised dough, the leavening phenomenon is a biological process involving living organisms, primarily the fungi called yeasts (*Saccharyomyces cerevisiae*). The production of carbon dioxide is a metabolic process in which yeasts and all other living cells convert carbohydrates into chemical energy that the cell uses for its growth and for the synthesis of all cellular constituents (Chapter 1). The over-all reaction can be summarized as:

$$\text{starch} \rightarrow \text{glucose (6-carbon sugar)} \rightarrow \text{2 pyruvic}$$
$$\text{acid} \rightarrow 2\ CO_2 + \text{2 ethyl alcohol.}$$

When baking bread, a small amount of sugar in the flour starts the fermentation process, enzymes in the flour and from the yeasts convert starch to sugar, and production of carbon dioxide is very rapid. Bakers may add extra sugar to the dough to speed up the fermentation and to give bread a slightly sweet taste. Because the enzymatic reactions of yeast are sluggish at low temperatures, temperature control during leavening is important—as any good cook can tell you.

Early leavenings were undoubtedly natural yeasts and bacteria that fell into the flour-water mixture. Bakers soon found that a piece of risen dough contained enough leavening agent to serve as a starter for the next day's dough. I Corinthians 5:6 says, "Know ye not that a little leaven leaveneth the whole lump?" indicating that the technique is probably 3000 years old. In Egypt, another method was used to obtain leavening. Crushed wheat was soaked in white wine for three days, and the mash was added to the flour-water dough. The brewing industry became a readily available source of yeasts. Brewers supplemented their income by selling pails of beer wort or skimmed beer foam for home baking. These made excellent leavenings, and the slightly beery taste of the bread was certainly not unpleasant. When Antony van Leeuwenhoek trained his primitive microscope on beer and rising bread in 1677, he was the first person to see yeast cells, and he remarked in his letter to the Royal Society of London that the organisms multiplied rapidly in sugar solutions. This observation allowed bakers to grow yeasts in sugar-flour solutions in order to have a stable type of yeast which gave rapid and even rising of the dough. By the middle of the eighteenth century, it was recognized that beer yeasts were not as good for breadmaking as those selected over time as bread yeasts. As bacteriology developed, Charles Fleishmann founded the yeast-making industry in the United States in 1869. Yeasts were grown in sugar solutions, the excess water removed, and the yeast with a starch binder was compressed into slabs that were sold to bakeries or to the home. Dried yeasts, a very stable source of active yeasts, were generally available after 1945.

THE BAKERY

Wheaten, raised breads were apparently invented in Egypt (Figure 2.47) and spread rapidly throughout the Mediterranean basin and the Middle East. Raised bread was used commonly in Greece, and unleavened barley bread was also eaten. Siphilis of Siphonos said in 100 B.C., "Bread made from wheat, as compared with that made from barley, is more nourishing, more digestible and in every way superior,"

and he was nutritionally and biochemically correct. Greeks were very imaginative about their breads. They made puff cakes mixed with honey, cinnamon breads containing dried fruits, egg breads, cheese breads, breads spiced with cloves, mace, and saffron; essentially all the kinds with which we are familiar. But they didn't make rye bread, because rye was unknown in Europe until about 150 A.D. The Romans, too, had dozens of kinds of rolls, twists, and even raised waffles glazed with Greek clover honey mixed with white wine. In Rome, baking, like milling, was a regulated trade, but the baker, unlike the hated miller, was a highly respected member of the community. Since he controlled the ovens, he not only made bread but also cooked meats, pastries, and fowl for the neighborhood. This esteemed profession was regularized by the formation of guilds, trade associations that controlled the quality of the bread and also set standards for the profession. Baker's guilds continued until the nineteenth century. A boy was apprenticed at about the age of eight, and by 18 he was ready to become a journeyman. This was exactly what he was; he was required to travel through several countries learning different techniques and upon his return was examined to determine his fitness to become a master baker.

Most daily bread was fabricated into a round loaf, much like that given to the Roman populace. The French name for the baker, *boulanger*, refers to the round loaves that came out of his oven (Figure 2.48). The quality of bread was of great concern to government. The English "Assize of Bread" in 1266 was the first set of regulations that had wide application, and equivalent laws in other countries quickly followed.

Bakers also were the makers of table service. Impervious dishes

Figure 2.48
Brick oven used in sixteenth-century Italy. It was called the peel oven for the wooden spatula with which bread was moved in or out of the oven. Taken from a sixteenth-century woodcut.

Figure 2.49
A fifteenth-century woodcut showing the use of slices of bread as trencher plates.

Figure 2.50
Wood-fired beehive oven.

did not exist before the thirteenth century, and "plates" consisted of a thick slab of stale, partly leavened brown bread about four inches by six inches. The absorbent plate, called a trencher, gave rise to our name trencherman (Figure 2.49), one who was able to soak through a number of trenchers at a single sitting. The grease-soaked trenchers were either flung to dogs fighting under the tables or, in moments of charity, to the poor who stood around the tables waiting for them. Only when glazed dishes became generally available in the sixteenth century were trenchers completely abandoned except for their vestigal placement under broiled lamb chops or soft-boiled eggs.

The American experience with bread paralleled its development in Europe of the seventeenth and eighteenth centuries (Figure 2.50). In pioneer days, bread was primarily cornbread with wheat bread made from imported and expensive flour. George Washington raised some wheat, milled it, and baked it for his personal table; other wealthy planters did the same. As wheat became more available in the middle of the eighteenth century, breadmaking was done in the massive fireplaces that were also the source of light, general cooking, and home heating. The bread oven was usually of brick or stone on the side of the fireplace, and only skilled bakers produced an evenly brown loaf without a soggy center (Figure 2.51). In the cities, baking was a commercial operation almost from the beginning of the eighteenth century. The first bakery in New York opened in 1645, and professional bakers were in business throughout larger communities by 1700.

The year 1850 can be taken as the watershed for baking in North America. At that time there were about 2000 bakeries in the United States employing 7000 workers who produced close to $14 million worth of baked goods. Urbanization increased in the United States as factories enlarged and as machinery began to supplant muscle power on the farm. Although home baking was still predominant in 1850, women began to form a significant proportion of the labor force, and they simply didn't have time or reserve energy to bake the family bread

Figure 2.51
A nineteenth-century American peel oven.

after a 12–16-hour factory day. By 1860, more than half the bread sold in the cities was baked outside the home. As demand increased, bread-making began to be mechanized. The deck oven came along in 1880. It carried bread into and out of ovens on a wheeled deck that eliminated the peel and permitted almost continuous operation of the ovens. A rotary oven invented at the same time looked like a Ferris wheel turned on its side. Mixing machines, dough dividers, rounders, and proofers were in use by 1895, and continuous tunnel ovens, "a tube with dough going in at one end and bread coming out the other," were constructed by 1913 (Figure 2.52).

As people became more and more crowded into cities, retail grocery stores began to replace the bakery as the place to buy bread. Large bakeries could produce loaves faster and cheaper (albeit not better tasting) than could the baker, and wagons, and later trucks, could place

Figure 2.52
An integrated cracker-baking factory. Courtesy Nabisco, Inc.

many loaves on the grocer's shelves. At the same time, the American fetish of supercleanliness was being promoted. Wrapped bread, "untouched by human hands," was, it seemed, more sanitary. Packaging machines that enclosed bread in impervious wrappers which negated the virtue of a crust came along at the end of World War I, and the automatic slicing of bread developed in 1928. Families were asked, "Can you make better clothes than a tailor? Then why try to bake your own bread?" Brand identifications and increasing real income of the working man and woman combined to make it almost a crime to place home-baked bread on the table when company was invited for dinner. The bread companies then asked, "Aren't your children and your husband as important as your guests?" As sales of "store bread" increased, breadmaking became a fully automated process, and mass production techniques were applied to it as they were being applied to the construction of automobiles.

BREAD IN RELIGION

Light, leavened bread could not be baked on hot slabs of rock, and the bee-hive oven was invented, heated by a wood fire and then brushed clean to hold the leavened dough. So important was the loaf and so mysterious was its rising that even the oven was invested with mythic importance. Bread came from a womb wherein was born the staff of life. Our expression "home and hearth" originated in Egypt, whose inhabitants despised nomadic herdsmen as barbarians who had no hearth in which to bake their bread. As long as the Israelites were nomads during the biblical 40 years, their bread was unleavened gruel, quickly baked and quickly eaten. This is memorialized in the Passover matzos: "Seven days shall ye eat unleavened bread" in Exodus 13. Only when they became settled agriculturists was it again possible to leaven their bread, and they invested bread with an aura of mystery. Each household had its own oven, and only women could tend it; thus woman was both the life-giver and the life-sustainer. As holy food, strangers and travelers were offered God's bounty, and the Hebrews broke bread as a symbol of God's peace. Nevertheless, when offered on God's altar, the bread was unleavened, uncorrupted by the unclean (from the earth) leaven. In medieval England, from whence our word bread comes, it meant "that which is fermented."

The role of bread in Christian ritual is based on two verses in the New Testament. Matt. 26:26 and Mark 14:22 state: "Now as they were eating, Jesus took bread and blessed and broke it, and gave it to the disciples and said, 'Take, eat. This is my body'." Luke 22:19, however, renders this differently, "And He took bread and when He had given thanks, He broke it and gave it to them saying 'This is my body which is given for you. Do this in remembrance of me.'" Early Christians were of two minds about this discrepancy. Were they, following Matthew and Mark, to take literally the idea of "body" or were the words, as given in Luke, to be taken symbolically? Did Christ mean that in order to be saved, one must eat human flesh—a concept that has parallels in other Mideastern cultures? Or, as St. Paul suggested in I Cor. 11:23, was the passage in Luke to be essentially a recommendation or parable of remembrance? Tertullian, Augustine, Origenes, and other early Church fathers agreed that bread represented the body of Christ but was not, in fact, the substantive body. St. Gregory of Nyassa in Asia Minor (331–394 A.D.) argued that once bread is eaten, it becomes part of the body, and thus the Eucharist is representational. St. Chrysostem of Constantinople (347–407 A.D.) disagreed, asserting that man does in fact eat Christ's body transformed as the blessed bread. The matter smouldered for a thousand years until, in the Lateran Edict of 1204, Pope Innocent III declared *ex cathedra* the Dogma of Transsubstantiation. The Roman Catholic Church now said that at the Last Supper, Christ required His followers to eat of His Flesh. A priest had the ability to transform bread into the fleshly body of Christ to be eaten by the faithful, thus following St. Crysostem, who said, "Why should we, the faithful, shrink from eating human flesh?" The Dogma of Transsubstantiation became a time bomb that exploded 300 years later—when the role of the priest was questioned during the Reformation.

In the meantime, there were practical matters that required attention. Eucharistic bread was, for the first 500 years, a large round, raised loaf of wheat and barley shared by the congregation. From about 650 A.D., each communicant was given a small wheaten loaf or roll containing some barley or other cereal grain—essentially the common bread of the people elevated by transsubstantiation. The Church decided, about 1000 A.D., to take again the Jewish injunction regarding the corruption of leaven and ruled that Eucharistic bread should be unleavened. It was also agreed that while in Christ's miracle of the loaves and fishes the bread was undoubtedly a mixture of wheat, barley, and probably millet flours, the passover bread was probably of wheat flour. St. Thomas Aquinas stated that it was almost blasphemy to assume that his Lord would have eaten anything but the best of wheat bread, and so the whitest possible flour was specified for the Host. Because even the prospective Host was almost holy, it had to be guarded carefully. Witches could steal it, or pious communicants might take some for their much loved domestic animals, feeling that their cows and sheep might profit from eating it.

The safe storage of the wafers presented several problems. In unheated, damp churches, the host could soften or even become covered with mold and have to be discarded with reverence. Much more seriously, the wafers occasionally seemed to be covered by drops of blood. The blood of Christ appearing on bread was a horrible thing to contemplate. Obviously it was caused by the Devil, and who among the residents of the community could be the Devil's agents? Why, it was the Jews, of course! In 1253, the members of the Jewish community of Beelitz near Berlin were burned—the Host had been bloodied. Paris in

1290, Vienna in 1298, Crakow, Posen, Breslau and Prague in 1300, all saw the defiled Host, and human blood ran in the streets. Over 2000 Jews were killed in Brussels in 1370, and uncounted thousands died in smaller towns. As late as 1510, 38 Jews were pilloried in Berlin. Not until 1848 did Professor Christian Ehrenberg find that the "blood" was a bacterium that produced a red pigment. In spite of this knowledge, at least two pogroms in Russia in the late nineteenth century were started by rumors of bloodied hosts.

The matter of transsubstantiation, although dogma, was still being questioned. Bishop Berenger called the pope *pulpifex* (flesh eater) instead of *Pontifex* (head of the pontifical college) and was consigned to the flames. One unnamed churchman asked why the Church wished to enclose the Savior in a loaf of bread and was flayed alive for his impertinence. By the early sixteenth century, transsubstantiation became part of the larger debate leading to the Reformation. Huldreich Zwingli, a Swiss reformer, spoke of the Eucharist as a swindle and exclaimed, "If Jesus is present in the bread, we would shudder to eat of it." Martin Luther, Zwingli's mentor, was angered at this. Luther wanted not to destroy the Roman Catholic Church, but to purify it. He recognized that attacks on dogma could damage the Church, and debates over parable, metaphor, and biblical interpretation had to be avoided.

For three years, the battle of words raged until the Landgrave Philip von Hessen offered his castle at Marburg—midway between Luther in Saxony and Zwingli in Zürich—as a neutral debating place where the two could defend their theses. Debate started on Friday, October 1, 1529. Its mood was set when Luther chalked on the conference table the words, "This is My Body." A week later, the debate was ended with mutual recriminations and thus was destroyed one of the last bonds that held churchmen together. Zwingli was wounded in 1531 in a regional battle between Catholic and Protestant cantons in Switzerland and was captured by Catholics, who beat out his brains, had his body drawn and quartered, then burned, and his ashes mixed with swine manure. Luther's response to all this was, "We see the judgment of God, who will not long suffer these mad and furious blasphemies."

The Swiss concept of the Eucharist as metaphor was accepted by the Church of England under Elizabeth I: "Transsubstantiation is repugnant to scripture and is the occasion of many superstitions." By 1700, a law was on the books stating that only those Englishmen who pledged allegiance to the king as head of the Church and who also declared against transsubstantiation could buy or inherit land. All others, that is, Catholics, were barred from the civil service or from living within ten miles of London. American colonists did not include the Eucharist in their religious observations, and at least some of the anti-Catholic feeling in the young United States can be traced to this schism. The French writer, Montesque (1689–1755) was, in his youth, something of a radical. In his book, *The Persian Letters*, he had Usbek, his fictional Persian diplomat, write home about an important magician called the Pope who had persuaded the people "that bread is not bread and wine is not wine." The Chevalier LeFebre was formally indicted in Paris in 1775 for saying that he could not understand why anyone would "worship a God of dough" and was decapitated for having a heretical tongue in his head. Voltaire (1694–1778) said that Catholics profess that they eat God and not bread, Lutherans eat both bread and God, and Calvinists eat bread but not God. He was excommunicated—burning was out of fashion. Even the devout Leo Tolstoy was excommunicated from the Russian Orthodox Church in 1901 for presumably mocking the Eucharist in his novel *Resurrection*. In 1976, the pope declared that a serious obstacle to the rejoining of Christian sects was the failure of non-Catholics to accept the Dogma of Transsubstantiation.

ADDITIONAL READING

Ashley, W. *The Bread of Our Forefathers.* New York: Oxford University Press (Clarendon Press), 1928.

Ashton, J. *The History of Bread.* London: 1904.

"Bread in the Ancient World." *Baking Industry Magazine* 47 (April 12, 1952):30–40.

Dunlap, F. L. *White Versus Brown Flour.* New York: Wallace & Tiernan Company, 1945.

Jacob, H. *Six Thousand Years of Bread.* Garden City, N.Y.: Doubleday, 1944.

MacEwan, G. *Harvest of Bread.* Saskatoon: Prairie Books, 1969.

McCance, R. A., and E. M. Widdowson. *Breads White and Brown.* Philadelphia: Lippincott, 1956.

Panscher, W. G. *Baking in America.* Evanston, Ill.: Northwestern University Press, 1956.

Rose, A. H. "Yeasts." *Scientific American* 202(1960):136–146.

Steen, H. *Flour Milling in America.* Westport, Conn.: Greenwood Press, 1963.

Storck, J., and W. D. Teague. *Flour for Man's Bread. A History of Milling.* University of Minnesota Press, 1952.

CHAPTER 3
Plants in Religion

Plants of the Bible

And God said, let the earth bring forth grass. . . .
GEN. 1:11

The Judeo-Christian Testaments are part of the foundation of Western civilization and are quoted and misquoted frequently. We can view the Testaments as the literal Word of God, as inspired allegory, as a fairy tale, or as an historical record. For botanists, the Bible provides a record of plant use by those peoples of the Middle East who gave us our religious heritage. The Bible is not a *de novo* source of plant symbolisms; accretions from other faiths are intercalated and have become incorporated into our attitudes.

BIBLICAL SCHOLARSHIP

In order to understand these symbolisms, we should have some idea of the history of biblical literature. The Old Testament is composed of books written 2000–3000 years ago, some written shortly after the Tribes of Israel entered the now Holy Land, and others written as contemporary records of the prophets and kings. Ancillary records and surviving fragments such as the Dead Sea scrolls validate many passages in the King James and subsequent English-language versions. These versions are based on translations from the Hebrew into Greek, the first one authorized by Ptolemy Philadelphus (d. 246 B.C.), patron of the library at Alexandria.

The history of the New Testament is less direct. In the first period of the new Church, the Good News (Gospels) was spread orally. The Gospel message was colored by the locale, with different plant imagery used to point up the message in Egypt than in Asia Minor or Greece. Written versions appeared in the second century A.D., but not until the fourth century was the New Testament collected and collated. The language of the early New Testaments was Greek, taken from oral depositions and fragmentary records in Aramaic, the common language of the peoples of the area, as well as from depositions in Greek, Syriac, Coptic, and Ethiopic. Several centuries passed before versions in Latin were prepared, the first being the Vulgate of St. Jerome in the fourth century. The first English-language version appeared at the beginning of the seventh century, and subsequent versions in English during the eighth to tenth centuries were prepared from the Vulgate. An Old English translation in the tenth century was followed by several Middle English versions and all of these culminated in the King James version of the seventeenth century. The American Standard version appeared at the end of the nineteenth.

This abbreviated history points up the difficulties in being sure of what plants and what plant imageries were used by Hebrews and early

Figure 3.1
Adam, Eve, serpent, apple, and Tree of the Knowledge of Good and Evil. Taken from Hans Sebald Beham's woodcut, *The Fall of Man.*

Christians. Many translations were inaccurate, and these errors were perpetuated in succeeding versions. Since the Gospels were bent to fit different cultures, names of plants were altered beyond our ability to be sure of the imagery of biblical peoples themselves. We have clues to these images, among the best being our knowledge of what plants were either native to the area or introduced by waves of immigration and conquest which were the fate of these lands. Modern Israelis grow maize, tobacco, and squash, plants that did not enter Europe from the Americas until the fourteenth and fifteenth centuries and did not reach Palestine until at least 1700. To blithely speak of the apple (*Pyrus malus*) as being Eve's temptation (Figure 3.1) doesn't make botanical sense; the apple could not develop in the present or past climate of Holy Land. This image was used in Europe only because it was a common fruit that people knew. We know that many plants that did exist in biblical times were either wiped out by man's activity or are now inconspicuous parts of the flora due to conversion of natural areas into arable land. There are probably no more than 200 endemic plants in the area—plants known only from Palestine and nowhere else—and these rarely figured in biblical writings. Most plants of symbolic interest or economically important as food or fiber either originated in the Middle East or were imported eons before the area became a cradle of Western religions. These, for the most part, are the ones mentioned in the Bible.

THE GRASSES
AND GRAINS

One group of native plants, the grasses, are frequently mentioned in both Testaments, the first in Gen. 1:11, "And God said, let the earth bring forth grass." Most subsequent references to grass are allegorical: "All flesh is grass" from Isaiah and I Peter; "My heart is smitten and withered like grass" from the 102nd Psalm; or the statement in the 103rd Psalm, "as for man, his days are like grass." For nomads, the fleeting nature of pasturage involved their very survival and the figurative image is perfectly understandable. Nomads of central Mongolia have the same image. References to hay (Prov. 27:25; Isa. 15:6) are later interpolations, since the sparse growth of natural grasses would have precluded haying. In Revelations 8:7 where "all the green grass was burnt up," it is likely that a more accurate translation would be "all the green plants"—plants used as forage—because tough native grasses will dry out, but not collapse as would broad-leaved plants.

The cereal grains, true grasses all, have figured many times in both Testaments. The first of over 30 references to barley (*Hordeum vulgare*) appeared in Exod. 9:31 (Figure 3.2). Barley was probably a more widely grown grain crop than wheat, because none of the other cereals can grow in as many different climatic conditions. It was planted as soon as man became agricultural and Hebrew invaders of Canaan found it in cultivation. As seen in Deut. 8:8, "A land of wheat, and barley . . ." was synonymous with fertility, and a land where "thistles grew instead of barley" (Job 31:40) was a symbol of infertility. Two barley grains made a finger's breadth, 16 made a hand's breadth, 24 a span, and 48 were the biblical cubit—about 41 cm. Although sprouted barley grains have for centuries been the preferred malt used in brewing, there is no statement in the Bible of this fact. Since wine was not forbidden, it is unlikely that beer was proscribed, but the absence of any references to malting or to beer is surprising.

The bread of the peoples of the Bible was primarily an unleavened loaf prepared from ground barley mixed with millet, pea meal, spelt, and other seeds, but the lighter, whiter, and more stable bread prepared from wheat flour was preferred. Most references to "corn" are to wheat. Reuben, a successful farmer, grew wheat (Gen. 30:14) and the Psalms use "the corn" as a symbol of a fruitful harvest of grain and of souls. Matthew, Mark, and Luke mentioned corn fields in historical context. John 12:24, "Verily, verily, I say unto you, except a corn of wheat fall into the ground and die, it abideth alone; but if it die, it bringeth forth much fruit," is a powerful image which made good sense to the hearer. Matthew used the image of a good harvest in several places (13:3–8 and 13:24–30). The sower who threw grains in stony

Figure 3.2
Seven-rowed ear of barley. Taken from a coin struck ca. 500 B.C.

ground and some in fertile soil and the comparison of the Kingdom of Heaven to a man who sowed good seeds in his field was immediately recognized as an object lesson for the faithful. Famine was an ever-present terror, and the concept of a granary as a hedge against real or religious famines is in many verses of both Testaments.

VEGETABLES AND FRUITS

The Bible is ambiguous about the use of fresh vegetables. Most would require more water than was generally available, and many cultures in the area had little use for them. Onions and garlic were, however, used extensively. Food was dressed with spices, some grown locally, like capers (*Capparis sicula*), bay leaf (*Laurus nobilis*), saffron (*Crocus sativus*) and coriander (*Coriandrum sativum*). Cinnamon and other exotic spices arrived on the caravans from the Indies and China, and others, like cloves, came from Africa.

Many fruits are celebrated in both Testaments. The pomegranate (*Punica granatum*) was grown in Egypt from at least 1600 B.C.; and it was grown in the Holy Land before the hegemony of the Jews. It was a luxury item, praised for the beauty of the plant, its flowers, and the handsome red-skinned fruit. Because of the many seeds surrounded by sweet, thirst-quenching juice, it became a symbol for abundance and especially for fruitfulness (Num. 13:23 and 20:5). Its sexual connotations appear twice in the Song of Solomon (6:7 and 8:2) and the same symbolism appeared in Europe. The fig (*Ficus carica*) was also a sexual and fertility symbol because of its large number of seeds, but because this fruit could be dried for storage, its role as a food staple was well established (Numbers 13:23). Rabbinical scholastics suggest that the fig was the Tree of Knowledge; this suggestion has as much merit as any of the other plants suggested for this role. As a plant used in allegory and parable, figs appear many times as a symbol of peace (1 Kings 4:25), good vs. evil (Jer. 24:1–8), literary analogy wherein "All thy strongholds shall be like fig trees" (Nah. 3:12), and parable (Luke 13:6–9; Matt. 21:18–19; Rev. 6:13).

The date palm (*Phoenix dactylifera*) was used for many purposes (Figure 3.3). Fronds were used as thatch, trunks as building supports (later to be stylized as stone pillars), and the fruit could be eaten directly or made into date honey to be fermented into "strong drink." *Tamara*, a common arabic given name for women, is translated as the palm, recalling the graceful and elegant "posture" of the tree. Lev. 23:40 and Neh. 8:14 report that palm branches, as a symbol of prosperous harvests, were used to decorate the booths erected to celebrate the Feast of Tabernacles. As a symbol of triumph, it has figured in several cultures. Judas Maccabee celebrated his military successes (1 Macc. 13:53; 2 Macc. 10:10) with palms, and John 12:13 reported that palm branches were waved with joy upon the entry of Jesus into Jerusalem. Romans, too, used palms in symbolic fashion. When the victorious African le-

Figure 3.3
The wondrous date palm, *Phoenix dactylifera*. Taken from Johnson's revision of Gerarde's *The Herball* published in London, in 1633.

Figure 3.4
Cross with palm fronds and olive leaves. Taken from an embossing found in the Christian catacombs of Rome.

gions returned, cohorts waved fronds of palm as they marched through triumphal arches. It is from these usages that early Christian martyrs chose the palm as a symbol of eventual triumph (Figure 3.4). On all Souls' Day, the second of November, palm leaves are burned, and their smoke rising unto the sky is taken as proof of the victory of souls leaving purgatory for heaven. Legends of the miraculous properties of the plant abound in Christian and Moslem cultures, but are not supported by scriptural reference—a phenomenon common for many plant legends.

Although temperance groups believe that biblical references to wine are really to grape juice, few scholars agree with them. The grape (*Vitus vinifera*) is first mentioned in Gen. 9:20 wherein Noah planted a vineyard and got drunk on the end product. Wine and raisins were articles of commerce and trade, as evidenced by records from Egypt and Mesopotamia. The names of specific places in Palestine are linked to the vine (Figure 3.5), including Abel Kramim (Plain of the Vineyard) in Judg. 11:33, Mount Carmel (Hill of the Vineyard of the Lord) in 1 Kings 18:19, and Nahal Eshcol (Brook of the Cluster) in Num. 13:23. Many references to vines, grapes, and wine are clear statements of fact or are horticultural directions (Isaiah). Most, however, are allegorical and

Figure 3.5
Caleb and Joshua bringing in the wine grapes. Taken from *Biblia Germanica Decinquarta* published in Strasbourg in 1518.

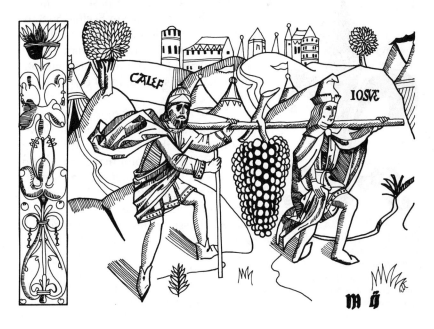

symbolic. The "vine brought out of Egypt" and the "fruitful vine" referred to the Hebrews themselves. John (15:1–6) observed that Jesus referred to himself as "the true vine and my Father is the husbandman" or again, "I am the vine, ye are the branches." The Parable of the Vineyard (Luke 20:9–16) is another example. Peace and tranquility can have no clearer definition than that in 1 Kings 4:25 where every man was under his own vine, or the vision of the prophet Joel (2:24) where "the vats shall overflow with wine. . . ." "The Battle Hymn of the Republic" by Julia Ward Howe includes the powerful metaphor, "He is trampling out the vintage where the grapes of wrath are stored." Rev. 14:18–20 speaks of the "great winepress of the wrath of God," and in Lam. 1:15 is written, "the Lord has trodden the virgin . . . as in a winepress." The image of devastation is, in Psa. 78:47 and Isa. 5:1–6, given as the uprooting of vineyards, an act of destruction more long-lasting than virtually any other act of violence.

THE LILY

"Consider the lilies of the field, how they grow; they toil not, neither do they spin. And yet I say unto you that even Solomon in all his glory was not arrayed like one of these." This quotation from Matt. 6:28–29, repeated in Luke 12:27, is only one of many references to lilies. There are only two native species of *Lilium* which could have existed in biblical times, but neither were nor are abundant. It seems unlikely that an image of such importance would be to a plant that most people would not be familiar with. This has provoked discussions as to which plant was the biblical lily. It is generally agreed that no single plant meets all the criteria imposed by the passages in which the word appears. Thus, reference in Ecclus. 50:8 to "lilies by the rivers of waters" implicates the iris, of which several species are native to Palestine. The modern Arabic word "Shushan" seen in Nehemiah, Esther, and in Psalm 60 also fits well with this concept. A woman who doesn't like the given name Iris could, in all etymological accuracy, change her name to Susan.

The iris (Figure 3.6) is a common spring-flowering plant found in moist habitats. Its blue and purple flower colors lend themselves nicely to the purple that early churchmen formally associated with the Passion, and the blue color that clerics standardized as Mary's color. Its use in France's heraldry stems from King Clovis (d. 511 A.D.), founder of the Merovingian dynasty. Clovis was pagan, but had a Christian wife, later to be St. Clotilde, who fervently wanted to convert her lord and master. Possibly after some nagging, the king agreed—so the story goes—that if he won the battle of Tolbiac against the Huns in 474, he would embrace the faith. As a symbol of his pledge, he painted the flower of the Trinity, the iris, on his battle standards and, once victorious, adopted the iris as a symbol of his reign. The crusade of 1137 under Louis VII used a somewhat stylized iris flower on its standards,

Figure 3.6
The iris. Top drawing taken from *Commentarii* by Mattioli published in 1579. Bottom drawing of the fleur-de-lis taken from a fifteenth-century engraving.

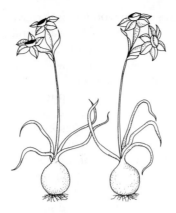

Figure 3.7
The narcissus. Taken from *Commentarii* by Mattioli published in 1579.

Figure 3.8
Anemone hortensis, a candidate for the "lilies of the field" in the Old Testament. Taken from *Rariorum Plantarum Historia* of Carolus Clusius published in Antwerp in 1601.

and by the end of the twelfth century, the flag of France was a blue field (Mary's color) sprinkled with white iris, the fleur de Louis. This became the fleur de luce—the flower of light—and then the fleur de lis. In 1364, the number of flowers reduced to the Trinity number. When England held part of France, the iris was included in the coat of arms of the Plantegenet kings, disappearing only in 1801, many years after France asserted its hegemony over these lands. Although the Massachusetts Bay Colony would brand an adulterous woman with an "A," the French in both the mother country and in New France branded a fleur de lis on the woman's cheek. The male partner was not marked.

In Isa. 35:1, we read that "desert shall rejoice, and blossom as the rose," but the image is more basic than the botany. The Hebrew word used here was *chanatselet* translated as rose, but it appears to mean bulb. Since roses don't develop from bulbs, a more adequate plant would be the narcissus (Figure 3.7) which grows in profusion on the Plains of Sharon and in fields surrounding Jerusalem. When, in Ecclus. 39:12, the prophet exhorted, "Hearken unto me, ye holy children, and bud forth as a rose growing by a stream of water," one may think of the oleander (*Nerium oleander*). The Lake of Galilee and brooks running into the Dead Sea and the River Jordan are lined with these poisonous, but graceful and fragrant-flowered plants. The Rose of Sharon (Song of Sol. 2:1) may have been the mountain tulip (*Tulipa montana*) or even the hyacinth. There are, however, several true roses found in Palestine, but they are uncommon and none are as showy as twentieth-century roses.

The lily in Matthew and Luke was not likely to have been a lily, and either the windflower (*Anenome coronaria*) (Figure 3.8) or a crocus fit the passages. Both are spring-flowering, tending to validate our admittedly unsubstantiated impression that the teachings were made in the spring. Even today, both genera grow in uncultivated areas of the Holy Land and both have showy flowers in many different colors. The plants develop quickly after the drab days of winter and were undoubtedly appreciated, then as now, as harbingers of renewed fertility of the land.

With these various interpretations of the botanical nature of the

Figure 3.9
Madonna lily of purity and motherhood. Taken from a fourteenth-century Italian woodcut.

lily and the rose, one wonders where we acquired the white lily as one of the most important of Christian symbols. White flowers were symbols of purity and chastity long before Christianity; white lilies were floral representations of Diana and Juno. Greeks and Romans prepared chaplets of white lilies for brides, and, when intertwined with sheaves of wheat, the benediction of a pure and fruitful married life was obvious to all. Exhibiting purity and grace, the white lily (*Lilium candidum*) was a symbol of several goddesses of the Fertile Crescent and parts of southeast Asia where the plant apparently originated. The lily was formally adopted by the Roman Catholic Church by the fifth century and, by the early Middle Ages, was firmly associated with the Virgin (Figure 3.9). Renaissance paintings of the Annunciation almost always include the lily, and this was formalized in 1618 when a papal edict laid down rules "as to the proper treatment of certain sacred subjects of art and the necessity of the introduction of the white lily into pictures of the Immaculate Conception." By the fifteenth century, the white lily was a traditional feature of the altar decorations in Europe. In addition to being Mary's flower, it is the floral attribute of several female saints. This device of using a common plant to facilitate the recognition of the holy in everyday life and to have symbols for saints for people who could not read can be found not only in Christianity, but in all religions.

In the eighteenth century, as plant explorers began bringing plants back to Europe from the Orient, the White Trumpet lily (*Lilium longiflorum*) was introduced from Japan and Korea where it was cultivated as a garden plant and religious symbol. The plant is easier to grow than is the Madonna lily, and its perfumed flowers last many days. It can also be forced into bloom in time for Easter. In North America, the Trumpet Lily has almost completely replaced the Madonna lily as a potted plant. In the prurient Victorian era, it was common practice to remove the yellow stamens from lilies used as altar decoration so that the flowers would "remain ever virgin." Can you imagine the distaste (or pleasure?) with which this task was performed?

ADDITIONAL READINGS

Goor, A. "The History of the Grape Vine in the Holy Land." *Economic Botany* 20(1966):46–64.
The Holy Bible, Revised Standard Version. New York: Collins, 1973.
Moldenke, H. N., and A. L. Moldenke. *Plants of the Bible*. Waltham, Mass.: Chronica Botanics, 1952.
Schonfield, H. J. *A History of Biblical Literature*. New York: New American Library, 1962.
Walker, W. *All the Plants of the Bible*. New York: Harper & Row, 1957.

Plants of Superstition, Myth, and Ritual

He loves me, he loves me not. . . .
PICKING DAISY PETALS

With a few exceptions, large urban areas are a recent development in social history. Ancient Rome, Athens, and Constantinople were within a short distance from farms and forests, and people had direct and immediate ties to the land. Plants were intimate parts of life, and it was natural for religious and secular leaders to use them for imagery, for parables, and for exhortation. So, too, did plants serve the people themselves; they were not merely for food and shelter, but were intertwined with the rhythms of the seasons. A child was born "just after the corn harvest," or a woman married "when the apricot trees were blossoming." When we remember that natural occurrences were mysterious at best or evil at worst, emphasis on the role of plants in human affairs was perfectly logical. Accidents and disease were caused by evil spirits or the anger of God, and one could be whole again by driving off Satan or placating Deity with plants. The child, to be trained in the mores of the group, was presented with the myths, legends, and superstitions of the group and was provided object lessons based on attributes and properties of things that formed his or her immediate environment. These lessons to the young also served to reinforce the tenuous grasp of adults on a world which impinged so ominously upon the people. Here again, plants—a vital ingredient for survival—were woven into the fabric of belief. We don't know where most of the legends and symbolisms of plants arose. Indeed, comparative religion research demonstrates that cultures which could not have been in contact had similar if not identical constructs and ideas about plants. Some of these have been transmitted down to us and form part of modern, twentieth-century rationalism.

PLANTS AND SAINTS

The association of a plant with a saint has been formally authorized since at least the sixth century. Its roots are in the pre-Christian attribution of plants to members of the pantheons of deities in all cultures. Many plants of economic importance to man were said to have been the visible symbol of a deity who, in kindness, sympathy, or generosity, presented the plant to man; Athena became the patroness of Athens for giving the olive to the citizens. Honoring Christian saints with special flowers on their days served to reinforce the flow of religious faith throughout the year. The choice of a particular plant varied with the saint. St. Patrick's shamrock (actually an *Oxalis*) was a druidic

mystic symbol associated with the Celtic sun wheel. The Celtic name *seanrog* (little clover) became *shamrog* and finally *shamrock*. St. Patrick 390–464) took this pagan symbol, pointed out that it really represented the Trinity, and subverted its symbolism to the new religion. Canterbury bells (*Campanula medium*) cannot be associated with other than St. Thomas à Becket (Figure 3.10), and the attribution of spring wild flowers to St. Francis of Assisi is a concept easily impressed on even young communicants. The birthday of a saint—the nameday of many children in Latin countries—can be equated with a flower that blooms at about that time of year. The Michaelmas daisy (*Aster spp.*) for St. Michael's Day (September 29) and the Christmas rose (*Helleborus niger*) for St. Agnes's Day (January 21) are examples. For St. Peter:

> The yellow floure, called the yellow coxecombe which floureth now in the fields is a sign of St. Peter's Day whereon it is always in fine mettle in order to admonish us of the denial of Our Lord by St. Peter; that even he, the Prince of Apostles, did fall from feare and denied his Lord. So we too are fallible crethures, the more likely to a similar temptation.

At the time of the Spanish conquest of the Americas, many plants entered Europe and were given religious significance. Pissaro saw a large yellow flower in Peru venerated by the Incas as an image of the sun god. It was of equal interest to note that the Peruvians wore hammered gold breast plates embossed with representations of this flower. Pissaro took both seeds and breastplates back to Spain, and while the sunflower (*Helianthus annuus*) assumed only minor importance as a symbol of the Glory of God, the gold was used to finance the running battle with the heretical English (Figure 3.11). The marigold (*Calendula*) became Mary's gold (Figure 3.12) after it was imported from Mexico. A Mexican vine was asserted by sixteenth century Jesuits to be the plant that St. Francis of Assisi saw in a dream. It was named *flor de las cinco llagas* (flower of the five wounds), and by the seventeenth century each

Figure 3.10
Canterbury bells, *Campanula medium*, assigned to St. Thomas. Taken from *Plantarum sen Stirpium Icones* by Matthias Lobel published in Antwerp in 1581.

Figure 3.11
Sunflower, Helianthus annuus. Taken from *Hortus Floridus* by Crispean van de Passe published in Arnheim in 1617.

Figure 3.12
Mary's gold, *Calendula officinalis*, one of many plants assigned to the Virgin.

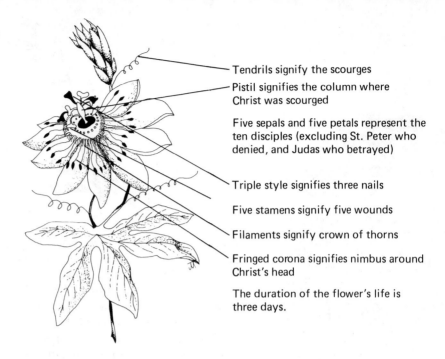

Tendrils signify the scourges

Pistil signifies the column where
Christ was scourged

Five sepals and five petals represent the
ten disciples (excluding St. Peter who
denied, and Judas who betrayed)

Triple style signifies three nails

Five stamens signify five wounds

Filaments signify crown of thorns

Fringed corona signifies nimbus around
Christ's head

The duration of the flower's life is
three days.

of the parts of the flower were given associations with aspects of the
crucifixion (Figure 3.13). We now call it the passion flower (*Passiflora
incarnata*).

DOCTRINE OF SIGNATURES

Imputation of religious significance to plants took some strange
paths, one leading to the doctrine of signatures. With disease a manifestation of evil and with medicine tied to the ancient Greeks, the need
for curatives—"simples" that anyone could obtain—was a necessity.
But unless there was some outward sign that a plant was useful and
some visible indication as to its powers, how could one know which
plant to use? This problem and its answer is best summed up by a quotation, given here in modern English, from Nicholas Culpepper's *English
Physician* of 1680.

> Though sin and Satan have plunged mankind into an ocean of infirmi-
> ties, yet the mercy of God which is over all His works has made the grass
> to grow upon the mountains and herbs for the use of man and He has not
> only stamped upon these a distinct form, but also has given them partic-
> ular signatures whereby men may read, even in legible characters, the
> use of them.

Figure 3.14
Walnut as a doctrine of signatures plant.

In Ecclus. 38:4, it is written: "The Lord created medicine from the earth.
And a sensible man will not despise them." The kidney bean looks like
a kidney and would cure urinary afflictions. A walnut (Figure 3.14)
looks like the human brain and should cure madness or even migraine

Figure 3.15
Dutchman's breeches, *Dicentra cucullaria*, as a doctrine of signatures plant.

Figure 3.16
Birthwort, *Aristolochia clematitis*, as a doctrine of signatures plant.

headaches. Dutchman's breeches (*Dicentra*) was clearly a man's plant (Figure 3.15) used to cure venereal diseases, and Dutchman's pipe (*Aristolochia*) has recurved flowers that look like a womb (Figure 3.16)—useful in difficult births. The red sap of the bloodroot (*Sanguinaria canadensis*) was a signature that the plant was specific for blood disease. The yellow inner bark of the barberry (*Berberis*) (Figure 3.17) and the yellow spice, tumeric, were specific for jaundice. The spleenwort fern (*Asplenium*) and *Hepatica* (Figure 3.18) were used in liver disease because their lobed fronds or leaves resembled the liver. Strap-shaped leaves were boiled into a tea to ease sores on the tongue. The list goes on and on.

Although the doctrine of signatures developed in Europe in the Middle Ages, the idea is far older. Theophrastus, an ancient botanist, reported that the root of polypody (a fern) is rough and has suckers like the tentacles of a polyp; if worn as an amulet, one would not get rectal polyps. The American Indians, too, had an equivalent doctrine system (Figure 3.19). To eliminate worms, eat a wormlike plant part such as the tendrils of a squash; to promote lactation, use a plant with a milky sap; to control convulsions, place a gnarled piece of wood next to the patient.

Figure 3.18
Liverleaf, *Hepatica triloba*, as a doctrine of signatures plant.

Figure 3.17
Barberry, *Berberis vulgaris*, as a doctrine of signatures plant.

Figure 3.19
Snakeweed, *Polygonium bistorata*, as a doctrine of signatures plant. Taken from *De Historia Stirpium* by Leonhard Fuchs published in Basel in 1542.

In India, gatherers of medicinal herbs say, "Blessed be thou, plant of virtue. I pick thee with the good will of Vishnu," and in Greece they would say, "I pluck thee with good fortune, the good spirit of the gods at the lucky hour of the day that is right and suitable for all things." This parallels the sixteenth century English incantation, "Haile be thou holie herb, growing in the grounde. All in the Mount Caluarie first wert thou found. Thou are good for manie a sore and healest manie a wounde. In the name of the sweet Jesu, I take thee from the grounde."

Not all physicians were happy about the doctrine of signature plants, but caution was advisable; one could not offend the sensibilities of the people, nor provoke the anger of a clergy dedicated to any idea that linked sanctity and health. Gerarde, author of *The Herball* of 1591, ridiculed many signature plants, but he did observe that Solomon's Seal (*Polygonatum multiflora*) roots will take away "any bruise gotten by falls or woman's wilfulness in stumbling upon their hasty husband's fist."

PLANTS OF GOOD AND EVIL

As a corollary to plants of virtue, there were plants of evil or of contraevil. The barberry, with its thorns and fruits like drops of blood, was symbolic of the Crown of Thorns, and a chaplet over a door was a sign to the evil one that only true Christians were within. Cloves look like nails . . . nails of the Cross . . . Christ's Passion for mankind . . . salvation and a traditional Easter cookie in Greece has a whole clove embedded in it. Most plants associated with holidays have traditional antievil associations that predate Christianity. Evergreens, flourishing when all else is brown and sere, were symbols of enduring life, and their association with midwinter festivals was obvious. Holly (*Ilex spp.*) is a plant that was hated by witches because its berries, once white, became red with Christ's blood. The Druids used it to build tiny huts as homes for woodland spirits that would protect humans, and even earlier the Romans included it in Saturnalian festivals to ward off evil. The mistletoe, too, has significance buried in pagan belief, and its non-Christian symbolism gave title to one of the most important studies in comparative religion, *The Golden Bough* of Sir James Fraser. Christian legend says that it was originally a large tree used to make the Cross, that it shrunk to its present size, and was doomed to live off the strength of other trees. Norse legend says that the mother of the god Balder asked all living things to protect her son, but the mistletoe wasn't asked. Balder went around bragging that nothing could harm him, and evil Loki, hearing of the mistletoe's failure to promise protection, dared Balder to stand up and prove his invulnerability. Loki gave the blind Hodur an arrow made of mistletoe which killed Balder. The goddess of love restored Balder with a kiss, as we cheerfully restore each other in the same way.

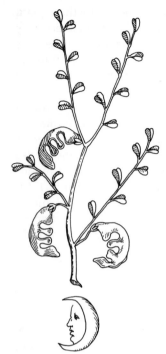

Figure 3.20
One of the "lunar" medicinal herbs. Taken from *Phytogono-mica* by Giambattista Proto published in Naples in 1591.

Plants of evil were rare, but included several members of the tomato family (Solanaceae) and a few other hallucinogenic and poisonous plants. Many otherwise useful or innocuous plants could be used to cast spells. St. Luke's flower, the marigold (*Calendula*) figures in one.

> Take marygold flowers, a sprig of marjarom, thyme and a little wormwood, dry them before a fire, rub to a fine powder, sift through a piece of fine lawn. Simmer with a small quantity of virgin honey in white vinegar over a slow fire. With this, anoint the stomach, lips and breast while lying down. Repeat these words three times: St. Luke, St. Luke, be kind to me. In my dreams, let me my true love see.

In an age when Christian physicians based their skills on the old writings of Galen, Aristotle, and Dioscorides (pagans all!), many were held in ill repute and nonphysicians turned to doctrine of signature plants or to other means of curing. Nicholas Culpepper published *A Physical Directory* in the fifteenth century and used his book to attack the doctors. He called them "a company of proud, insulting Doctors whose wits were born about 500 years before themselves." Culpepper espoused a combination of signature plants used with botanical astrology. Since, he reasoned, people were under the influence of a particular star, it followed that plants and animals were also affected similarly. Thus, the potency of plants as medicines depended upon the celestial bodies, and the healer should use that herb under the sign of that planet; "the eyes are under the luminaries" and eye troubles are due to the moon and plants picked at full moon would be effective (Figure 3.20). Philippus Aurelius Theophrastus Bombastus von Hohenheim, city physician of Basel and known throughout Europe as Paracelsus said:

> each plant is to be under a terrestrial star, and each star is a spiritualized plant. Each plant is under the influence of some particular star and it is this influence which draws the plant out of the earth when the seed germinates.

This confirmed the even older belief as to the importance of the phase of the moon in planting and in harvesting.

Up to the beginning of the eighteenth century, plants were known to possess means of communicating with people and of rectifying wrongs done to their vegetable persons. They could shriek, speak, or sing. The American Indian apologized for stripping a birch of its bark to make a canoe, and woodcutters in cultures as disparate as fourteenth-century Bavaria and seventeenth-century Japan would never enter a forest after dark for fear that the spirits of trees they had cut down would wreak a horrible revenge. Children were passed through holes in trees (Figure 3.21) to "rebirth them and make them well." One could pull petals from a daisy and know whether one was truly loved . . . something that no one would ever think of doing today! From the beginning of the twelfth century, people spoke of the Tartarian, or Sythian, lamb, the plant-animal. Described by Sir John

Figure 3.21
The birth of Adonis from a tree. Taken from a fresco by Bernardino Luini of Lombard painted ca. 1500.

Maundiville in 1360, it was a large shrub bearing a full-sized lamb fixed at its navel (Figure 3.22). It fed by stretching to reach the grass surrounding the rest of the plant, but since most of the vegetable lamb plants were eaten by wolves, few people saw them. Medieval clerics argued that it grew in paradise, for the vegetable lamb was the True Lamb of God. It is likely that it was a cotton plant (Chapter 8), seen by European travelers to India. The legend continued until at least the middle of the seventeenth century, along with the goose tree on which grew barnacles which gave birth to geese.

Tracing the history and symbolism of virtually any plant is a fascinating exercise in comparative religion, sociology, philosophy, and history. A case in point is the rose. Many of the hundreds of cultivated roses are from the Orient, and hybridizations and selections over the past several hundred years has obscured the botanical and geographical origin of the roses we now grow in gardens. True roses have been cultivated for centuries and have contributed to mythology for about the same length of time. In Greek legend, Chloris, goddess of flowers, found a dead nymph and asked the other gods to transmute her into a special flower. Aphrodite gave beauty, the Graces bestowed brilliance, Apollo gave breezes to waft its perfumes, and Dionysius, god of wine, gave nectar. Thus was born the rose. Cupid gave the new flower to Harpocrates, god of silence, and this resulted in our term *sub rosa* or "under the rose." Until the seventeenth century, a rose was hung from the ceiling of rooms in which silence and secrecy were to be observed. Ceiling paintings in council chambers still use the rose in their designs, but modern legislators apparently don't know Greek mythology.

Specific flower colors were mandated by the Council of Trent in the late sixteenth century, but the use of the red and white rose in Christian ritual predates this by centuries. Plutarch spoke of the rose "garded in modest bud" and noted that the sun is born each day from the flower bud. White as purity (remember the white rose in "Beauty and the Beast"?) and red as martyrdom were color associations before Christianity. The rose was adopted in the third century as a religious symbol to point up stages in the life of Christ.

Figure 3.23
Heraldic use of the rose. Red rose of Lancaster (left), white rose of York (center), and the combined red and white rose of the Tudors. Taken from Parker's *Annales* of 1855.

The rose has been a heraldric device since the tenth century. The War of the Roses in 1455–1485 between the English Houses of York and Lancaster was named for the battle devices of these families, white for York and red for Lancaster (Figure 3.23). The Tudors came out on top, and, to placate the opponents, they combined the two symbols. Because of the association of roses with the Virgin, presentation of a single rose to a woman was to bestow a blessing upon her. North Americans have corrupted this practice in our penchant for extravagance. We rarely give a single flower, but often present a dozen or more. In point of fact, giving more than one rose was, during the Renaissance, a symbol of extreme and not particularly holy passion. Acceptance of a bunch of roses signaled the sealing of an unspoken bargain that the passion would be reciprocated at the earliest possible *sub rosa* moment.

ADDITIONAL READINGS

Crow, W. B. *The Occult Properties of Herbs.* New York: Weiser Press, 1970.

Eliade, M. *Patterns in Comparative Religion.* New York: New American Library (Meridian Books), 1963.

Fergeson. G. *Signs and Symbols in Christian Art.* New York: Oxford University Press, 1973.

Fraser, Sr. J. *The Golden Bough.* London: Macmillan, 1911.

Jacob, D. *A Witch's Guide to Gardening.* New York: Taplinger Publishing Company, 1965.

Kirk, G. S. *Myth, Its Meaning and Functions.* New York: Cambridge University Press, 1970.

Koch, R. "Flower Symbolism in the Portinen Altarpiece." *Art Bulletin* 46(1964):70–77.

Lehner, E., and J. Lehner. *Folklore and Symbolism of Flowers, Plants and Trees.* N.Y.: Tudor Publishing Company, 1960.

Skinner, C. M. *Myths and Legends of Flowers, Trees, Fruits and Plants of All Ages and in All Climes.* London: privately printed, 1911.

The Tree of Life
and the Rod of Aaron

Thy rod and thy staff they comfort me.
PS. 23:4

In Gen. 2:8–9 it is written, "And the Lord God planted a garden in Eden to the eastward and there He put the man He had formed. And the Lord God caused to grow out of the ground every tree that is pleasant to the sight and good for food. And the Tree of Life in the midst of the garden and the Tree of Knowledge of Good and Evil." The Tree of Life appears only three times more in the Testaments: in Gen. 3:24 where cherubim and a flaming sword guard the way to the Tree of Life, as an allegory in the tale of Jesse (Isa. 11:1), and in the Apocalypse (22:1–2) where the Tree is on the banks of a river. With both Testaments replete with plant symbols, one wonders why this powerful concept was used so sparingly. To evaluate this question, we must go back to the period before the Israelites.

The Levant of the Middle East—Iran, Iraq, Syria, Palestine—were a shifting series of nation-states including Mesopotamia, Phoenicia, Chaldea, Sumeria—lands of shifting sands, peoples, and gods (Figure 3.24). They were harsh lands, where water was literally life, where obtaining food from the soil was difficult, and where natural phenomena like lightning, dust storms, disease, and death had to be explained to give meaning to life itself. All these pre-Israelite cultures had pantheons of gods, and each was venerated in forms which the people could see and feel. Rocks, solid and unchanging; water, the source of life; and trees, visible signs of renewal, growth, and fertility, were immortal objects that "lived forever." A tree, solitary and tall against a desert sky, was endowed with will, strength, and perception. It epitomized death in

Figure 3.24
Winged bull unicorn kneeling before tree of life composed of lotus buds. Taken from an Assyrian stone relief.

Adapted from "Maypoles and Earth Mothers," *Natural History* 85(1976):4–8. Copyright American Museum of Natural History.

Figure 3.25
Chinese version of a tree of life. Taken from a tomb of the Han dynasty, ca. 200 B.C.

Figure 3.26
"God's deer" bearing a Japanese tree of life. Taken from the Kasuga Mandara in Nara drawn ca. 700 A.D.

the dry season and resurrection when the spring rains came to the parched land. Trees were sacred symbols—visual manifestations or epiphanies—of powerful forces, and their desecration or removal was sinful. Modern women who interpose baby buggies between a shade tree and a bulldozer are repeating an act of faith which almost predates history.

Specific trees were assigned to specific deities. The sycamore (*Ficus sycomorus*) was sacred to Egypt's Ra, Hathor, and Nut; the date palm (*Phoenix dactylifera*) to Marduk of Babylonia; and oaks (*Quercus spp.*) to Zeus. Zeus's oak groves at Dodena, rustling leaves of which were the voice of the thunderer, were paralleled by those of the Celtic Druids. Hera was made manifest in willows (*Salix spp.*) that grow by springs, and who was to say that the tree was not the cause of the water? Apollo was the laurel, whose leaves contain enough prussic acid so that His oracles could eat the god and become intoxicated with Him. As deities, trees often gave birth to humans. Yggradsil gave birth to the first Norseman, Ask. Queen Maha-Maya gave birth to Prince Siddartha—He who became Gautama Buddha—through strength provided by the Bo tree (*Ficus religiosa*). The Yurucase of Bolivia know that Tiri opened a tree and brought forth the people. In east Africa, the first man was born within the baobab and fed on its breast-shaped fruits. In Persian mythology the tree of life stands in the sea where its seed, falling into the water, maintain the fertility of the earth. The Chinese, Japanese, Aztecs, and American Indians had such trees and developed virtually identical mythologies about them (Figures 3.25–3.27).

TREES OF LIFE

Trees of life were the highways traveled between earth and heaven. Ra of Egypt started His daily journey through the sky from the branches of the sycamore fig (Figure 3.28). Indeed, Ra became the sun

Figure 3.27
Aztec tree of life on which perches the parrot of wisdom. Taken from a stone relief in Mexico.

Figure 3.29
Pacal, ruler of the Aztecs, is captured in death by the cosmic tree. Taken from a bas-relief carved on his limestone sarcophagus dated 683 A.D.

Figure 3.30
Baobab, *Adansonia digitata*, in west Africa.

god by climbing the tree of life, and His earthly counterparts, the pharaohs, reached their home in heaven by following Ra. Rulers of the Aztecs, too, moved between earth and heaven via tree ladders (Figure 3.29), as did Jacob (Gen. 27:10–14). Jack used a beanstalk. The tree-ladder symbolism is almost universal, and one doesn't walk under such a powerful symbol. Egyptians solved the riddle of why the sky doesn't fall by asserting that the universe was a huge box whose lid, forming the sky, was held up at its corners by branches of the tree of life. In some cultures, the tree was upside down. The Rg-Veda of the Hindu says: "The branches grow towards what is low, the roots are on high and all worlds rest on it." If one sees a tree leafless against a sky, it is easy to believe that this is true (Figure 3.30).

All this was a source of embarrassment for the writers of the Testaments. Ancient Israelites invaded the lands of Canaanites and Moabites who believed in trees of life. Their deity, Asherah of Canaan, was symbolized by a tree stripped of branches, and Old Testament writers commented most reprovingly of such worship. The Israelites had no doubt about where they originated; they were made by Yahweh in His own image and He did not have to show Himself to his followers; his transcendence was a canon of faith. Success in agriculture, in procreation, and in battle was controlled by a single, male God with whom they had a covenant. Placation of the abode or the epiphany of a deity was not only unnecessary, but as evidenced in the first two commandments, sinful. Yet, as Israelites married women of subjected nations, they absorbed the deities represented by trees and adopted and reshaped myths, symbols, and concepts of the conquered peoples. The tree of life, so basic to other Middle Eastern faiths, was just one of these symbols and, somewhat altered, it could not fail to have been included in Genesis. Nor could the symbol fail to have influenced those peoples of pagan Europe who came in contact with cultures who accepted and revered trees of life.

Implicit in tree of life motifs is the belief that it was at the center of the earth, that it was the *axis mundi* about which the earth turned. Virtually forgotten in our mobile Western society, this concept was central to beliefs of many cultures. As the tree of life was permanently wedded to the earth in the people's homeland, so do we, today, "put down roots" and the wanderer is "rootless." The Cross (Figure 3.31) and the Menorah are not only trees of life, but they are the center of the believer's world. Cemeteries in the Western world are planted as parks, not only for esthetic considerations, but because the plantings are symbols of the tree from which life and death spring. The annual resurrection of leaves in the spring, the changes in leaf size and color through the summer, and the fall of leaves in autumn parallels the life of man. The white cedar (*Thuja occidentalis*) of the Northeast and its Asiatic equivalent (*Thuja orientalis*) are both called arbor vitae—tree of life. They, like other conifers, are symbols of everlasting life.

"Paradise" is the Persian word for garden, a rare thing in that dry land, and our concepts of the heavenly paradise involves trees of life. Rev. 22:1–2 said that those victorious over evil "shall eat of the fruits of the Tree in the heavenly city." According to the *Mishkat al-Masabih* (a Persian commentary on the Koran), the Tooba tree in the Persian heavenly garden not only gave everlasting life to those who ate its fruit, but also gave men the strength and vigor to have connection with many women, "the power of a hundred men given to a single man." When Mohammed journeyed to Heaven, he saw on God's right hand the Lote tree, made of gold and laden with jewels, with the river of Paradise flowing from its roots. Rabbinical scholastic scholars suggest that the biblical Tree of Life was forbidden Adam and Eve because it conferred immortality (Gen. 4:22), a concept that is in the belief structure of cul-

Figure 3.31
Combination of Tree of Life and the Cross. Taken from a seventeenth-century Romanian house decoration.

130   *Plants in Religion*

tures all over the world. Trees, and what they stood for, could not be other than sources of awe, desire, fear, and wonder.

TREES OF GOOD AND EVIL

A tree with an unusual shape or one riven by lightning acquired special powers of malignity. Those that failed to grow fruit could be inhabited by evil spirits. Indians believed that the banyan (*Ficus benghalensis*) could be dangerous under certain circumstances, and, since one could not know the circumstances, it was prudent to be careful; one didn't sleep under this sacred tree or carelessly knock off a branch. Predating Greek mythology, it was known that trees were the dwelling place of spirits, elves, or sprites who could cause suffering and had to be placated. Ornaments on Christmas trees were memorials to sacred animals and spirits that lived in that tree and were placed there to ease the insult of having their homes removed to human habitats. Witches used bits of the local tree of life to cast evil spells, but these could be countered by keeping branches of the tree on one's door. A baptized man could kill a witch with such a branch or by driving a stake of the tree through the heart of warlock or werewolf.

Trees could also be benevolent. A fine looking tree or one used for food could be the home of a protective spirit. Sick children could be cured by passing them through the hole of a tree—essentially resurrecting them. Boy Scouts plant trees on Arbor Day because our ancestors planted a tree for each baby, knowing that the tree would become the home of a spirit who would, in gratitude, watch over the child. As a plant with many seeds, trees became fertility symbols. The pineapples on four-poster beds are stylized pine cones (Figure 3.32), the manifestation of several goddesses of fecundity.

THE STAFF AND ROD

Staffs, rods, and wands are also part of this complex of symbols. In order to demonstrate that he was the envoy of a powerful Force, Moses had his brother Aaron cast his staff upon the ground whereupon it became a serpent (Exodus 7). In the same chapter, the staff is used again, this time to turn water into blood—one of the plagues visited on Egypt. When the children of Israel thirsted, Moses was commanded to touch a rock at Horeb with his staff, and a spring gushed forth. We remember that Noah's dove brought back the branch of an olive, whose wood formed the staffs of many Mideastern deities. Later (Num. 17:17–26), Yahweh commanded that the staffs of each of the tribes be collected before His shrine. The staff of Aaron, elder of the Tribe of Levi, budded, flowered, and produced ripe almonds, testimony that priestly functions were to be assumed by the Levites. The budding staff appears in legends of St. Christopher and in the fable of the Glastonbury Thorn (*Crategus oxyacantha praecox*) in England where it budded

Figure 3.32
Cones of the fir tree, *Picea,* were called pineapples and were fertility symbols. Taken from *Hortus Floridus* by Crispean van de Passe published in Arnheim in 1617.

Figure 3.33
Zeus with staff and bundle of
lightning. Taken from a
Greek amphora of 400 B.C.

from the staff of Joseph of Arimathea. Charles I, in an effort to "root
out, branch and all" the Roman Church, had the tree cut down. Representations of deity frequently involved a staff. Baal of the Canaanites
carried a sceptre of cedar; Zeus a staff of oak (Figure 3.33); Anu (sky
god of Sumeria) carried Kiskanu, the tree of life, and Dionysus, originally a Babylonian harvest god, had His staff of the tree of life twined
about with ivy and surmounted by the pine cone of fertility. Mortal
rulers use a staff, rod, or sceptre to make clear that they possess the
powers of the deity; the royal umbrella of the Kings of Siam was a
frond of their tree of life, and the Christian bishop's crook is descended
from the Zoroastrian priests of Persia who carried the tree to protect
their "flocks."

Possession of a branch of God's tree conferred extraordinary
powers on mortals. In Scandinavia, the ash (*Fraxinus*) and the rowan
(*Sorbus aucuparia*) were sacred to Odin and Thor. The word rowan is
derived from *rune* (to divine), and twigs, each marked with an oracular
message, were selected by priests as prophetic. The ancient Judaic judicial system included having a judge choose a stick from a bundle,
each stick containing the name of a litigant. Our custom of choosing
long or short straws is derived from this custom. The children's game of
jackstraws and the divination of tossing sticks in the Chinese I-Ching
have similar origins. The magician's wand is a magic staff made of the
tree of life. Modern magicians use a black rod to add the dark powers
of the underworld to enhance their spells. Circe, daughter of Helios,
used a magic wand to transform lustful sailors into pigs (Figure 3.34).
Medea used her wand to assist Jason (and little good it did her!). The
traditional broomstick of witches is a profane use of the tree of life.
Touching with a wand or rod can either confer power—as when a notable is knighted—or bring power—as when an American Indian "counts
coup" in battle by touching his enemy with a feathered stick.

THE CADUCEUS

Although the medical caduceus was not formally adopted as a
symbol until the 1956 General Assembly of the World Medical Associations, it has symbolized the healer for 3000 years. Hermes, messenger of Zeus, conducted souls of the dead to Charon's boat which
transported the righteous to the Elysian fields where they were reborn
as immortals and transported the unrighteous to Hell. His staff was
originally a forked stick, and only later was it depicted with the forks as
wings. Hermes' staff was an extension of Zeus' power, given the demi-
god both for its inherent power and as a symbol of Zeus. Carrying the
king's staff was continued by royal envoys until the eighteenth century,
and passing the baton in a relay race is similarly related. Another
bearer of the Caduceus was Asklepios. He was raised by Cheiron the
Centaur who taught the boy the healing arts. Since recovery from seri-

Figure 3.34
Circe turning one of Odysseus' sailors into an animal.
Taken from a Greek
amphora.

Figure 3.35
Asclepius, god of healing. The sacred rod is intertwined by the serpent of renewed life.

ous illness was equated with return from the dead, Asklepios liberated people from the thrall of the underworld with a branch of the tree of life (Figure 3.35). The snake, depicted as coiled about one leg, was later moved to the staff as a symbol of rebirth since snakes regularly shed their skins. The wings on the medical caduceus are a later addition (Figure 3.36).

Although we use magic staffs to search out water, their original purpose was to locate metals. In 1556, Georgius Agricola, physician of Bohemia (modern Czechoslovakia) published a *De Re Metallica* in which he gave directions for using the *Virga mercuralis*, the rod of Mercury, to locate ores (Figure 3.37). As late as 1710, Jonathan Swift noted a "certain magic rod that, bending down its top, divines when o'er the soil has golden mines." Dowsing or water-witching dates from the late sixteenth century when the Baroness de Beausoleil popularized the art in *La Restitution de Pluton*. The rods are, of course, branches of sacred trees—oak, rowan, hazel, willow—and their users are viewed with awe, superstition, and suspicion. Does it work? It all depends upon who you ask.

Figure 3.36
The Sigillum Chirurgorum Dresdensium, a crest used in a book on diseases published in Dresden in 1663.

Figure 3.37
Dowsing for minerals with rods made from hazel, *Corylus spp.* Taken from *De Re Metallica* by Georg Agricola published in 1556.

THE MAYPOLE

On the first day of May, tall poles are erected in public parks and are hung with colored ribbons. Young girls in frilly dresses are crowned with flowers and dance, first clockwise and then counterclockwise, wrapping and unwrapping the ribbons about the maypole (Figure 3.38). It is a happy, innocent amusement, a symbol of the joy of spring, but few recognize that the maypole and its dance is one of the oldest and most sexual of public ceremonies involving the tree of life. The maypole originated in the lands of the Mesopotamians and Syrians who all believed in an Earth Mother. Ishtar of the Assyrians became Astarte of the Chaldeans, Asherah of Canaan, Cybele, Demeter, Ceres, Diana, and many others. It was to Earth Mother that people prayed for good crops, human and animal fertility, and, through Her, intercession was asked to the male God for rain, gentle breezes, and pure, ever-flowing springs. If Earth Mother was with child, so would be the fields, the women, and the ewes; trees long dead during the dry season would burst forth with new leaves, and flowers and fresh pasturage would appear in the fields.

But gods were forgetful and had to be reminded annually of their responsibilities to the faithful, and the people simultaneously reminded of their duties to the gods. Cybele-Astarte-Demeter was symbolized by a tree, shorn of branches and carried in solemn procession to the temple where She was erected before the awed gaze of the people. Decked with flowers, the tree was ritually worshipped with dances that included sexual orgies. Minoan Cretans painted an olive branch bright pink and planted it in a bed of grass as the imminent symbol of Isis.

Figure 3.38
Children dancing around a maypole in Central Park, New York. Etching of 1877.

Old Testament writers commented on such worship in most reproving terms. Thus, the maypole was not accepted in Judeo-Christianity, and we must look to the Greek and Roman roots of our Western heritage for information.

The Hellenes probably invaded Greece from the Near East, changed the names of the gods, and made both gods and goddesses more "human," accessible, and understandable. Cybele of Phrygia became Artemis, and Demeter of Phoenicia became Maia—daughter of Atlas, mother to Hermes (courtesy of Zeus) and the month of May. We know only that the mayday pole dance was retained and modified, but little else.

Roman rule was marked by changes in the names of the gods, but not in their function. Demeter became Ceres, and Cybele-Artemis became, in part, Diana. The spring ritual began in Rome on March 22 when a pine tree was cut down, debranched, decked with violets as a symbol of the male God, Attis, and borne to Diana's temples. March 23, the Day of Blood, started with dancing around the erected maypole. The music of cymbals, drums, and flutes became wilder as the day progressed. Frenzied by the dance, participants lacerated themselves and dripped their blood upon Diana's epiphany. The dance culminated in self-emasculation of young men wishing to become Diana's priests. The excised organs were thrown at the tree to hasten the resurrection of the earth and its impending fertilization. The ceremonies ended when a procession, led by a human representative of the goddess and her male consort—the King of the Woods—mated to signal the fertilization of the earth. The King of the Woods, originally the male god, also had different names in different cultures—Attis, Marduk, Virbius, Zeus, Jupiter—and was symbolized by a tree. It was usually a forest giant like the oak or the ash, one that annually lost its leaves, an indication that Earth Mother and Her conifer were immortal, but a mere male was replaceable.

With minor variations, the festival was similar throughout Europe. Young men and women went to the woods, cut down a tree, stripped off the branches, and carried it triumphantly into the village. It was decorated with flowers and other symbols of fruitfulness and set up in front of the church—to the discomfiture of the clergy. A May Queen was selected. She and the King of the Woods—called Green George, Father May, Bark King, Grass Lord, or Leaf Man—presided over the dancing. As night fell, the queen and king mated in the fields to emphasize the point of the ceremony, and this act was affirmed by the faithful. Celts of Scotland, Ireland, Normandy, and parts of Scandinavia included a fire ceremony in their May Day rituals. These beltane fires were lit from a flame started with a firebow whose spindle was a shaft of oak (Zeus) fitted into the slot of a board of softer wood such as willow (Hera). Tinder was dried mushrooms and puffballs. In Sweden, the Maj Stäng is dressed with crossarms bearing hoops of willow wrapped with flowers. Hoop rolling on the quadrangles of many colleges is derived from these garlanded May hoops except that, in the sixteenth century, the hoop stick of oak was thrown through the hoop instead of being used to propel it.

It is to England that we must look for the North American version of the maypole ceremony. Before the seventeenth century, the mating of the queen and Jack o' the Green was not pretend, and the midnight cutting of the tree was accompanied by the normal exuberance of lusty youth. Phillip Stubbes, in his *Anatomie of Abuses* of 1583 called the ritual an "act of Satan," the maypole itself "that stynking ydol," and noted that after the revels, "scarsely the thirde part of ye girls returned unde-filed." In 1626, Governor William Bradford of the Plymouth colony appointed Thomas Morton as administrator of Merrymount, a suburb of the main colony. Morton, apparently taking the community's name seriously, introduced the maypole to New England, thereby creating a scandal. Bradford noted that the celebrants invited Indian women to participate in "dancing and frisking together . . . and worse." John Milton wrote of the dallyance of Zephyr and Aurora "as he met her once a-Maying." The Morris dances of England were May dances, and May Queen Mary only later became Robin Hood's "Maid Marion."

By the beginning of the eighteenth century, Mayday celebrations had become just celebrations and were no longer spring fertility rituals. By 1833, Tennyson could write: "You must wake and call me early, call me early Mother dear./For I'm to be Queen o the May mother, I'm to be Queen o the May." Kipling remembered the old ways in a bit of doggerel: "Oh, do not tell the priest our plight or he would call it sin./But we have been out in the woods all night conjuring summer in."

One is struck by the universality of concepts which flow around the tree of life. Life, death, and the perils of everyday existence must somehow be understood, for if they are not explicable, reality is without meaning. The psychologist Carl Jung seized upon this as one of the arguments to support his theory of the collective unconscious, an attempt to integrate the universal symbols human beings use to relate themselves to their environment. Whether Jung was correct in believing that homologies among symbols in all cultures represents an archetypical and universal mind, or whether the choices and beliefs were identical because the biological facts of life are identical is an open and most intriguing question.

ADDITIONAL READINGS

Butterworth, E. A. S. *The Tree at the Navel of the Earth.* Berlin: W. de Gruyter, 1970.
Campbell, J. *The Masks of God*, 4 vol. New York: Viking Press, 1964.
DeWaele, F. J. M. *The Magic Staff or Rod in Graeco-Italian Antiquity.* Amsterdam: 1927.
Eliade, M. *Patterns in Comparative Religion.* London: Sheed & Ward, 1958.
Hastings, J. *Encyclopedia of Religion and Ethics.* New York: Scribner, 1916.
Holmberg, U. "Der Baum des Lebens." *Ann. Acad. Sci.* Fennicae B. 16 (3) (1922):1–146.

Jung, C. G. *The Archetypes and the Collective Unconscious.* Princeton, N.J.: Princeton University Press, 1959.

MacMulloch, J. A. (ed.). *The Mythology of all Races.* London: Marshall Jones Company, 1928.

Porteus, A. *Forest Folklore, Mythology and Romance.* London: G. Allen, 1928.

Thiselton-Dyer, T. F. *The Folklore of Plants.* London: Chatto & Windus, 1889.

Peyote

A white man uses prayers out of a book; these are just words on his lips. But with us, peyote teaches us to talk from our hearts.
CHEYENNE PEYOTE LEADER

In the American Southwest and Great Plains, the post-Civil War nineteenth century was, for the American Indian, a period of turmoil. When the overcrowded East viewed the land beyond the Mississippi, it was considered manifest destiny to open the country for families devastated by the war. Increased immigration, a rising birth rate, railroads, and the Gatling gun made it clear that the Indians of the plains were doomed. The Indian Agency of the Great White Father was resettling hunting tribes on reservations selected to include the worst lands of the Oklahoma Territory—lands so deficient in water and good soil that they couldn't be tilled, particularly by Plains Indians whose culture was not agricultural. The Indians could either starve or become completely dependent upon federal welfare. More insidiously, the government was supported by missionaries whose task, as they saw it, was to make the savage into a Christian content with his or her lot—for the meek shall inherit the earth. Conversion was facilitated by violating the U.S. Constitution and channeling much of the welfare through the missions. Food was distributed with covert and overt statements that the people's gods were impotent—demonstrated by the success of the Seventh Cavalry—and that their customs were savage and immoral. Not unexpectedly, the world of the American Indian fell apart and disease, sloth, and degradation became the lot of the Indians. Deprived of their culture, swindled by federal authorities, seeing their children kidnapped into mission schools which specifically denied their heritage, and buying firewater at prices inflated by the alcohol prohibition act passed by the combined efforts of the missions and the bootleggers, the people began casting about for some mechanism which would provide a modicum of dignity, hope of survival as Indians, and some spiritual sustenance.

The *Deus ex Machina* was a cactus botanically known as *Lophophora williamsii*, the peyote (Figure 3.39). The plant itself is an inconspicuous grey-green hemisphere with few spines, sparse white hairs, and a carrotlike tap root. When dried into *mescal* buttons, it is a hard, medium-brown disk less than a centimeter in diameter. Peyote is a modification of the Aztec *peyotl* which means "caterpillar's cocoon" referring to the

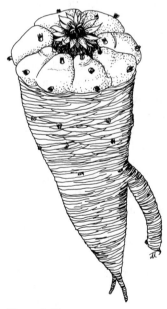

Figure 3.39
Peyote cactus, *Lophophora williamsii.* The top is about 3 cm in diameter.

white, wooly interior of the plant. It was in Mexico where Western man first learned of it. In 1560, the Spanish Jesuit Bernardo Sahugan wrote, "there is another herb called peiotl; it is produced in the north country; those that eat it see visions either frightful or laughable; this intoxication lasts two to three days." Francisco Hernandez, physician to the King of Spain, noted that the Chichimecas imputed to it wonderful properties including the ability to foresee the future, to uncover lost articles, to predict the weather, "and other things of like nature." Peyote use among the Aztecs was restricted to the priesthood whose members used it to reach a state in which they could communicate directly with the gods. As the cross accompanied the sword across New Spain, peyote was ruthlessly suppressed and the shamans put to torture. From 1570 until the middle of the nineteenth century, peyote was the property of *brujos* (warlocks) and *curanderos* (healing women); note that it was not a generally used hallucinogen, but retained its mystical attributes.

PEYOTE AS A CULT OBJECT

Peyote's physiological effects are due to at least nine alkaloids (Figure 3.40), divided into two major categories. The first group produces a mild euphoria, accompanied by dilation of the pupils, increased blood supply to the face and body, and diminished kinesthetic responses. There is a tendency for the participant to talk a lot. This phase is relatively short-lived and is followed by lightheadedness and nausea. Only after this phase do the other peyote alkaloids begin to be effective. The participant is not sleepy, but finds it more comfortable to lie down. The pupils dilate excessively, motor coordination is impaired, reflexes are more pronounced, but the sense of touch and minor physical aches and pains are reduced. Muscular twitching, particularly of facial muscles is pronounced. In addition, there are profound mental changes, the most dramatic being over-estimation of time and an inability to focus one's attention for any length of time. Sensory hallucinations, particularly those involving vision, are noted, with the participant experiencing brilliant colors and patterns. He or she may hear the voices of ancestors, protective spirits, or gods. In some cases, unusual odors or tastes may be perceived as integral parts of hallucinogenic visions. Hallucinations are short-lived, and the participant may experience many visions during the period of maximum responses.

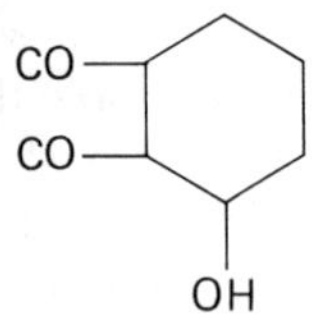

Peyote made its first appearance north of the border around 1870 when Mescalero Apaches obtained and used it ritually. It entered the Oklahoma Territory through a notable figure, Quanah Parker, son of the Comanche war chief, Nokoni, and Cynthia Parker, a white woman who had been captured as a child. Himself a war chief who fought the whites, he capitulated in 1875 and attempted to bridge the gap between cultures. In 1884 at the age of 40, he became seriously ill, was treated with peyote by a Mexican curandera, and recognized that it could be a useful factor in binding his people together.

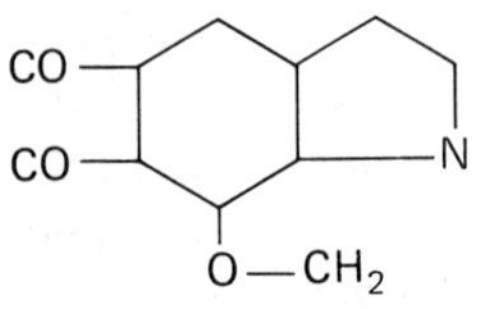

CO
CO
OH
Mezcaline

CO
CO
O—CH$_2$
Mescaline

CO
CO
OH
Anhalamine

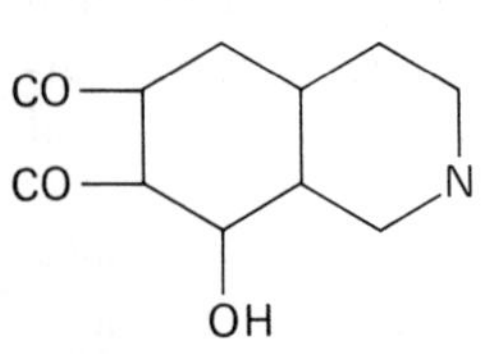

Figure 3.40
Major hallucinogenic alkaloids in peyote.

The Comanche were the right tribe at the right time. Powerful, warlike, prone to visions, and with a long history of religious ritual including self-torture, they were not only among the more mystical Indians of the Southwest, but as hunters on an agricultural-welfare reservation, they were completely demoralized. Taking cultural elements from Comanche, Kiowa, Apache, and parts of Christianity—notably (and ironically) the concept of love—Parker fashioned a series of ceremonies which were neither at variance with the beliefs of the tribes, nor destructive to the cultural mix that was developing into pan-Indianism. Parker's prestige, his genetic bridge between white and red, and his superb organizational ability were focused on the development of a ritual system to provide the self-esteem that Indians desperately wanted.

Between 1890 and 1900, the peyote ritual system was restricted to Oklahoma and part of New Mexico, but between 1900 and 1910 it spread rapidly and widely. The Tonkawa, Lipan Apache, Cheyenne, Arapaho, and Navaho embraced it by 1891, the Sac, Fox, Shawnee, Mescalero, and Witchita were added by 1895, and the Algonquin, Omaha, and Pawnee by 1900. The Peyote cult was introduced to the Winnebago in 1900 by John Ray. Returning to Nebraska he converted his family, then a few friends, and eventually so many of the Winnebago and Omaha that the tribe was rent with a bitter dispute between the pan-Indian peyotists and the Christianized conservatives. Eventually, peyote reached up to Hudson Bay and Alaska. Considerable impetus was supplied the cult by the graduates of Carlisle, an Indian college whose students were effective radicals and innovators. With accretions of elements from Christianity and the inevitable cross-cultural modifications attendant upon its adoption by different tribes, tribal leaders were attracted to the cult. Church buildings were erected, and business and professional members provided respectability and a formal infrastructure. The Navaho, members of an agricultural, settled, and sophisticated group, could now make common cause with Florida Seminoles or the Potlach culture of the Columbia River Basin. With less scholarly nitpicking than attends Judeo-Christianity, peyote practitioners held a fairly uniform set of beliefs and rituals which include the concept that God put some of his holy spirit into the peyote and gave it to the Indian. By eating this sacrament, God's spirit is absorbed. Thus, although central to the system of beliefs and practices, peyote is a means and not an end.

OPPOSITION TO THE PEYOTE CULT

As peyotism spread on reservations, opposition to it came not only from conservative tribal members, but from the white establishment exemplified by the Indian Bureau and the missions. Insinuating that the peyote ceremony was a black mass, and that mescal buttons inflamed ignorant savages, the missions demanded and got, in the person of William "Pussyfoot" Johnson, an Indian Bureau officer who, they believed, could crush the cult at its center in Oklahoma. Johnson was a U.S. marshall and a nondenominational fundamentalist preacher. Ostensibly, his job was to enforce laws preventing alcohol consumption by the Indians, but he stretched the law by stating that peyote was an intoxicant. His favorite trick was to raid a peyote ceremony, jail the celebrants, and release them only after they had signed a temperance pledge which he then enforced by jail sentences for breach of contract. The real objections to peyote were clearly stated by a minister, Walter C. Roe, who wrote in 1908, "one of the worst results of [peyote] use is that it creates a very strong barrier in the way of presentation of the Christian religion. . . ." Reverend Robert D. Hall, then in charge of Indian work for the YMCA, demonstrated to the psychology department of Yale University that peyote was dangerous to health since, after eating several buttons, he vomited. What Hall didn't tell these Easterners was that vomiting is part of the ceremony; the peyotist believes that he or she is thereby ritually cleansed. A 1914 conference of missionary leaders and the Law and Order Section of the Indian Bureau resolved: "It is now well known that the increasing use of the mescal bean, or peyote, is demoralizing in the extreme. We recommend accordingly that the Federal prohibition of intoxicating liquors be extended to include this dangerous drug." This was an attempt to include peyote in the Harrison Narcotic Act of 1914. The Internal Revenue Agency ruled otherwise. Although concerned with the "ravages of peyote," the IRA, charged with enforcement of the Narcotics Act, published in 1916 a detailed attack on peyote as a drug and on peyotism as an immoral and anti-Christian cult: "Its followers seem to abandon Christian teachings and in their frequent nightly gatherings indulge in excesses through the midnight hours in which men and women participate. It is claimed that in the nocturnal debaucheries there is often a total abandonment of virtue, especially among the women."

Yet, as early as 1891, James Mooney, an ethnologist with the Smithsonian Institution, had published a series of reports on the culture of Indian tribes on Oklahoma reservations. Mooney was vilified when he noted that peyotists were rarely drunk although surrounded by alcoholism, that they were clean, self-aware, and emotionally stable. As an observer and participant in peyote ceremonies, he wrote of the total absence of orgy, the dignity of the ceremony, and the reverent attitudes of the celebrants.

In 1918, Congressman Carl Hayden of Arizona introduced a bill mandating total prohibition of sales to Indians of alcohol and "other intoxicants" including peyote. The hearings were held by the Committee on Indian Affairs, and it was then that the Indians finally learned how powerful were the forces arrayed against them. They realized that if the Hayden bill passed, harassment would quickly become suppression, not only of the peyote cult, but also of virtually every facet of non-Christian Indian life. Testimony on both sides was marked by dramatic rhetoric, but that coming from the antipeyotists was notable for fabrications. Sexual immorality, drug addiction, and demoralization were charged, and denigration of reports of anthropologists, ethnologists, and pharmacologists duly inscribed as facts in the records of the hearings. Christianized, conservative tribal members were coached by missionaries, and the testimony of such solid citizens far outweighed that of the peyotists, many of whom could scarcely speak English or were long-haired college students. The hearings boiled down to two major points of contention. Is peyote a harmful, addictive, and intoxicating drug? Is peyotism a cult formed to foster a drug habit or is it a genuine Indian religion? For both questions, the committee found for the antipeyotists, reported HR 2614 out of committee, and it passed the House of Representatives. It did not get out of committee in the Senate, but the Indians knew how fragile was their victory. Clearly, protection could only come under the wings of the Constitution; the peyote cult had to become a legally recognized and established church.

THE NATIVE AMERICAN CHURCH

With the assistance of James Mooney, provisions of a charter were drawn up. It was at first decided to call the church the "First-Born Church of Christ," but this was rejected overwhelmingly, and the name "Native American Church" was chosen to emphasize intertribal solidarity. It was pan-Indian and also defined its participants as Americans. The articles of incorporation stated that it was a Christian religion, adding ". . . with the practice of the Peyote Sacrament as commonly understood in the several tribes of Indians." Incorporation was in the new state of Oklahoma in October 1918. For his efforts, Mooney was permanently expelled from Oklahoma by the Indian Bureau and died in 1921 without finishing what promised to be a classic study of Indian life. Opposition to the Church continued, with bills introduced (but never passed) into four Congresses. At the state level, peyote was outlawed in Kansas, Montana, the Dakotas, Arizona, Iowa, Wyoming, and New Mexico—all states with large reservations and active missions. The Indians responded by chartering the Church in these states and by 1925 legal harassment had almost stopped.

Peyote contains at least 15 psychoactive alkaloids, the most prominent and effective being mescaline. When used alone, as is usual in psychological experimentation, mescaline is a powerful hallucinogen, but when it is only one of the complex found in the cactus button, responses are milder, somewhat different, and usually less extended. Mescaline and related alkaloids are not restricted to *Lophophora*; at least seven other cactus genera contain them and one, the San Pedro cactus of Northern Peru, is employed by folk healers to diagnose and cure illness. Here, as with peyote, elements of Christianity are part of the ceremony, but the social and political structure found in the Native

American Church have not developed—possibly since Peru is officially Roman Catholic and ties between government and Church are stronger.

In defense of the missionary's use of the term "intoxication" throughout the debate on the peyote cult, at least part of the problem was a failure of communication. The Aztec word *mexcalli* was given the *Agava* or century plant from which the term *mezcal* developed to refer to the alcoholic drink tequila, made from fermented and distilled *Agava* sap. In addition, there is a shrub in the American Southwest, *Sophora secundiflora*, having dark red, beanlike fruits called mescal beans, which contain a very toxic compound causing nausea, hallucinations, and occasional death from respiratory failure. It was a custom of tequila drinkers to add crushed *Sophora* seed to their mezcal to render it more potent, and the seeds formed a basic part of the red bean dance—an oracular ceremony of the Mexican and Texas Indians. *Sophora* seeds are used by peyotists as decoration on ceremonial dress. This terminological confusion led whites to assert that the Native American Church fostered drunken orgies (the mezcal of tequila), ritual poisonings (the mescal of *Sophora* beans), and the drug habit (use of the ill-named alkaloid of peyote). Since the red bean dance also involved use of the poisonous, hallucinogenic, and possibly aphrodisiac *Datura* plant (Chapter 4), sexual debaucheries were included under the mescal rubric. This confusion was not cleared up until the 1930s, when it didn't matter any longer.

Whether the Native American Church is effective as a way of life and as a source of dignity and solace to the practitioner is difficult to determine. Its quarter of a million members know that it has reduced endemic alcoholism and has provided them with a rational set of beliefs. But, if it serves to make them content to suffer the outrages that are still being imposed on them by the white man, peyote will not do what Quanah Parker hoped that it would.

ADDITIONAL READINGS

Aberle, D. *The Peyote Religion among the Navajo.* Chicago: Aldine, 1966.

Furst, P. T. (ed.). *Flesh of the Gods.* New York: Praeger, 1973.

Hertzberg, H. W. *The Search for an American Indian Identity.* Syracuse University Press, 1971.

LaBarre, W. *The Peyote Cult.* New Haven, Conn.: Yale University Press, 1938.

LaBarre. W. "Twenty Years of Peyote Study." *Current Anthropology* 1(1960):45.

Lanternarri. V. *The Religions of the Oppressed.* New York: New American Library (Signet), 1963.

Myerhoff, B. G. *The Peyote Hunt.* Ithaca, N.Y.: Cornell University Press, 1974.

Opler, M. E. "The Influence of Aboriginal Pattern and White Contact on a Recently Introduced Ceremony; The Mescalero Peyote Rite." *Journal of American Folklore.* 49(1936):271.

Safford, W. E. "An Aztec Narcotic (*Lophophora williamsii*)." *Journal of Heredity* 6(1915):291.

Schultes, R. E. "Peyote and Plants Used in the Peyote Ceremony." Harvard University Botany Museum Leaflets, 1937.

Slotkin, J. S. *The Peyote Religion.* Glencoe, Ill.: Free Press, 1956.

Underhill, R. M. *Red Man's Religion.* University of Chicago Press, 1965.

The Lotus

Some of the plant symbolisms in religion are almost universal: the tree of life, the resurrection symbolism of new, green leaves, and the falling of leaves as a death symbol are understood by all people. But each religion has symbols unique to that faith, ones that aren't in other religions. The Jewish use of the *ethrog*, a citrus fruit, is not found in Christianity, and the lily has no significance in Judaism. Yet, both these faiths sprang from the same roots and both drew on common sources. If we look at religions different from Judeo-Christianity, we find the universal symbolisms and, in addition, some whose roles are not in our Western experience. The lotus is such a plant. It, more than any other plant, seems to have acquired symbolic importance among many cultures which may not have had direct contact with each other. Whether this cultural difference is more apparent than real is of anthropological importance, for as we learn more about trade routes and amazing sea voyages, we realize that there may have been extensive economic and cultural intercourse among peoples whom we thought were unconnected in ancient times.

The botany of the lotus is rather confusing since many plants have been given the same common name. The botanical family to which all of these plants belong is the Nympheaceae, the water lily family, containing eight or nine genera and about 50–60 species. All are aquatic plants with a rootstock (rhizome) buried in the mud at the bottom of lakes and ponds. Leaves grow up to the surface of the water on long petioles, and the flowers, too, arise from the rootstock. Flowers are large, white, or brightly colored. In most species, the flowers open at dawn and close again at dusk. On a bright sunny day, small waves and light breezes cause the leaves and flowers to dance gracefully on the water's surface. It was thus that Linnaeus chose the family name, "flowers of the water nymphs."

The most familiar genus is *Nuphar*, the common pond lilies, widely distributed throughout North America. A second genus, *Nymphaea* includes the fragrant water lily (*N. odorata*) and the tuberous lily (*N. tuberosa*) whose flowers run the color spectrum from white through blue, yellow, pink, and red. There is also the blue lotus of the Nile (*N. caerulea*) and the white lotus of the Nile (*N. lotus*). One species (*N. mexicana*) is found in Central America, one is native to Europe (*N. alba*), and another is the blue lotus of India (*N. stellata*). To compound the confusion, a third genus, *Nelumbo* (Figure 3.41), contains the sacred lotus of Persia and Asia (*N. nucifera*) and the lotus of East India (*N. nelumbo*). Members of these three genera are widely distributed throughout the world, and all are sufficiently similar so that all have been called lo-

Figure 3.41
Opened flower of the sacred lotus, *Nelumbo nelumbo*.

Figure 3.42
Horus, god of the rising sun, seated on a lotus. His finger at his lips signified silence. Taken from an Egyptian wall painting.

tuses. In the same way, your Christmas tree this year may be a scotch pine, next year a balsam fir, and the following year a white spruce—yet they are all Christmas trees, and all have the same symbolism.

THE LOTUS IN THE NEAR EAST

In order to get a feeling for the place of the lotus in non-Judeo-Christian faith, Egypt is a good place to start. Both white and blue lotuses are found in the Nile and the tributary streams that feed the Nile on its journey from Lake Victoria. The lotus was, for the Egyptian, a symbol of the nascent life of this sacred river, a reminder of the mysterious yet bountiful fertility given yearly as spring floods covered the land with rich, black mud in which crops grew luxuriously. The Egyptian gods themselves partook of the wonder and worry of birth and death, day and night, and those other dualities that trouble humankind. Horus, the new-born sun, is born daily from a lotus flower expanding at dawn under His influence (Figure 3.42), and both Horus and the day die each evening as the petals close. Osiris, father of Horus and husband of Isis, as well as king and judge of the dead, wears a crown of lotus blossoms in the closed position. When His flowers open, the entire world will be reborn. To reach the realm of Osiris, the pilgrim soul grasps a lotus as it embarks on the boat of the sun which will carry it to the throne of judgment and eventual resurrection, Isis, in her role as Earth Mother, was manifested by young flower buds of the lotus just emerging from the waters of the deep. Atum, the primaeval deity under whose influence the world took shape, is pictured as arising from a lotus bud floating on a sea without form. Heaven, Egyptian priests proclaimed, was supported by the flower stalks of the lotus and, to make this image concrete, the roofs of temples were upheld by columns whose capitals were composed of the lotus (Figure 3.43). The hippopotamus, a sacred animal living in and emerging from the equally sacred Nile, was imbued with the aura of the lotus (Figure 3.44). There is a

Figure 3.43
Egyptian column and lotus capital.

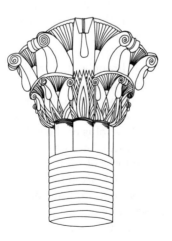

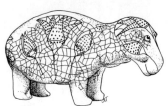

Figure 3.45
Homage to the winged globe representing the Deity with stylized lotus as the tree of life. Taken from an Assyrian stone relief.

bust of Pharaoh Tutankhamen sculpted ca. 1500 B.C. in which the head of this god-king emerges from the lotus flower; pharaoh himself is a god.

Lands in the Middle East, whose cultures are almost as old as those of Egypt, used the lotus in much the same way. In Assyria, the moon god is figured as sitting on a disk composed of lotus petals from which wings extend (Figure 3.45). Phoenician seals depict the moon and stars arranged around a lotus flower which, in turn, is supported by wings bearing the plant to heaven. In Mesopotamia, the lotus is a symbol of conquest, the triumph of the gods of the people against those of lesser faiths. Persians venerated the opened lotus flower as a manifestation of the sun and hence of light and life. The closed lotus was a moon symbol, evidence of darkness and of death. Since the flower will again open in the morning, life has meaning because of this resurrection.

The Greek gods did not directly employ the lotus, but some of its symbolism came from Egypt. In one legend a nymph, betrayed and then deserted by Hercules (the cad!), threw herself into a pool and was drowned. The other gods had pity on her and changed her into a lotus to provide a perpetual reminder to Hercules of his philandering ways. It apparently was ineffective; Hercules continued to be the ruin of many an innocent maiden. The stylized lotus developed into the anthemion where it adorned greek temples. In the Odyssey, Homer noted that Odysseus was acquainted with a land in which the people lived in languorous forgetfulness—the land of the lotus eaters—and we still speak of someone who is forgetful and unmindful of the world as a lotus eater. Since no part of the lotus is narcotic, it is likely that another plant had the same common name. Most authorities suggest that it was a shrubby member of the buckthorn family, *Zizyphus,* the jujube, whose fruit can be fermented into a potent wine. Etruscan art utilized the stylized lotus in architecture as did Greece (Figure 3.46), and traces can be found in Byzantine and Romanesque churches. The lotus appears in western art and religion only as a pretty plant.

Figure 3.46
Hellenic Greek decorative frieze of the lotus.

THE LOTUS IN INDIA

India, a land replete with interlocking cultures and religions, has employed the lotus in all of its many faiths. Prior to 2000 B.C., invaders came from the northwest, bringing with them barley—suggesting con-

Figure 3.47
Birth of Brahma. Taken from
a Hindu woodcut.

tact with peoples of the Middle East—and a mystery religion, Vedic
Hinduism, in which the lotus plays a prominent role. Creation legends
vary. In one version, Om, the spirit of creation, moved on the surface of
a vast sea which existed before the earth was formed. He quickened
into life a wondrous golden lotus, resplendent as the sun, floating on
the lifeless water. From this flower arose the Hindu Trimurti: Brahma
the Creator, Vishnu the Preserver, and Siva the Destroyer (Figure 3.47).
A mango tree arose from the lotus, and a seed of the mango gave birth
to the daughter of the sun. She mated with a god, and their children
populated the earth. In another version, the world itself is a lotus flower
floating in the center of a shallow pond which rests on the back of an
elephant who is standing on the shell of a turtle.

In Hinduism the lotus is a center of vitality, partaking of the
strength of the sun, and is therefore a source of tremendous power. Just
as the rays of the sun can be focused to cause fire, so can the lotus focus
man's thoughts upon universal concepts. In the Tantric tradition, there
are power centers (*cacra*) in the body from which radiate understanding.
Perception, bliss, sexual power, and pure thought are, accordingly,
called the lotuses. In the Puranic tradition, the cosmic tree was a lotus,
growing from the navel of Narayana (Om), and both gods and people
arose from this tree. Lakshmi, goddess of love, fortune, and fecundity
rests on a lotus leaf. Even Kamadwa, Hindu equivalent of our Cupid,
rides the sacred river Ganges on a lotus leaf, reinforcing the role of the
lotus as a symbol of physical love and child-bearing. It is said that souls
of the dead sleep in lotus blossoms until admitted to a paradise containing a sacred lake covered with the flowers.

In the middle of the fifth century B.C., Siddhartha Gautama was

Figure 3.48
Buddha receiving enlightenment leading to Nirvana. The tree is the Bo, *Ficus religiosa,* and the Buddha is seated in the lotus position on lotus leaves.

Figure 3.49
Hindu decorative frieze of the lotus. Taken from a stone carving in India dated at 250 B.C.

born into the Sakya clan near Nepal. At age 30, Siddhartha left home, wife, and family to obtain enlightenment. He achieved victory over the evil one and over himself after seven years under a Bo tree (Figure 3.48), and a new Buddha, fourth of the line, was born in the glory of the lotus whose opened flower revealed that which He wanted to know. Having pity on mankind, the Buddha gathered five disciples to whom he preached and whom he ordained. Soon there were many converts who were enjoined to go forth and proclaim the *dhamma.* The Buddha's fame as a teacher—Padma Sambhava (preacher of the lotus)—spread throughout northeast India. For many years, He moved from place to place organizing the expanding order until, full of years, he died. His body was cremated and relics given to the holy places where, in lotus-decorated reliquaries, they are venerated.

Buddhism, like Western faiths, showed both sectarization in which various paths or "vehicles to enlightenment and salvation" developed, and syncretism in which merging of philosophies and rituals occurred. Yet, the place of the lotus as a symbol was little altered (Figure 3.49). The Buddha and the *Bodhisvattas* (persons who attained enlightenment but chose to remain in contact with man) are figured on lotus pedestals just as are the Hindu Trimurti. Some bodhisvattas, such as Kwan Yin the goddess of mercy, may carry a lotus flower and are known as *Padmapani,* the lotus bearers or, more appropriately, bearers of the good news. Maitreya, Buddha of light and living kindness, is portrayed with the fingers of His uplifted hands forming a lotus bud about to open. Amitabha, Buddha of endless light and infinite compassion, is represented by golden fish from the sea of eternity and a lotus that grew in the sea. Symbols acquired from other faiths were combined with the lotus; many of the *mandalas,* schematized representations of the cosmos (Figure 3.50), are resting—as does the Buddha—upon the lotus. In some mandalas, the lotus may be conventionalized into a wheel (*Rimbo*) with the petals representing the spokes of the wheel which is the circle of birth, death, and rebirth—the rolling of time endlessly into eternity. The wheel became a rosette or was further simpli-

Figure 3.50
Mandala of Brahma resting on a lotus with the sacred swastika-in-triangle.

Figure 3.51
Korean Buddha on lotus pedestal. Taken from a painted and gilded wood carving of the twelfth century.

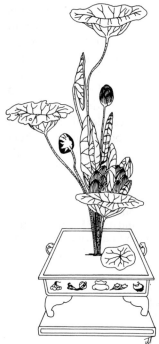

Figure 3.52
An eighteenth-century Japanese engraving of a flower arrangement using the lotus.

fied into an octagon. Mantras, words of praise to be sung or hummed, frequently incorporated the Indian word *padma* (lotus) as a symbol both of the Buddha and of the enlightenment. Yoga, a technique of bodily control to allow the unity of body (Avalokita) and the spirit (Amitabha) with the universe, uses specific body positions, including the lotus position in which the legs are crossed so that one's feet are resting on one's thighs (Figure 3.51). This derived from the Hindu Tantric tradition in which the lotus position signified a rebirth.

THE LOTUS IN CHINA AND JAPAN

China, Japan, and Korea used the lotus theme in some of the most beautiful art the world has ever seen. Lotus blossoms and leaves were translated, literally or after stylization, into every medium. Poetry, legend, and prose involving the miraculous lotus have passed through generations of people. In the visual arts, calligraphy, painting, sculpture, metal-working, wood-carving, and other techniques were used to portray the lotus. Even after Buddhism in China lost its prominence, artists continued to use the flower. No single plant, with the possible exception of the lily attributed to the Madonna, has so captured the imagination of artists.

Throughout this intense emphasis on the lotus as a symbol and epiphany of faith, the lotus was and still is a common food item. Lotus roots are grown in ponds, the rootstocks cut into rounds, dried, and used as a crunchy additive to soups. Lotus seeds are added to meat or vegetable dishes, and dried, roasted, and salted lotus seeds are eaten as we eat peanuts; there is no thought that the plant, as food, is to be equated with partaking of the godhead.

When Ch'an Buddhism was taken to Japan and metamorphosed into Zen Buddhism, the lotus lost much of its importance, although Buddha is still depicted on the lotus throne, and the lotus position of Tantric and Yoga meditation is used in Zen. Legend tells us that once, when the Buddha was at rest, a Brahman ruler came to Him and gave Him a golden lotus flower. The Enlightened One accepted the flower and spent many hours gazing in silence at the blossom. Finally, the Buddha smiled, and this moment of complete joy was passed on to his patriarchs until the Indian patriarch Bodhidarma came to China in 520 A.D. and taught the Ch'an concepts. This intense and complete riveting of attention upon some natural object became a cornerstone of Zen. The development of flower arranging and the tea hut garden by Zen tea masters drew upon the lotus whether it was used literally as a three-part flower arrangement (earth, man, cosmos) (Figure 3.52) or as carefully cultivated lotus blooms on a quiet temple pool.

There is only the most fragmentary evidence that the ancient civilizations of the New World were in contact with either Egypt or India, and yet Mayan freize decorations or religious symbols depict the water

Figure 3.53
Mayan decorative frieze of the lotus. Taken from a stone carving dated at 800 A.D.

lily (Figure 3.53). In Illinois, the American Indians of the Fox tribe, centered near the Fox River and Fox Lake, used the native yellow lotus, *Nelumbo lutea*, as a symbol of purity, of the sun and of death and rebirth.

WHY THE LOTUS?

In the cultures and religions examined here, there is an obvious concordance of the symbolic significances of the lotus. Some of these are, at least superficially, fairly obvious. The opening at dawn, the round disk shape of the flowers, and the number of white or yellow-petalled species suggest the sun's course through the sky. The closed blossom can easily be related to the moon. The lotus was not the only flower in which birth of souls occurred. The narcissus in Europe, the daisy in South Africa, and others, have been used in an identical manner. In North America, we still tell young children that they were born in a cabbage or in a rose or another plant. In several cultures, the opened, round, yellow or golden flowers of the sunflower, the marigold, or the cone flower (*Rudbeckia*) have been used as sun symbols.

Other aspects of the varied symbolism of the lotus are, however, not so obvious, and our attempts to rationalize them may merely reflect our Western "need" to put explanatory labels and names on concepts which really elude us. Most cultures have used the triple symbols of earth-water-sky, and the lotus partakes fully of all. Rooted in the primaeval mud and covered by endless water, the plant sends its leaves and, eventually, its sunlike flowers above the water into the sky. It betrays its humble, earthly origin in the mud, transcending this with leaves of bright green that are not wetted by the water upon which they float. Perhaps this characteristic, not being injured by water and rising above it, can symbolize the truly wise One who has achieved the state exemplified by the sun while still participating in and being a part of water and earth. The meanings of the lotus transcend man's ability to put these concepts into words or into more representational images. The symbols, and the concepts that underlie and magnify the symbols, can be variously interpreted by individuals. Were they made concrete, they would stand for concepts that could not be put into words with which all others could agree. Just as a nation's flag means different things to different people, the lotus and other plant symbolisms mean what we, as individuals, choose them to mean or want them to mean.

ADDITIONAL READINGS

Armstrong, R. C. *Buddhism and Buddhists in Japan.* London: Society for Promoting Christian Knowledge, 1927.
Edkins, J. *Chinese Buddhism.* London: Kegan Paul, 1893.
Jung, C. G. *Man and His Symbols.* New York: Dell, 1974.
May, R. *Symbolism in Religion and Literature.* New York: Braziller, 1966.
Soothill, W. E. *The Lotus of the Wonderful Law.* New York: Clarendon Press, 1930.
Thomas, P. *Hindu Religion, Customs and Manners,* 2nd ed. Bombay: Tara, 1910.

CHAPTER 4
The Drug Scene

Opium

An opiate is anything that causes dullness or inaction
or soothes men or emotions.
Random House Dictionary

The botanical family Papaveraceae contains 25 genera and 120 species of flowering plants. Most are herbaceous annuals, although a few die back to the ground each year and form new shoots from a perennial rootstock. The family originated in Asia Minor. The genus *Papaver* contains ten species, several of which, the Iceland poppy (*P. nudicaule*) and the oriental poppy (*P. orientalis*), are common garden plants. The California poppy (*Eschscholzia*), well known to most gardeners, is a member of another genus in the family. None of these plants produce alkaloids of medical interest. The one that does is the opium poppy (*P. somniferum*) whose specific name was chosen by Linneaus because of the sleep-inducing properties of the gum produced in the young seed capsule of the plant (Figure 4.1).

Raw opium has been gathered by virtually the same method for ages. When the red, pink, or white flower petals fall, usually in June or July, the immature seed capsule synthesizes opium, a mixture of over 20 alkaloids plus water, sugar, several plant gums, and latex. The alkaloid content is close to 30 percent of the dry weight of opium. Within a day after the petals fall, workers go into the fields with sharp, three-bladed knives that look much like a table fork with sharpened tines. They make several slits in the capsule, taking care not to puncture into the chambers where the young seeds are forming. Within a day, the opium oozes out and begins to dry down (Figure 4.2). A curved spoon is used to scrape the opium from the capsule, and another set of slits causes a second flow of opium. The collected opium is pressed into cakes or balls and allowed to dry to a rubbery-textured, yellow to brownish mass. From an acre of opium poppy (20,000 plants), an average of ten kilograms of opium can be collected. Because of the delicacy of the operation, opium collection is entirely a hand process.

The opium poppy can be grown in many parts of the world where the growing season is sufficiently long with warm, sunny weather. Because of the need for cheap labor, it is presently grown in relatively few countries of the world (Figure 4.3). India and Pakistan each produce about 100 metric tons of opium each year, most of it consumed locally. Afghanistan produces the same amount, but the crop is smuggled into other countries of the Middle East. Turkey has traditionally been the major supplier of legal (medicinal) opium for the West, with between 60 and 100 metric tons produced per year. Mexico has only recently become an opium-producing country, although its production is still under ten metric tons. Less is known about production in the area of southeast Asia called the Golden Triangle, an area embracing parts of Burma, Laos, and Cambodia, but it may well be the greatest producer.

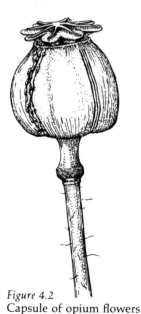

Figure 4.1
Opium poppy, *Papaver somniferum*.

Figure 4.2
Capsule of opium flowers with raw opium gum oozing from cuts made in its side.

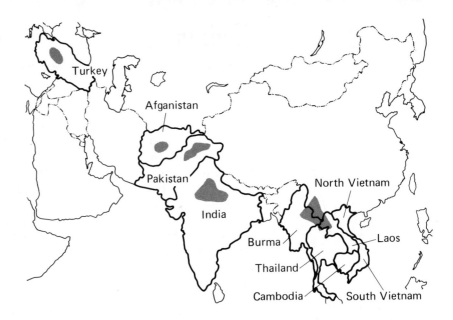

Figure 4.3
Major opium-growing areas
of the world. Smaller
amounts are grown in
Mexico.

Reports on production vary from 250 to over 500 metric tons per year. The world's production of raw opium is close to 1000 metric tons per year, of which less than 250 tons enters legal medical channels; the United States processes about 150 metric tons to isolate morphine and codeine.

Of the alkaloids found in raw opium, three are of medical importance. Close to 11 percent of opium is the single alkaloid morphine, while codeine constitutes about 2 percent and thebaine a bit less than 1 percent of the weight of the opium gum. Isolation of pure morphine and codeine is not a difficult or expensive process, and the legal cost of these alkaloids has been constant for many years. Raw opium is mixed with slaked lime, heated to remove the gums and latex, leaving an alkaloid base which constitutes about ten percent of the original bulk of the starting opium. Further purification of morphine, codeine, and thebaine involves dissolving them in weak alkali and partitioning the alkaloids into alcohol. Water-soluble alkaloids are made by adding concentrated hydrochloric acid to form the medical morphine-hydrochloride.

AN ESSAY ON ALKALOIDS

In addition to making proteins, carbohydrates, fats, and other compounds familiar to most people, plants synthesize a huge array of substances that are usually found in relatively low quantities. The general name of these substances is secondary plant products, which gives no idea of the chemical range of the materials or any idea of their importance. The medicinal products, the natural dyestuffs, products for

Figure 4.4
Chemical interconversion of
major opium alkaloids.

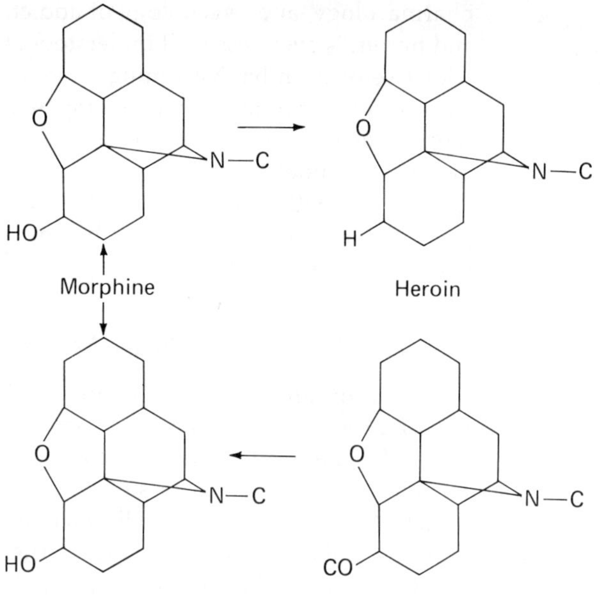

Morphine

Heroin

Codeine

Thebaine

the chemical industry (gums, resins, etc.), and a wide variety of common, everyday substances used as flavorings and essential oils (perfumes, peppermint) are secondary plant products.

Among the secondary plant products that have long played important roles in human life are the alkaloids. The word alkaloid is a purely chemical word defined as an organic chemical molecule—one containing carbon atoms—which also contains at least one atom of nitrogen. All can be crystallized and, when dissolved in water or in alcohol, they give an alkaline reaction to the solution. The nitrogen in an alkaloid is usually found in combination with a ring of carbon atoms, the so-called heterocyclic ring (Figure 4.4). When an alkaloid is mixed with an acid, such as hydrochloric, the very water-soluble hydrochloride is formed; this is the usual way of getting alkaloids into simple solutions. In solutions or as the crystal, they are colorless, are most soluble in alkaline solutions such as weak sodium hydroxide, and almost invariably have a bitter taste. The flavor of tonic water is due to quinine, one of the alkaloids.

In spite of a great deal of intensive research, the physiological mode of action of alkaloids in the animal body is poorly understood. Some, like the caffeine alkaloids in tea and coffee, are stimulants. Others, like the alkaloids in ergot, cause constriction of smooth muscle, and still others, like those in the opium poppy, are powerful pain-killers. It is likely that all operate on or in some part of the central nervous system and that the responses reflect alterations in control over cellular function by the brain and peripheral nerve network. The physiology, pharmacology, and psychology of addiction to alkaloids like morphine and heroin is even less well understood. Certainly, not all alkaloids are addictive or even habit-forming. One *can* get along without a morning cup of coffee without experiencing withdrawal symptoms. Even for those which are addictive, the nature of the addiction and its consequences are poorly understood, and the same can be said for the response to withdrawal from the substance. There are certainly psychological factors as well as biochemical and physiological factors which must be evaluated.

THE WESTERN OPIUM EXPERIENCE

Raw opium has been used medically for centuries. Sumerian tablets of 2500 B.C. note that when small balls of opium were eaten or taken after mixing with wine, the drug induced sleep and relieved pain. Homer spoke of *nepenthe*, a substance which will "lull pain and bring forgetfulness of sorrow." Hippocrates, Theophrastus, Pliny, Dioscorides, and other ancient medical writers recommended opium, and it was the most used analgesic up to the twentieth century. Thomas Sydenham, an influential British physician of the seventeenth century said of opium, "Among the remedies which it has pleased Almighty God to give to man to relieve his sufferings, none is so universal and efficacious as opium," a statement that many physicians have echoed. Morphine was first isolated in 1805 and replaced crude opium in medicine. Isolation of other alkaloids soon followed, but it was not until 1925 that the chemical structure of morphine and other alkaloids was fully known. Although morphine has, to a large extent, been replaced by synthetic analgesics, codeine is still used as a local anesthetic and as an effective cough suppressant.

In the latter half of the nineteenth century, pharmaceutical chemists started altering the morphine molecule in order to make a compound which would be more effective than the natural alkaloid and less addictive. Several German chemical companies found that simple chemical treatments would yield a compound called heroin which was more effective as a pain-killer than morphine and was a very effective cough suppressant. In the 1890s and up to World War I, heroin-containing cough medicines were openly sold in North America without prescription, and it is estimated that there were as many heroin addicts in New York City in 1900 as there are today, but most of them were children (Figure 4.5). For reasons that defy explanation, most of these addicts simply cured themselves suggesting that addiction has a large psychological component. Only after World War II did widespread addiction become a problem in North America, reaching epidemic proportions in the large cities of the United States and, to a lesser extent, in Canada. Addiction in the cities of European countries developed a few years later until, today, heroin addiction is a major problem in most of the Western world. During the 1950–1970 period, the bulk of the illegal

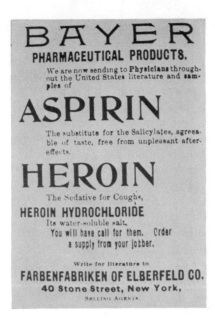

opium reaching the West was grown in the Anatolian provinces of Turkey, reduced to morphine base, and shipped to Marseilles, France, where it was converted into heroin. Profits were enormous. Ten kilograms of opium, which cost Turkish smugglers $25–30 per kilogram, could be converted into one kilogram of pure heroin, a full year's supply for 50 addicts. The "street" price for this kilogram in New York was, in 1970, close to $50,000. When diluted with quinine, lactose, or other adulterant and packed into "bags" of two grains each selling for $8, there was a gross profit of $600,000 per kilogram.

THE OPIUM BAN

In 1970, President Richard Nixon declared war on the international heroin traffic. He concluded that the best way to eliminate the North American supply was to prevent the growth of the opium poppy. Since Turkey was dependent upon U.S. military and economic assistance, the U.S. government forced the government of Turkey to place a total ban on opium production. The U.S. offered $35 million per year to Turkey to pay the farmers not to grow the poppies and, it was hoped, to give them time to develop another crop. Seen in retrospect, these decisions were disasterous.

In the opium-growing areas, the poppy is a multipurpose crop. The most important is opium itself, accounting for 90 percent of the cash value of the plants. After the capsules mature, seeds are used as a source of oil for the peasants or are sold for use in baking; most of the poppy seeds on rolls come from the Turkish opium poppy (Figure 4.6). The oil-free meal cake is fed to cattle or ground into a flour for cookies.

Figure 4.6
Mature capsule of opium poppy showing poppy seeds.

In early spring, young leaves are used in salads, and after the capsules mature, the stalks are used as silage. Since the peasants of Anatolia had grown poppies for centuries, they had no other cash crop. Wheat and other cereals grow poorly, the road system does not permit the development of a fresh fruit and vegetable economy, and the land receives too little rain for extensive pasturage. The impact of the ban on the peasants was immediate and profound. The *New York Times* reported that one village headman said, "We have heard that our opium becomes heroin and kills all the Americans. They are rich and that is why they use it. Their richness kills them and makes us poor." A well-to-do farmer is one who owns five hectares on which he normally planted two to the poppy. His cash income from the illegal market was about $350 for 10 kilograms. Had he planted wheat, barley, or tobacco, he would have realized about $150, assuming that he could get the crop to market. One such farmer was reported to have said, "The Americans use it and they die. We produce it and don't use it. Don't they have any brains?"

Keeping their part of the bargain, the Turkish government tried to eliminate the poppy plantings in 1972 and 1973. Soldiers set fire to plants, and helicopters sprayed remote poppy fields with herbicides supplied from the American stockpiles used in the Vietnam War. Afyon, a city 150 miles southwest of Ankara and the center of the poppy-growing area (afyon means opium in Turkish), reflected the depressed economy of the region. Debts piled up, bank loans remained unpaid, and a reasonably prosperous land was plunged into penury. During this period, thousands of young men and women, without jobs or economic resources, entered the European labor markets in Germany and France and kept their families solvent by sending home their pay. Separated from their families, lacking linguistic skills, freed from the stabilizing influences of their culture, and eating strange food, they were relegated to the dirty jobs that their affluent hosts no longer wished to do. Not unexpectedly, they caused "trouble" in Europe and were looked down upon and sometimes jailed.

Recognizing the unhappy fate of their youth and realizing that the American subsidy was both demoralizing and inadequate, the Turkish government asked the United States to reexamine the agreement. The Turkish foreign minister, Turan Gunes, said, "The sacrifices we have to bear for humanity cannot be placed on our shoulders only." The United States Senate voted 81 to 8 to cut off all economic aid to Turkey if it did not tighten the ban, and the U.S. State Department recalled its ambassador. The international situation became tense, and the North Atlantic Treaty Organization, of which Turkey was a member, appealed to the United States to relax its uncompromising stand. Turkey agreed to organize a poppy-growing system designed to reduce the possibility of opium reaching the illegal market. The Turks would also develop their own factories to process opium, license all growers, and would require that only unlanced, mature poppy capsules would be

handled. The mature capsule yields little morphine, but contains high concentrations of codeine needed by the medical profession. The United States government agreed to allow a trial of this system, and the first licensing and planting was done in 1975. Its long-term success is still unknown.

The question is, of course, did this attempt to dry up the source of opium alkaloids work, and the answer seems to be that it did not. The United States offended and antagonized one of its allies, and after a short period of time when "street heroin" was in relatively short supply, the availability of the drug actually seemed to increase. In part, this was due to the release of stockpiles of morphine base in Turkey and heroin in France. Equally important, supplies of opium were obtained from other growing regions. Mexican heroin began to enter via Texas, California, and New Mexico, some was obtained in Iran from opium grown in Afghanistan, and a supply in the Golden Triangle of Burma, Laos, and Cambodia (Figure 4.7) soon reached the North American market via Hong Kong and Japan. At this time, 80 metric tons of the 1000 metric tons produced each year can be made into enough heroin to saturate the illegal North American market; this amount of raw opium can be produced on about 20 square miles of suitable land. Addiction to heroin is not controllable by restricting the poppy, but must be handled by other means including alteration of the cultural and psychological stresses that society imposes on individuals.

THE OPIUM WARS

This contemporary political and economic contretemps is far from the only one involving the opium poppy. One of the most profound of these was the Chinese Opium Wars of 1839–1842.

Opium was used in China for medical purposes for centuries, but there was only a small amount of addiction among the people. Called *chandu*, raw or processed opium was smoked in special clay pipes, and the intoxicating smoke was inhaled to cause sleep with, it is assumed, pleasant dreams. The ash remaining in the pipe was mixed with other burnable material and smoked in turn by the less affluent, but this was highly toxic, and much of the death rate attributed to opium addiction was due to smoking this dross. Yet, until the nineteenth century, China did not grow or process opium, and the number of addicts was very small.

Although Europeans had obtained a foothold in islands off the Chinese coast, the Ming and early Ch'ing emperors restricted commerce with the big-nosed, round-eyed barbarians who, they believed, were their intellectual and moral inferiors. China had little use for Western goods and ideas; its society was stable, and the vast country supplied all its needs. Europe, however, was extremely interested in the goods of China. Its tea, silks, spices, and porcelain commanded high prices on a market fascinated by the Orient. Although Portugal was the first country to have much trade with China, the conquest of Bengal in 1773 by Britain and the development of the world's finest merchant fleet allowed England's East India Company to obtain first a foothold in the China trade and, by 1800, to monopolize it.

After delicate and, to the English, humiliating negotiations with the Chinese government, a small island off the shores of Canton was established as a base for trade. Even here, the intermediaries had to be Chinese, and the British traders were restricted in their movements. More importantly, the Chinese insisted that all goods be paid for in silver, since they had no desire for European products that could be used in trade. The China trade was so successful that by 1810–1815, China had acquired a good share of the silver of Europe which, because of scarcity, was rising rapidly in price, thus reducing the profits of British merchants. Obviously, there had to be something that the Chinese wanted and, with unerring intelligence, the British decided to sell opium to the Chinese. At first, the opium was given away, and as addiction spread, prices rose accordingly. Because of its absolute control over India, the East India Company subverted the agriculture of Benares, Baher, and other areas of India to the growing of the poppy and the production of opium. Poppy cultivation was compulsory, and since the production of food crops was limited, the peoples of these Indian provinces were reduced almost to starvation.

As addiction in China rose to astronomical proportions, silver began to move out of the country and back into Europe where its price fell and the profits of the East India Company soared. Since silver was the currency of China, taxes went unpaid and internal business was disrupted. Based on Confucian and Tao ideals, Chinese society was grounded in the philosophy of self-discipline and a hierarchical arrangement of duties to family and emperor. The addict sacrificed family, duty, and self-discipline, and the fabric of Chinese society and government began to collapse. Britain recognized this danger and forbade the legal importation of opium into England, but in spite of the pleas of a few humanitarian members of Parliament, the British government refused to order the East India Company to stop the opium trade; tax revenues from India, tea duties, and opium sales were simply too profitable.

In 1833, Emperor Tao Kwang, whose son had become an addict, ordered Commissioner Lin Tse-Hsuto to stop the opium traffic. On March 10, 1839, Lin issued a proclamation to the foreigners that asserted that it was not right to harm others for the sake of profits: "How dare you bring your country's vile opium into China, cheating and harming our people?" He demanded and obtained all opium stored in Canton and, to the horror of the British, destroyed it. The Houses of Parliament rang with the shouts of indignant members who demanded that Prime Minister Palmerston revenge this insult to the honor of Great Britain. Besides, this provided a rationale for taking over China, something that Britain had long desired.

A punitive force, assisted by the fleet of the East India Company and the merchant fleets of France and the United States, formally invaded China. The Chinese, long believing that gunpowder was for fireworks and that armies were to maintain the peace, were no match for the experienced Westerners. Britain lost 500 troops; the Chinese lost 30,000. The Treaty of Nanking, signed in August 1842, provided that all costs of this "punitive expedition" were to be borne by China, that an indemnity of $21 million was to be paid to Britain, and that the country was to be opened to Western trade. Hong Kong was converted into a British Crown Colony. France and the United States soon wrung equivalent concessions from a demoralized Chinese government. For practical purposes, China became a colonial outpost. No future Ch'ing emperor was more than a puppet; the unifying concept of "The Mandate of Heaven" upon which the dynasty was built, was lost.

Subsequent treaties extended the control of Britain. The opium tariff recognized the right of Britain to addict the population, Christian missions were imported with full protection, customs regulation was in the hands of the West, and additional indemnities were forced on the Chinese. The importation of "foreign mud" increased, and the number of addicts in China increased even more rapidly until, by 1900, it was estimated that close to ten percent of the population of China used opium regularly. More than half of Britain's trade with China was opium from India. A few Britons protested but were told, "If we don't sell it, someone else will. Besides Her Majesty's government needs the money for its expansion." In areas ceded to Britain, opium cultivation was started, and soon the Chinese themselves were growing opium for home consumption.

In 1906, the last of the Ch'ing emperors decided that China must give up opium, and a ten-year period was designated in which to accomplish this seemingly hopeless task. Britain reluctantly agreed to limit and then stop importation. Local production was restricted, but smuggling continued for many years. When, in 1911, the republic was organized and the Ch'ing dynasty came to an end, the detoxification of China was continued and even accelerated, but the problem was simply too large to handle. Opium addiction continued and was enhanced during the Japanese occupation, when it became imperial policy to

Figure 4.8
Goddess of the night giving opium poppy to man. Taken from a Roman cameo carved ca. 100 A.D.

subjugate the Chinese peoples by supplying opium. The Mitsui family had a virtual monopoly on opium and, with the money so obtained they used slave labor from China to dig their coal and work in their factories. Sony, Toyota, Sapporo, and Nishi-Nippon are companies whose financial base was opium addiction in China. They are not alone. A fair share of the wealth of Indian princes and rajahs came from converting their lands to poppy production, and some of the respected families in New England got their start by renting clipper ships to the India-China opium trade. Today, China is opium-free.

And yet . . . opium and the morphine and codeine it contains have eased more physical pain and saved more lives than the products of any other plant (Figure 4.8).

ADDITIONAL READINGS

Ball, J. C., and C. D. Chambers. *The Epidemiology of Opiate Addiction in the United States.* Springfield, Ill.: Thomas, 1970.
Beeching, J. *The Chinese Opium Wars.* New York: Harcourt Brace Jovanovich, 1976.
Fay, P. W. *The Opium War 1840–1842.* University of North Carolina Press, 1976.
Gates, M. "Analgesic Drugs." *Scientific American* November 1966.
Junnes, R. *The American Heroin Empire.* New York: Dodd, Mead, 1973.
Lidesmith, A. R. *Addiction and Opiates.* Chicago: Aldine, 1968.
Musto, D. F. *The American Disease.* New Haven, Conn.: Yale University Press, 1973.
Reynolds, A. K., and L. O. Randall. *Morphine and Allied Drugs.* University of Toronto Press, 1957.
Straus, N. *Addicts and Drug Abuse.* Boston: Twayne, 1971.

Pot—Grass of the Living Room

The only part of conduct of anyone, for which he is amenable to society is that which concerns others. In the part which merely concerns himself, his independence is of right, absolute. Over himself, over his own body and mind, the individual is sovereign.
JOHN STUART MILL, *On Liberty*

It has been suggested that hallucinogens permit people to enter the "real" world, closed off from childhood by the many layers of culture that surround us from birth. There are hundreds of plants containing mind-altering drugs which have been used at some time in human history, but aside from alcohol, none has had as much social impact in North America as the hemp plant, *Cannabis sativa*. Botanically, cannabis is a genus of several species in the Moraceae, a family containing hops used in beer, the mulberry, figs, and the Osage orange. It is a tall annual

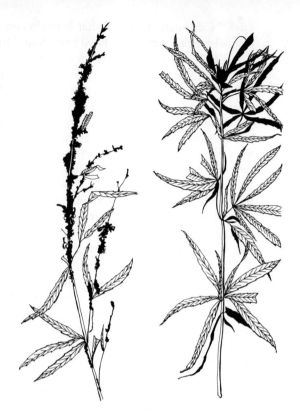

Figure 4.9
Hemp plants. *Cannabis sativa.*
Female flowers produce
hashish gum or resin (left).
Male plants do not form
hashish.

weed which sometimes reaches four meters in height and can grow in most temperate and tropical climates and in virtually any soil. Cannabis is dioecious, with both male and female plants which can be told apart even when young. The male plant tends to be taller, less prone to branching, and its leaf form is slightly different from that of the female plant (Figure 4.9). The genus is native to China, although it has been disseminated by man to all continents.

SOOTHER OF GRIEF

Although its use as a fiber and food crop dominated its production in the Orient for a long time, its medical uses were known and exploited. In a medical book ascribed to the legendary Emperor Shen-Nung in 2737 (but probably written in the Han dynasty about 100) B.C., ground, dried leaves of cannabis, known as *ma-yo*, were recommended for malaria, beriberi, constipation, as an anaesthetic in surgery, and for that disease of old age—absent mindedness. Shen-Nung noted that its primary medical value was in calming hysterical women. In India and throughout southeast Asia, cannabis was also used medically, and its capacity to ease anxiety was noted in both Hindu and, later, in Buddhist writings, where it was referred to as the "soother of grief." Curiously, it seems that cannabis entered India about 1500 B.C. from Assyria, since seeds have been found in archeological sites in Iran dated from the ninth century B.C., before any records of Indian use.

It was in India where cannabis was first formally exploited as an hallucinogen, although the ancient Chinese knew that it could be "the delight giver." In India, cannabis had its greatest use among those religious cultures whose norms include introspection, meditation, and bodily passivity. Indra, Lord of the World, was believed to drink it, and the plant thus became a gift of the gods, imbued with mystical properties; those who used it were partaking of the sacred. This attribution of drugs to the godhead is found in many societies. Certainly, anything that could deaden pain and "remove the noise of the world" had to be holy. To the Hindu peasant, it was "the joy giver, the sky flyer, the heavenly guide, the poor man's heaven" although to the Brahman it had to be used religiously, and even its fabrication into rope was forbidden as a sacrilege. Nevertheless, as with all other mind-altering drugs, it could not be restricted to magic-religious functions, and it quickly became a product of everyday use.

The refinements accompanying the use of cannabis and much of our basic nomenclature is derived from India. There are three basic grades. The poorest of these is *mj*, consisting of leaves and twigs of both male and female plants. Also called *kif* or *dogga*, it is used in India by the poorest of the poor—it is also the most common grade in North America. *Bhang* is a better grade whose name is derived from a supplication to Siva, the Hindu god who taught that the word should be prayerfully chanted over and over during the period of cultivation to ensure a good harvest. Bhang consists of the upper leaves and cut tops of female plants harvested before the flowers open. The plant material is sun dried, coarsely ground, and taken either as a decoction with milk or smoked. Better grades in North America are basically Bhang, although we have substituted the Portuguese word marijuana (marihuana) which came into North America via Mexico in the late nineteenth century. The third grade is *ganja*, composed entirely of the upper leaves of carefully cultivated female plants. In India it is smoked, mixed with wine or milk, or incorporated into candy. The Indian government considers cannabis cultivation a licensed agricultural industry and, until recently, licensing was solely to regulate quality; it was not taxed, for, it was asked, "who can dare tax a necessity of life?"

A special grade is called *charas*, consisting solely of the pure, unadulterated resin of cultivated female plants at the flowering stage. This resin, which we call *hashish*, is synthesized when the female flowers open. It was collected in many ways; at one time naked men ran through the fields and the resin stuck to their bodies. Since the odor of sweat offended the nostrils of the elite, the men were later clothed in leather vests and pants. In Persia, female plants were cut down and beaten onto carpets, the rugs were washed and the dissolved resins concentrated.

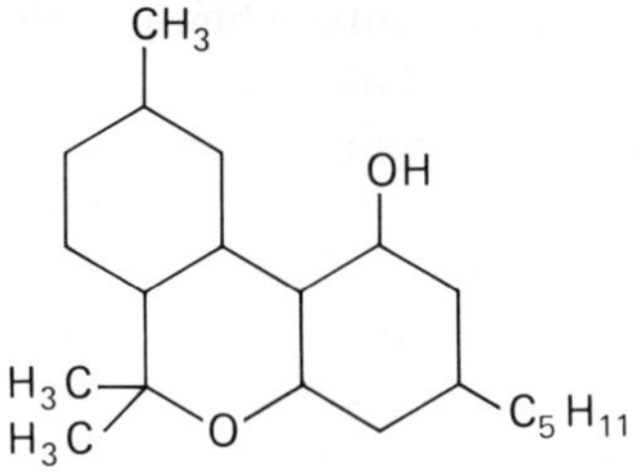

CH3

OH

H3C
H3C   O   C5H11

Δ¹ trans-tetrahydrocannabinol

Figure 4.10
Chemical formula for THC.

The active hallucinogen is delta¹-tetrahydrocannabinol (THC) (Figure 4.10), but several other closely related compounds are also present in resins, and these may interact with THC in the body. THC was first synthesized in 1965 and is of chemical interest because no other hallucinogen has a chemical structure even remotely resembling it. THC is synthesized in special gland cells (Figure 4.11) found on the leaves and in the female flower parts; male plants have much lower concentrations than do female plants. The THC content of stable genetic strains of cannabis, grown under good conditions, will vary from season to season; in spite of subjective judgments, chemical analyses indicate that THC levels are greatest in plants grown under high temperatures and with a long growing season, conditions not usually obtained in dormitory closets or in the colder parts of the United States and Canada. The dramatic, well-publicized burning of a field of cannabis in northern areas is basically an exercise in publicity. It is also fairly well established that the form in which the material is taken is a factor in the response, even with synthetic THC. Smoking provides the quickest and most long-lasting euphoria, but subjective attitudes, peer-group pressures, and other factors are clearly involved. Clinical, psychological, and pharmacological effects can be found in references cited herein and in many other reputable reports. It is, however, worth noting that use of cannabis does not lead to sexual orgies as detailed in many bad films, to murder, or to other lurid activities lovingly detailed by the communication media.

Examination of cannabis is an excellent way of looking at the flow of a plant through various cultures. Knowledge of its hallucinogenic properties spread from India throughout Asia and Asia Minor about 500–1000 B.C. Herodotus described its use by the Scythians in 500 B.C.; they made a woolen tent, burned hemp in a dish, and stood in a circle

Figure 4.11
Diagram of a *Cannabis* leaf showing structure of the glands that synthesize THC.

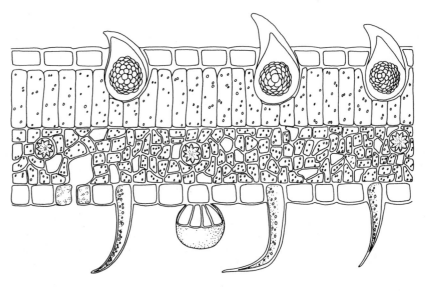

with their heads under the tent. The ancient Thebans mixed hemp with sweetened wine, and Galen reported that the Greeks baked it into little cakes. Under the influence of Mohammedanism, whose faith proscribed alcohol, cannabis spread throughout the Arab world. The word "hashish" is derived from the name of a Persian prince, Hasan-i-Sabbah, whose private army, the Hashishins, were paid off partly with resin; our word "assassins" is derived from these mercenaries whose avowed purpose was to kill all "enemies of the faith." The sect was not completely suppressed in Syria and Persia until the thirteenth century. At least part of the legends about the dangers of cannabis relate to this sect and the story (probably false) that the hashishins committed their ritual murders while under its influence has been cited as reason enough to ban it. It then entered the lands of Africa below the deserts. In the Zambesi valley, piles of fresh leaves are simply thrown on a fire and the celebrants inhale the smoke. When a Bantu tribe, the Baloubas, conquered much of what is now the Congo River Basin, their chief, Kalamba-Moukenja, wished to unite the various groups. To do so, he developed a cannabis ceremony now common throughout central Africa. Hottentots, Bushmen, and Kaffirs have used cannabis for at least 300 years. In the Middle East and in Africa a variety of reed tubes, pipes, and the famous waterpipe were developed. Preparations of cannabis have been mixed with tobacco, with other hallucinogens, or with alcohol in order to increase physiological responses.

HASHISH CANDY

Although it is likely that the hallucinogenic properties were known in Europe during the Middle Ages, we have no good evidence that it was commonly used. The introduction of cannabis to "polite" society came when Napoleon's army brought hashish back from conquered Egypt. In 1844 a private club, the Club de Hachichins, opened in Paris, dispensing hashish in candy or mixed with wine. Its founders included Gautier, Baudelaire, Dumas, and others of the socially elite, literary set. Other clubs sprang up and introduced smoking. At the same time, French medical authorities began recommending it as a calming agent for the hysterical, justifying the recommendation of Emperor Shen-Nung.

Marijuana entered the New World via Mexico when her French rulers introduced its cultivation for hallucinogenic purposes in the nearly ideal hot, dry lands around Mexico City. Its use spread very slowly to the native peoples because the peon preferred native hallucinogenic plants that were sanctified by usage dating back to Aztec times—marijuana was good for a mild smoke after a long hard day in the fields or mines. The product entered the United States through the Southwest and by 1860 was introduced to the Eastern Seaboard via the immigrants from the Caribbean Islands. In New York, it was used by the black poor and by the socially chic. *Harper's New Monthly Magazine*

Figure 4.12
"A New York Hasheesh Hell" wherein innocent young women of the upper classes were lured into a life of drugs and shame . . . and heaven alone knows what else. Taken from a cartoon published in the *Police Gazette* of the 1890s.

reported in 1883 that "there is a large community of hashish smokers in this city . . . and I can take you to a house uptown where hemp is used in every conceivable form and where the lights, sounds, odors and surroundings are all arranged to intensify and enhance the effects." Not unexpectedly, lurid journals and newspapers, like the *Police Gazette*, had a field day (Figure 4.12)! In the late nineteenth century, Tilden's Extract of Cannabis was medically prescribed for nervousness and even for lockjaw, and marijuana cigarettes, made by Grimault & Sons, New York, were a prescription item for easing asthma. Marijuana entered the American heartland via the Mississippi River, along with New Orleans jazz, and was popular among the black poor who were then moving to Chicago, Kansas City, and St. Louis. From all these centers, it moved into the major Canadian cities.

Immediately after World War I, as crime and violence seemed to take a quantum leap upwards, political and police authorities were quick to blame marijuana rather than to evaluate shifts in population and other social changes—it was a convenient scapegoat. In 1937, prodded by the same temperance groups who had outlawed alcohol through the Volstead Act, Congress passed the Federal Marijuana Tax Act, ostensibly to limit and control distribution, but which actually made cultivation or possession a felony punishable by a $2000 fine and up to five years in federal prison. Most states passed their own marijuana acts with penalties about as severe as for narcotics. A Canadian commission, on the other hand, stated, "If we confuse the issue of whether an individual should use a substance with that of whether society should forbid his doing so, we may be laying the foundation for tragic errors in public policy."

In recent years, there has been considerable agitation to modify or repeal aspects of the marijuana proscriptions. In part this is due to the increasing amounts of clinical and pharmacological data suggesting that tetrahydrocabinnol is not a narcotic, is not addictive, and does not cause sociologically adverse behavior. That cannabis is now being used by people in those socioeconomic strata who are articulate and have sufficient influence to force legal change (in addition to the poor and the young whose political clout is minimal) is also a factor in altering the statutes. Relatively few of the available clinical studies are sufficiently comprehensive, adequately controlled, or conducted with statistically significant populations to provide reliable information upon which to base political and legal decisions. Thus, the question of adverse short-term or long-term consequences to the user is still open. The hemp plant is likely to be newsworthy for some time to come.

ADDITIONAL READINGS

Andrews, G., and S. Vinkenog. *The Book of Grass*. New York: Grove Press, 1967.
Bloomquist, E. R. *Marijuana. The Second Trip*. Beverly Hills, Ca.: Glencoe Press, 1971.
Brecher, E. M. *Licit and Illicit Drugs*. Boston: Little, Brown, 1972.
Gamage, J. R., and E. L. Zerkin. *A Comprehensive Guide to the English-Language Literature on Cannabis*. New York: Stash Press, 1969.
Grinspoon, L. "Marijuana." *Scientific American* 221(1969):17–25.
Mendelson, J. H., A. M. Rossi, and R. E. Meyer. *The Use of Marijuana: A Psychological and Physiological Inquiry*. New York: Plenum Press, 1974.
Nahas, G. *Marijuana—Deceptive Weed*. New York: Raven Press, 1973.
Simmons, J. *Marijuana: Myths and Realities*. North Hollywood, Ca.: Brandon House, 1967.
Solomon, D. (ed.) *The Marijuana Papers*. New York: New American Library, 1966.

Mushroom Madness

♗ She stretched herself up on tiptoe and peeped over the edge of the mushroom and her eyes immediately met those of a large blue caterpillar.
LEWIS CARROLL, *Alice's Adventures in Wonderland*

In addition to Magna Carta, common law, Shakespeare, and over-cooked vegetables, many of us living in North America inherited the Anglo-Saxon mycophobia. Many of us will pick at broiled mushrooms garnishing a steak, but few will eat them raw and even fewer have the courage to taste one of the four or five edible wild species which can be

Adapted from "The Mysterious Mushroom," *Garden* 1(1977):18–19, 30. Copyright The New York Botanical Garden.

Figure 4.14
The chanterelle, *Cantharellus cibarius.* This yellow-to-orange mushroom is readily identified and is excellent in flavor.

unequivocally identified (Figures 4.13–4.16). There are exemptions to this blanket statement; some Scandinavians and Germans avidly search for *pfifferlings* in the spring and *steinpilz* in the summer, but the common attitude is summed up by a statement from a British farmer, "If my caows won't eat 'em, Oy ain't agon'ta."

There is a long history of mushroom distrust. Except for mention of "leprosy of a house," referring to dry rot and the notion that leavened bread is not ritually clean, neither of the Testaments speak directly of fungi. There is a Latin saying: "If you awaken in the morning after an evening meal of wild mushrooms, you know they were good ones." Pliny asked, "What great pleasure can there be in partaking of a dish of so doubtful a character?" and Dioscorides in the first century A.D. stated: "Of fungi there is a double difference, for either they are edible or they are poisonous," and he considered that even edible sorts were hard to digest. If they are served at a banquet, Dioscorides recommended that prophylactic vomiting should be induced by swallowing

Figure 4.15
The gem-studded puffball, *Lycoperdon gemmatum.* Puffballs are edible when their interiors are pure white and firm.

Figure 4.16
The shaggy-mane, *Coprinus comatus.* Easily identified and very tasty when young.

Figure 4.17
A bolete, *Boletus edulis*, one of the prized mushrooms of the Romans.

hen's manure mixed with vinegar. Peter Treveris, author of an herbal, stated most forcefully:

> Fungi ben mussherons. They be cold and moyst in the thyrde degre and that is shewed by theyr vyolent moysture. There be two maners of them, one maner be deedly and sleeth them that eateth of them, and be called tode stooles, and the other dooth not.

"Tode" is a German—a Saxon—word for death, and this concept pervades our mycophobia.

On the other hand, mycophiliac cultures esteem them highly. Egypt's pharaohs reserved them for their personal use. Wealthy Romans vied with one another to create new and different recipes, and Cicero noted that "since these elegant eaters which to bring into high repute the products of the soil, they prepare their fungi with such high-seasoned condiments that it is impossible to conceive of anything more delicious." The favorite in Rome was the *cibus deorum*—the divine food—and in the area near Rome, this mushroom was first offered to the emperor. Julius Caesar did, however, issue an edict that those found on foreign soil could be eaten by captains of cohorts. The boletes (Figure 4.17) were prepared in special cooking vessels, *boletaria*, that could not be used for less noble purposes. Emperor Tiberius (r. 14–37 A.D.) awarded one Asellius Sabinus the sum of 200,000 sesterces (about $15,-000) for a recipe containing boletes, small whole birds (*beccafios*), oysters, and boned thrushes. *De Re Coquinaria*, a cookbook written by Caelius Apicus about 200 A.D., contained the following recipe:

> Boil truffles, sprinkle with salt and thread them on twigs. Roast them over an open fire and then place them in a cooking pot with wine, pepper, boiled wine, oil, greens and honey. Boil, and while simmering, add flour to thicken. Arrange them around a pig while it is roasting.

Galen, however, wrote that fungi must be counted as among insipid things for "they yield clammy juices and it is far safer to have nothing to do with them."

Medically, mushrooms are of little value. Hippocrates recommended that dried, powdered fungi be used to cauterize wounds, a practice still used in Lapland, Sikkim, and a few other out-of-the-way places. Dried fungi were powdered, mixed with saltpeter, placed on the wounds, and set afire. Although imputed with magical properties, the fungi serve only as tinder. One fungus, a bracket type found on trees, contains laricic acid, a powerful purgative that can cause death and hence is rarely employed. In Solzhenitsyn's book, *Cancer Ward*, he refers to a fungus called *chagi* which grows on birch trees in Siberia and has a local reputation of curing cancers. Several such fungi grow on trees and one, *Polyporus hispidis*, contains a compound that Soviet scientists are evaluating as a cancer cure. Most edible mushrooms are about 90 percent water, 4 percent protein, 5 percent carbohydrate, and 0.4 percent fat—they are as nutritious as watermelon.

In North America, it is almost impossible to purchase any except

Agaricus bisporus, the least tasty of the edible species. It has the sole advantage of ease of cultivation and ready recognition. Cultivation of *A. bisporus* developed in England in 1779 and was introduced into the United States about 1865 by immigrants to Pennsylvania. They are usually grown in caves (Figure 4.18) where light will not darken the pristine whiteness we prize and where temperatures are constant. Horse manure mixed with wheat or rye straw is composted for several weeks and then placed in the bedding area where it is allowed to pasteurize naturally at the high temperatures caused by the metabolism of microorganisms. This "sweating-out" process makes the bed disease-free. Beds are inoculated with either a pure laboratory culture of the fungus or with mushroom spawn—a compressed block of the threads (mycelium) of the fungus. Small buttons begin to appear in a month, and a crop is ready to be picked in about two months. A well-cultivated bed will continue to produce for about two or three months with yields of five kilograms of fresh weight per square meter of bed area.

ꙮ AN ESSAY ON THE FUNGI

The familiar fungi are those whose fruiting, reproductive structures are big enough to be visible to the human eye. Most prominent are those evolutionarily advanced forms, the basidiomycetes, including the mushrooms and toadstools. The fungi are, however, among the most varied of living organisms, with well over 80,000 species divided into several hundred families. The vast majority of fungi are micro-

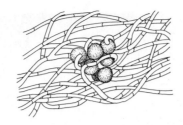

Figure 4.19
Diagram of the association of fungi and algae to form a lichen. Hyphal threads of the fungus utilize sugars formed in algal photosynthesis and provide the algae with water and minerals.

scopic in size, with their vegetative structures either as single cells or as threadlike strands called hyphae. The fluffy, cottony masses of hyphae that reach visible size are referred to as a mycelium or, when they become twisted into ropelike masses seen under rotting bark of trees, as rhizophores. None of the fungi contain chlorophyll, and hence all are dependent on other organisms for their organic food supply. Some can obtain this food from dead plants and animals; the mycelium surrounding a dead fly on a windowpane or the fungi that cause rot in wood are examples of this *saprophytic* mode of nutrition. Other fungi live in association with algae, obtaining sugars and other compounds from the photosynthetic forms and, in return, providing water and minerals to the algae (Figure 4.19). Such associations are termed *symbioses*—living together. Others are parasitic, living on or in plant or animal cells. Many diseases of plants and of animals are caused by fungi whose activities result in large economic losses and considerable numbers of human deaths each year.

The range of habitats in which fungi are found is enormous. Some species live in the depths of the oceans, others in virtually every soil type on earth, and some can even grow in jet fuel, where they clog engines. Leather, some plastics, paint, asphalt, soap and, other inhospitable substrates can each harbor one or more species of fungi. Fungi excrete enzymes that break down organic substrates into soluble compounds which are then taken in by the hyphal cells and converted into more fungus.

The reproductive capacities of fungi have been of scientific interest for several hundred years. Many fungi produce microscopic asexual spores, or conidia, which can be carried long distances in the air. When they land on a suitable substrate, they germinate and again form a hyphal thread. The blue, green, and black powders formed on rotting oranges are conidia of *Penicillium*, a fungus which also produces the antibiotic penicillin (Figure 4.20). The red streaks which form on wheat leaves, giving the name wheat rust to the disease, are the spores of a pathogenic fungus, as are the black powders that form in the ears of corn as corn blight disease develops (Figure 4.21). One hyphal fragment can produce millions of conidia, and the spread of plant and animal diseases "like a prairie fire" is due to this efficient asexual reproduction.

Fungi, like most other organisms, can also reproduce sexually, that is, by the fusion of genetically dissimilar nuclei. Many of these

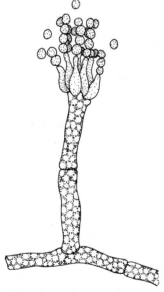

Figure 4.20
Conidial branch of *Penicillium spp.* The brushlike form of the branch was responsible for the genus name.

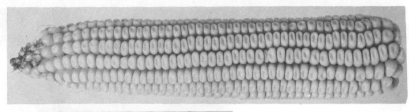

Figure 4.21
Effects of southern corn blight caused by *Helminthosporium mydis*. Courtesy U.S. Department of Agriculture.

sexual processes are extremely complicated, involving several stages. The wheat rust fungus can only complete its life cycle with a total of five distinct kinds of spores and two different hosts—the wheat plant and the barberry. A disease of white pines also involves the raspberry plant as an alternate, obligatory host. The taxonomic separation of the fungi into classes is based on their modes of sexual reproduction. In the mushrooms, the visible entity is the reproductive structure, the terminal point in their life cycles. The mushroom itself is composed of hyphal threads closely appressed and organized into the typical form for each species. The mycelium that supplies the nutrients to the fruiting structure may ramify through soil or through trees and is rarely seen except when one digs carefully and finds rhizomorphs in the soil or in the decaying wood. Picking mushrooms does not injure the mycelium, and new fruiting structures will develop when conditions again become favorable.

Most emphasis has been placed on those fungi that cause the rots, blights, murraines, rusts, smuts, bunts, etc. of plants, or the skin, lung, or systemic diseases of humans and other animals. This should not obscure the fact that the vast majority of fungi play irreplaceable roles in the ecosystems of the world. They, and the bacteria, are responsible for the return to the soil of the accumulated organic and mineral constituents that comprise plants and animals. Without this decay, the cycling of minerals and of carbon-containing compounds would not occur, and the biotic world would soon come to a grinding halt. People have exploited the phenomenal capacity of fungi to synthesize compounds of use to them. The bulk of the world's antibiotics, many of our cheeses, all of our bread and fermented beverages, and compounds for the chemical industry (acetone, citric acid, etc.) are produced by fungi.

Mushrooms and toadstools have been part of legend and myth in many cultures, and have inspired the work of poets, mystery writers, and other artists. Many of these stories are charming, not the least being those about the Irish elves that sit on, or in rainy weather under,

Figure 4.22
Puffballs, *Lycoperdon pyriforme*.
Courtesy Dr. Kent H.
McKnight, Agricultural Re-
search Service, U.S. Depart-
ment of Agriculture.

giant umbrella-shaped toadstools. In England, one type of fruiting mushroom, the puffball, is called pucksball, from Puck, the English equivalent of the Greek satyr Pan (Figure 4.22). This indirect reference to male virility led to their being considered aphrodisiacs, but this is not supported by any scientific evidence. The stinkhorn fungus (*Phallus impudicus*) is also called the devil's penis; both common names are accurate, for it has a horrible smell and its shape is anatomically correct (Figure 4.23). Sir John Maundiville wrote in 1335 about the fungus named *Hirneola auricula-judea*, stating that it could summon the Devil from hell. Maundiville declared it was first found on the tree upon which Judas hanged himself after betraying Christ. Thus, from Judas's ear to Jew's ear. Edible, but not especially tasty, it was used for several centuries as one of the doctrine of signature plants to cure ear inflammations (Chapter 3).

The rapidity with which the fruiting body can appear has been a source of superstition. Even in temperate zones, these can appear within a day after a rain, a fact that tended to emphasize their other-worldliness. Since the vegetative threads of the fungus can grow out equally in all directions from a single center, it is possible to find an almost perfect circle of mushrooms, a circle with a diameter of up to 30 m (Figure 4.24). People believed that these "fairy rings" were the result of the round dances of the wee people on moonlit nights. A Scottish proverb has it, "He wha tills tha fairie green, nae luck again sall hae." Young lassies wishing to improve their complexions need only rub grass dew on their faces, but if the dew came from within a fairy ring, their faces would be disfigured. Dew from within the circle could, however, bewitch a loved one.

Associations of fungi with the supernatural, with evil, rot, decay, and death pervade English literature from at least the fourteenth century (Figure 4.25). The deep woods, the gloomy, wet days following prolonged rainstorms, their rapidity of appearance, fairy rings, the luminescence of several species as a faint glow in a dark night, and, of

Figure 4.23
Stinkhorn fungus, *Phallus impudicus*.

Figure 4.24
A fairy ring of mushrooms.
Courtesy U.S. Department of
Agriculture.

course, the convulsions and death resulting from eating a toadstool have reinforced our legacy of general revulsion. Alfred Lord Tennyson, for example, wrote "As one that smells a foul-fleshed agaric . . . ," and Thomas Hardy evoked visions of "an afternoon which had a fungus smell." And remember that in order to consolidate his nefarious schemes against the lost boys in *Peter Pan*, Captain Hook "sat down on one of the enormous forest mushrooms, [for] in Never-Never Land, mushrooms grow to gigantic size."

A good deal of the Anglo-Saxon avoidance of mushrooms is related to the inadvertent eating of poisonous species. Not only have many people died by innocently eating them, but some have died by deeds most deliberate and foul. When Agrippina, second wife to Emperor Claudius, decided to kill him so that her son, Nero, could become imperial caesar, she is said to have added the juice of *Amanita phalloides*—the death angel—to his favorite dish of cibus deorum (*Amanita caesarae*). Nero, when safely on the throne, was asked whether this mushroom was really the food of the gods and replied, "Yes, since it led to the deification of my father." Among the poisonous species none are more deadly than *Amanita phalloides* and *Amanita virosa*. Their toxins are complex cyclopeptides which, depending on their chemical struc-

Figure 4.25
Poisonous serpents and toadstools. Taken from *Commentarii* by P. Mattioli, edition of 1560.

Figure 4.26
Psilocybe spp.

ture, will kill in 6–12 hours (the phallotoxins) or 15–20 hours (the ama-toxins). By causing lesions in the liver, mortality can run higher than 60 percent. A classical antidote was a mixture of the stomachs and brains of rabbits . . . eaten raw; this was recommended because rabbits eat *Amanita* with impunity. Even modern treatments are ineffective once the toxins begin their work on the liver and kidneys.

HALLUCINOGENIC MUSHROOMS

Our knowledge of hallucinogenic mushrooms in the Americas began with the Spanish conquest of Central America. Bernardo Sahugun wrote in his *Historia General* that "the Chichimecas possess great knowledge of plants and roots. . . . They discovered harmful mushrooms which intoxicate like wine. There are some small mushrooms which are called teonanacatl. . . . only two or three need be eaten. Those that eat them see visions . . . that provoke lust and even sensuousness." The Aztec *teonanacatl* is best translated as "flesh of the gods," and contrary to the Spanish belief that they were orgiastic stimulants, the evidence is overwhelming that they were used in as holy a ritual as the Eucharist and, indeed, they served the same function as does the Body and Blood. King Philip's physician and sometime reporter on New Spain, Francisco Hernandez, piously noted that priests banned the use of the flesh of the gods by the conquered Indians because it contributed to pagan behavior, promoted idolatry, and mocked Holy Communion.

Apparently two different groups of fleshy fungi were involved in what has come down to us as mushroom cults. The teonanacatl group includes several genera especially *Psilocybe* (Figure 4.26), of which 15 distinct species have been named; *Psilocybe mexicana* is the most important in ritual. The genera *Paneolus* (one of the Orient's laughing fungi), *Psathyrella*, *Conocybe*, and *Strophoria* are used as hallucinogens. All contain psilocybine (Figure 4.27) and other tryptophan derivatives, alkaloids that cause visions, muscular relaxation, hilarity, alterations in perceptions of time and space, and a feeling of total isolation from one's environment. Although they may be used by ordinary people, most of the modern rituals involve the eating of the fungus by the descendants of the Aztec priests, the male or female shamans. Under the influence of the hallucinogen, the shaman will chant the "truth" about health and disease, success or failure, and how to remedy the affairs of everyday life. For the Indians of Central America, the psychedelic experience awakens the forces of creation, and partaking of the flesh of the gods is a serious act, not taken lightly. It is not a hedonistic but rather a deeply religious act.

The other hallucinogenic mushrooms of the Incas, Mayans, and Aztecs is *Amanita muscaria*, the fly mushroom (Figure 4.28), so named because water extracts of it were used by North American pioneers to kill flies. Our first record of its use came from reports of Cortez, who

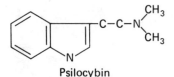

Psilocybin

Figure 4.27
Chemical formula of psilocybin.

observed the mushroom being eaten during the coronation of Montezuma. *Amanita muscaria* contains at least five psychoactive compounds related to those found at the junctions of nerve cells in body and brain and responsible for excitation transmission along nerves. These compounds, in ratios and combinations present in the mushroom, affect the central nervous system producing mental confusion, abnormal behavior states, visions, and, if too much is taken, convulsions and death by respiratory failure. Indians of Mexico believe that the magic mushroom came from drops of God's sperm, that it grows from the vulva of Earth Mother, that it appears above ground as a small penis that becomes more and more aroused, and that the cap (woman) is penetrated by the stipe-penis (Figure 4.29). In Guatemala, *Amanita* is identified with the thunderbolt and is believed to appear when lightning strikes the earth (Figure 4.30). The sacredness of the mushroom is attested by glyphs seen in the Madrid Codex dating from the middle of the sixteenth century in which an animallike deity is shown presenting the mushroom to a figure seated on a throne (Figure 4.31). Even today, mushroom cults gather the fruiting structure just before dawn, at the time of Earth Mother's new moon.

Figure 4.29
Stages in the growth of an *Amanita*.

Figure 4.30
Guatemalan mushroom stone dated from 1000 B.C.

175   *Mushroom Madness*

Figure 4.31
Messenger from the gods offering the sacred *Amanita* to a ruler. Taken from a wall painting in Mexico and copied in the Madrid Codex.

Mushroom madness is not restricted to "less civilized" portions of the world. A fresco in a Roman Catholic church in Plaincouralt (Indre), France, depicts Adam and Eve on either side of a tree of knowledge that is unequivocally a branched *Amanita muscaria*. The Viking Berserks, were, according to one authority, under the influence of *Amanita*, as evidenced by their capacity for tremendous feats of strength and their own reports of visions and high sexual activity. In western Siberia, among the Finno-Ugrian peoples and among the Lapps, hallucinogenic use of this fungus is widespread. Women of the Koryak tribe in Siberia prepare the mushrooms by chewing them (without swallowing) and then pass the softened pellets to the men who are allowed to swallow them. The hallucinogenic alkaloids are excreted in urine which the women are permitted to drink. Their urine, in turn, is taken by the children. Russians refer to these hallucinogenic fungi as *paganki* [little pagans] and believe that they have Powers of the Dark. There are scattered reports in the English literature of the late nineteenth and early twentieth centuries that hallucinogenic properties of *Amanita* were exploited by upper-class opium eaters, and one wonders whether the Rev. C. L. Dodgson of Christ Church College in Oxford was aware of this when, in *Alice's Adventures in Wonderland*, he wrote about our heroine nibbling at a mushroom to change her size (Figure 4.32).

SOMA

When the Aryan peoples entered India from the North about 3500 years ago, they brought with them a cult involving a plant called *soma*. The *Rg-Veda* (X, 119:7–9) glorified soma in the following couplets:

> Heaven above does not equal one half of me.
>   Have I been drinking soma?
> In my glory I have passed beyond earth and sky.
>   Have I been drinking soma?
> I will pick up the earth and put it here or there.
>   Have I been drinking soma?

Figure 4.32
Alice speaking to the green caterpillar who is seated on a mushroom. Taken from the drawing by John Tenniel in the first edition of *Alice's Adventures in Wonderland*.

Botanists and anthropologists have suggested many plants as being soma, but none seem to meet all the criteria found in the ancient writings, poems, and legends. The possibility that soma is a drink made by steeping *Amanita muscaria* in milk or wine has been suggested by several people. They argue that in primitive societies a psychedelic experience is a journey into the realm of the supreme, a transcendent opportunity to become part of the supreme. Soma, according to the ancient writings, could send one's soul to the abode of deity and, equally important, could allow the traveler to return—something which death does not permit. John Allegro has extended this argument in *The Sacred Mushroom and the Cross* and concluded that *Amanita* was a focal point in the Christian tradition. The distinction between the mycophobe and the mycophile may be basically religious, based on acceptance of the divine mushroom concept. These speculations are not generally accepted by other scholars and are most assuredly not in favor with organized Western religions.

In a book on hallucinogenic plants, *Flight from Reality,* the late Norman Taylor asked why human beings have searched so avidly for psychedelic plants. Since use of hallucinogens has great religious ritual significance, was religion itself a flight from reality? Admittedly, this is a cold, hard, cruel world we live in, and it was much colder, harder, and more cruel in the past. Today we are exhorted to face life, to stand up to reality, to become masters of our fates and captains of our own souls. In our Western European milieu, we condone the use of only one drug, alcohol, as our vehicle to flee reality and have interdicted virtually all others, some with many fewer physical and mental consequences. Did the Aztec noble, the Siberian peasant, and the Vedic priest know something that we don't know?

ADDITIONAL READINGS

Allegro, J. *The Sacred Mushroom and the Cross.* Garden City, N.Y.: Doubleday, 1970.

Gray, W. D. *The Use of Fungi as Food and in Food Processing,* parts 1 and 2. Cleveland, Oh.: CRC Press, 1971.

Grover, J. W. *Edible and Poisonous Mushrooms of Canada.* Canada Department of Agriculture, Ottawa, 1962.

Harner, M. J. *Hallucinogens and Shamanism.* New York: Oxford University Press, 1973.

Marshall, N. *The Mushroom Book.* Garden City, N.Y.: Doubleday, 1974.

Puharich, A. *The Sacred Mushroom.* Garden City, N.Y.: Doubleday, 1970.

Rolfe, R. T., and F. W. Rolfe. *The Romance of the Fungus World.* London: Chapman & Hall, 1925.

Savonius, M. *Mushrooms and Fungi.* London: Octopus Books, 1973.

Schleiffer, H. *Sacred Narcotic Plants.* New York: Macmillan Press), 1973.

Schules, R. E. "Teonanacatl: The Narcotic Mushroom of the Aztecs." *American Anthropologist* 42(1940):429–443.

Singer, R. *Mushrooms and Truffles, Botany, Cultivation and Utilization.* New York: Wiley, 1961.

Sterling, D. *The Story of Mosses, Ferns and Mushrooms.* Garden City, N.Y.: Doubleday, 1955.

Wasson, R. "The Hallucinogenic Fungi of Mexico." *Botanical Museum Leaflets* 19(7)(1961):137–162 Harvard University

Wasson, R. G. *Divine Mushroom of Immortality.* New York: Harcourt Brace Jovanovich, 1972.

Wasson, R. G., G. and F. Cowan, and W. Rhodes. *Maria Sabina and Her Mazatec Mushroom Velada.* New York: Harcourt Brace Jovanovich, 1975.

The Tomato Family

℞ Give me to drink mandragora
That I might sleep out this great gap of time
SHAKESPEARE, *Antony and Cleopatra*

Solanaceous plants—potato, tomato, eggplant, bell peppers, that stinking weed tobacco, hot chili peppers, and their milder cousins the pimentos and paprika—are used throughout the world. Among the ornamentals, the Jerusalem cherry is obviously tomatolike, but the Chinese lantern, butterfly flower, *Salpiglossis, Browallia, Nerembergia,* and *Petunia* are not usually thought of as being related to the tomato. Within this family are four genera which have been involved in murders most foul and devious, in witchcraft and demonology, in military campaigns, in sly invitations to seduction, and in sexual orgies.

These literally unholy four are the jimsonweeds (*Datura*), deadly nightshades (*Atropa*), henbanes (*Hyoscyamus*), and mandrakes (*Mandragora*). The entire family is a chemist's delight; its members containing several hundred chemically distinct alkaloids. The family name itself, *Solanaceae,* is an allusion to the sedative or hallucinogenic properties of its genera. The terrible four genera, however, contain three alkaloids not found in more familiar genera. Atropine, hyoscyamine, and scopolamine are representatives of the tropane series of alkaloids, similar in structure to cocaine. Their biosynthesis is imperfectly known, but a good deal is known about their physiological action in human beings.

TROPANE ALKALOIDS

Atropine is the most medically useful of the tropanes, serving as a stimulant of the central nervous system and a depressant of the parasympathetic nervous system. The central nervous system includes the brain and spinal cord nerves which monitor many direct body functions and most of the actions of the minds of men and women. As a central nervous system stimulant, atropine in moderate doses causes loss of motor coordination and at higher concentrations hallucinations, delirium, stupor, and even death. The parasympathetic system, part of the autonomous network, regulates function of internal organs (heart, blood vessels, intestinal tract) over which there is little voluntary control. Thus, atropine can increase the heart rate to 150–180 beats per minute (tachycardia) with no increase in cardiac output. It increases parastaltic movements of the intestines and interferes with the normal regulation of breathing. Atropine blocks the normal transmission of signals across synaptic junctions between nerves. Most people have experienced one parasympathetic action of atropine when an ophthalmologist puts drops in the eyes to prevent the autonomous closing of the pupil in bright light. Atropine eases the wheezing of severe hayfever victims and effectively dries up drippy noses.

Scopolamine is a depressant of the parasympathetic system, but can depress the central nervous system. It is a mild analgesic and soporific, and an effective antidote for seasickness. Scopolamine as a truth serum may or may not still be used depending on whose national intelligence agency is being asked.

The third tropane alkaloid, *hyoscyamine,* is similar in action to atropine, but the clinical responses are sufficiently different to suggest that the receptor sites are not the same. The amounts used medically are very small; practitioners know that even slightly higher dosages, particularly when administered internally, have grave consequences and may lead to death.

DATURA

Of the four genera, we have more information on the history, anthropology, and ethnobotany of *Datura* than on the others (Figure 4.33). The genus is found world-wide, with *Datura innoxia* and *Datura metaloides* in Mexico and the American Southwest; *Datura metal* in India and southeast Asia; *Datura sanguinea* in tropical and subtropical South America, and *Datura stramonium* in both Europe and North America. There are 50 described species; ten have been reported from Europe, and an equal number are found in Africa. Most are annuals, growing to two meters in good soil with a long, warm growing season. The coarse-textured leaves are usually ovate and irregularly toothed or notched. Flowers are trumpet shaped, superficially resembling a white- or green-streaked petunia. The fruit is a spiny capsule, accounting for one common name, thornapple. The unpleasant smell of leaves and fruit accounts for another name, stinkweed. Alkaloid content, both tropane and nontropane, may be very high in tropical climates; a goat or even a child may die after eating a few leaves or a single fruit. The genus, named by Linnaeus, is derived from the Hindi word *dhatura* which means poison.

Datura has been used ritually in India as far back as records have been kept. Seeds were reverently offered to Siva the destroyer, and His priests ate seeds to induce hallucinogenic, oracular states. Sanscrit manuscripts recommended leaves, stems, roots, or fruit for pneumonia, mumps, heart failure, toothworm, asthma, and sexual perversion. Widows destined to die by *suttee*—burning alive on their husband's funeral pyres—were given a potion made from datura leaves to sedate them to the point that they could be led calmly to the pyre, and would not feel the fire. According to Christoval Acosta, who visited India in 1678, Indian prostitutes added ground-up seeds of datura to their patron's drinks to increase sexual excitement, and "these mundane ladies [were] adepts in the use of the seed that they give in doses corresponding to as many hours as they wish their poor victim to be unconscious or transported." In the East Indies, women fed datura leaves to beetles, and then fed the resulting highly poisonous dung to faithless lovers.

Figure 4.33
Thornapple, *Datura spp.*
Taken from a woodcut by
Acosta published in 1578.

Datura was the original knockout drops, and thieves in India and in Europe used it for centuries. Up until at least the beginning of this century, juice of datura leaves was added to milk given young Indian girls who were to be initiated into prostitution. The drink was narcotic and, it was asserted, aphrodisiac, so that the victim was believed to have actively contributed to her own downfall. The Chinese, believing that datura was a sexual stimulant, administered it to brides on their wedding night to calm their nerves and to make them more sexually receptive.

The European attitude towards datura parallels that in the Far East. Apollo's priests drank datura to achieve sedated, prophetic, and oracular states. The sacerdotal plant of Delphi was undoubtedly datura; the mumbling speech, trance states, and known fears of overdosage are consistent with datura intoxication. Greek physicians knew it as *nux-metal* or *neura*, a reference to its sedative action, and when extended unconsciousness was desirable, as during surgery, datura was mixed with opium. Rome followed Athens's lead in ritual and in medicine and added the drug-induced orgy in which datura mixed with wine was used to induce hallucinogenic states and to heighten sexual activity. Avicenna, a tenth-century Arabian physician, recommended *jouz-methal* not only for surgery, but as an excellent treatment for anxiety. From Arabia, the medical and aphrodisiacal use of datura spread to Spain and to western Europe; northern Europe was too cold to support the growth of datura with high tropane content.

Much of the information on the use of datura in the Middle Ages and into the Renaissance has been lost, because its use became so intertwined with witchcraft that it was suppressed. Certainly no love potion worth paying good money for was without datura, usually supplemented with extracts of other solanaceous plants and, for good measure, attar of roses, marjoram, other herbs, and a newt's tongue. Witches's sabbats, the infamous black masses that so intrigued prurient Victorians, utilized ointments and unguents containing datura. Whole leaves were inserted into the rectum or vagina where tropanes are quickly absorbed; the broomstick developed as a symbol of this part of the ritual. Shortly before World War I, Professors Karl Wieswetter and Will-Erich Peukary compounded a datura ointment from an old witch recipe and rubbed it on their arms. They reported that they soon felt as if they were whirling around the laboratory. If two sober-minded German scientists thought they were flying, one can appreciate the sensations of less erudite men and women of the fourteenth century. The belief that datura was aphrodisiac was an important component in witchcraft and demonology cults. This belief flows through the drug structure of all cultures that have used datura. Our "winging it" and our "flying" are derived from the concept of hallucinogenic states.

Datura was an important ingredient in the poisonings that pervaded southern Europe from 1400 to 1700. Nobles and merchants sent members of their family to schools teaching the art of poisoning for much the same reasons as we send our children to graduate schools of business administration. In 1650, the widow Toffana was convicted of poisoning 600 people in Palermo with *aqua toffana,* an alcoholic extract of opium and datura. In her defence, Signora Toffana asserted that she was not the only professional poisoner in Palermo or, for that matter, in Italy, and it was pointed out in her favor that the deaths caused by datura are painless. Reputable physicians of the Renaissance, still under the influence of Galen and Hippocrates, used datura as a sedative, anaesthetic, and soporific until the nineteenth century. John Gerarde, in his famous *Herball* of 1597 reported that "the juice of this plant, boiled with hog's grease . . . cureth all inflamations whatsoever . . . as meiself hath found by daily practice to my grete credite and profite." Nevertheless, the potential danger of internal use was well known. One herbalist-physician reported,

> He who partakes of it is deprived of his reason; for a long time laughing or weeping or sleeping and often times talking and replying so that at times he appears to be in his right mind, but really being out of it and not knowing to whom he is speaking nor remembering what has happened.

Datura was listed in official pharmacopoeias, codexes, and dispensatories for several hundred years with specifications for extraction, formulation, and dosages. Only when the tropanes were isolated and characterized were the recommendations altered to specify the purified alkaloid. In 1905, Dr. Carl Gauss of Freiburg brought down the wrath of the Lutheran church when he used datura extracts and morphine to induce *dammerschlaf,* the twilight sleep treatment for women experiencing difficult births. The Church fathers—none of them mothers—declared that Dr. Gauss had fallen from grace and was near unto heresy because the Old Testament says (Gen. 4:16) that women were to bring forth in pain.

Datura has had great importance in puberty rites of passage. In the western Amazon basin, initiates are required to drink a datura-beer mixture presented successively by each elder of the tribe. When the boys drink as much as they can hold, they fall into trances, and additional *maikoa* is ritually inserted into the rectum until the child is comatose, remaining barely alive for several days. In the Darien and Choco regions of Central America, seeds are fed to young children. They run around wildly for a short time before falling down in a stupor, whereupon tribal elders roll the child aside and dig beneath the body to find buried treasure. The Colombian Jivaro, a notably stoic people who shrink the heads of their captured foes, punish unruly children by forcing them to eat datura seeds in the hope that they will be visited by their ancestors who will admonish them for their social infractions; it is a rare child who does not reform after one experience. Infanticide in the Andean highlands involves smearing mother's nipples with datura juice mixed with wild boar's fat, providing a peaceful death that also ensures the baby's safe passage to the spirit world.

Datura is, as readers of Carlos Castaneda's books know, a powerful and respected hallucinogen. Mexican boys will eat datura leaves to acquire the strength of the evil sorcerer Kieri Tewiyara (datura person) under the watchful eyes of the *brujo,* who makes sure that they do not become permanently bewitched; the power of datura can be for good or evil and those who use it must constantly be on guard.

The British colonists in North America were introduced to datura very early. Robert Beverly's *History and Present State of Virginia* discusses the incident which led to our common name for *Datura stramonium.* In 1676, British troops landed at the Jamestown colony to put down a tobacco tax rebellion led by Nathaniel Bacon. Since the war office mandated inclusion of either citrus fruit or fresh vegetables in the diets of all troops to prevent scurvy, obedient regimental cooks prepared a "salide" of native plant leaves and inadvertently included datura. Beverly reported:

> The effects [were] a very pleasant Comedy, for they turn'd Fools. . . . one would blow up a feather in the air, another would dart straws at it with much fury; and another stark nakid was sitting in a corner like a monkey, grinning and makeing Mows at them. . . . a thousand such simple tricks they play'd and after eleven days return'd to themselves again not remembering anything that had pass'd.

Hence Jamestownweed, or Jimsonweed (Figure 4.34). They were not the first troops to be rendered *hors de combat;* Mark Anthony's African legions made the same mistake, delaying Julius Caesar's timetable for the conquest of Egypt.

With the brutal and total repression of Indian culture by the Spanish, the no less effective repression by northern European settlers and the less physical but equally effective repression by nineteenth century missionaries, ritual use of datura went underground, leaving a

Figure 4.34
Jimsonweed, *Datura stramonium.* Courtesy U.S. Department of Agriculture.

residue of secularized use. The people seized upon datura as a vehicle to escape, if only for a little while, the pain of spiritual and physical degradation. Ritual significance was never entirely lost; the cuarandas and brujos of the American southwest perpetuated and guarded the ancient mysteries inherent in datura.

BELLADONNA

Three sisters born of the mating of Erebus and Clotho became the Three Fates of the Greek pantheon, who spin out the thread of every man or woman's life. Atropos was physically the smallest of the sisters, but was the most feared, for it was she who snipped the thread at the appointed time and not even the major gods could alter her decision. The deadly nightshade (*Atropa belladonna*) was given her name by Linnaeus who simply took one of the common names for the plant (Figure 4.35). The species name, *belladonna,* dates from the late Middle Ages. Mattioli, a famous herbalist of the late sixteenth century, reported in 1560 that Italian ladies put drops of diluted nightshade sap in their eyes to induce *mydriasis*—the deep, dark mysterious look one gets when the pupils do not close in bright light. The pallor of Renaissance ladies as seen in portraits by Botticelli and Raphael appears, alas, to have been cosmetic, since it was brought about by rubbing their faces, arms, and shoulders with sap from the belladonna root. Like datura, deadly nightshade contains all three tropane alkaloids. The action of nightshade on the parasympathetic nervous system was known and exploited by ancient physicians; Dioscorides and Theophrastus both commented favorably on its sedative action.

Belladonna plants are usually perennial with a thick, woody rootstock and a branched, purplish stem bearing large leaves with a dull green sheen. Inconspicuous flowers with dull purplish petals are formed in midsummer and given rise to a small berry, green when young, turning red, and finally almost black when mature. In full flower, belladonna has a pleasant scent which attracts pollinating moths in the early evening. It can reproduce by seed, by stolens arising from the roots, or even from pieces of the rootstock. The plant is common throughout Europe, southwest Asia, and parts of North Africa. It is not native to North America, but was imported by early colonists for medicinal use, where it escaped cultivation and now grows in meadows, along marshy ponds, and even in tilled fields.

Greece and Rome knew it as a sedative and an hallucinogen. Bacchanalian orgies utilized new wine spiked with small amounts of the sap, and it was assumed that the frenzy with which women threw themselves at the male guests was due to the aphrodisiacal qualities of the plant juices, but it must be noted that these women were professional, paid entertainers. Jocasta, a professional prisoner of imperial Rome who specialized in the art of removing unwanted wives and mis-

Figure 4.35
Belladonna, *Atropa belladonna,* the deadly nightshade. Taken from an old print.

tresses, was devoted to belladonna, administering it in wine to disguise the bitter taste of the alkaloids. Arabian poisoners, as adept as their trans-Mediterranean counterparts—but usually male rather than female—knew the plant as *bu-rénguf*, the sorcerer's herb. The norse called it *dwale*, the devil's trance plant, and believed that it was especially beloved of Satan. On Walpurgisnacht, Satan had to go into the mountains to officiate at witch's sabbats, and the plant was transformed into a beautiful enchantress (belladonna) whose evil beauty would kill any man who saw her.

With the spread of Christianity, bacchanalian orgies became a lamented aspect of the golden age of Rome, but belladonna's star rose again as witchcraft and demonology captured people's imagination. One recipe guaranteed to project one into the astral sphere was an ointment of "baby's fat, juice of water parsnip, aconite (wolfsbane), cinquefoil, belladonna and soot," and another was compounded of water parsnip, sweet flag, cinquefoil, bat's blood, belladonna, and sweet oil. In his *"Masque of Queens,"* Ben Jonson quotes one adept as saying

> And I ha been plucking plants among,
> Hemlock, henbane and Adder's Tongue
> Nightshade, moon wort, leppard's bane.

Dryness of the throat and mouth, difficulty in swallowing, great thirst, impaired vision, and a strange rolling gait, all symptoms of tropane poisoning, are reminiscent of the symptoms of rabies, and it is no wonder that those who took these brews or rubbed on these ointments believed that they had become werewolves. A seventeenth-century coven of witches in Somerset's rolling and wild land confessed that they would anoint themselves "under their arms and in other hairy places" to be transported to the appointed place of a sabbat. They further confessed that they recommended to women who wished sexual satisfaction that before intercourse they rub their private parts with unguents containing nightshade or with fresh leaves of the plant. John Gerarde said that "women with child . . . do often long and lust after things most vile and filthie" following the use of nightshade, and Francis Bacon in 1605 noted that women who used nightshade were prone to think they flew and indulged in forbidden acts. John Gerarde was violently opposed to the nightshade. He reported:

> It came to pass that three boies of Wisbich in the Isle of Ly did eate of the pleasant and beautiful fruit thereof; two whereof dyed in lesse than eight houres. The thirde childe had a quantitie of honey and water mixed together given him to drinke, causing him to vomit often; God blessed this meanes and the childe recovered. Banish therefore these pernitious plants out of your gardens and all places near to your houses where children or women with childe do resort.

He did suggest that nightshade had some use in the treatment of epilepsy, "but if you will follow mei councill, deale not with the same in any case."

MANDRAKE

Jacob, son of Isaac and brother of Esau, served Laban, son of Nahor, for seven years and was tricked into marrying Leah. She bore him four sons while Jacob labored another seven years for the hand of his beloved, but barren Rachel. Distraught, Rachel chose in her desperation to employ the feared fertility plant of the Middle East, the awesome mandrake.

> In the days of the wheat harvest, Reuben went and found mandrake in the field and brought them to his mother Leah. Then Rachel said to Leah, "Give me, I pray, some of your son's mandrakes." But Leah said to her, "Is it a small matter that you have taken away my husband? Would you also take away my son's mandrakes from me?" When Jacob came from the field in the evening, Leah went out to meet him and said, "You must come in with me, for I have hired you with my son's mandrakes." So he lay with her that night . . . and she conceived and bore Jacob a fifth son. (Gen. 30:14)

Apparently without the magic of the mandrakes, Rachel did conceive and gave birth to Joseph.

Long before the Hebrews, the mandrake (*Mandragora officinarum*) had been associated with sexuality and sins of the flesh (Figure 4.36). The Elbers medical papyrus of 1500 B.C. listed it as *dudajm*, the fruit that excites love; Pharaoh Tutankemen was buried with 11 mandrake roots in the sixth row of his floral collarette to ensure his potency in the next world. The Greeks named it *circeium* after Circe who, Homer reported, lured men to her and changed them into swine, that is, into sexual pigs. They referred to Aphrodite as *Dios Mandragoritis*. Mandrake roots carved into big-hipped and -breasted fertility figurines have been unearthed at Antioch and Damascus and from tombs in Constantinople and Mersina. The Songs of Solomon noted: "The mandrakes give a smell, and at our gates are all manner of pleasant fruits, new and old, which I have laid up for thee, my beloved." Pliny said that those roots which resembled penises were love charms, and Galen echoed this: "If a root falleth into the hands of a man, it will ensure woman's love." By the second century A.D., everyone knew that female elephants gathered mandrakes and presented them to male elephants. Shakespeare was very aware of the sexual power of the mandrake. Justice Shallow was described in *Henry VI* (II) as "lecherous as a monkey and the whores call him mandrake," and Falstaff cursed his page in Henry VI (I) exclaiming, "Thou whoreson mandrake, thou are fittest to be worn in my cap than to wait at my heels." Contemporaries of Shakespeare were equally impressed; in *Polyolbion* Michael Drayton wrote: "The power of mandrake in philtres to procure love and worn about the body to correct barrenness was unduly recognized," and again, "The fleshy mandrake was, par excellence, the love-compelling agent." Nicolò Machiavelli wrote a comedy, *La Mandragora* about 1513 in which mandrake root permitted a noted seducer to bed down the virgin wife of

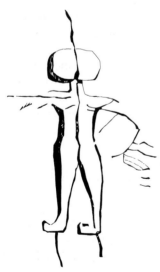

Figure 4.36
Prehistoric rock carving found in France and presumed to be of mandrake.

another man, and La Fontaine's fable, "La Mandragora" written about 1680 was based on the same theme. In the seventeenth century, scrapings of the root were believed to be effective in curing the French disease—Syphilis.

Any plant that could ensure potency and fertility must be gathered with appropriate respect and all due caution. The first written set of instructions on collecting mandrake was provided by Theophrastus who said, "Around the mandragor one must make three circles with a sword and dig, looking to the west. Another person must dance about in a circle and pronounce a great many aphrodisiacal formulae." A more complete method was given by Flavius Josephus, scribe and historian of the Roman Jews in the first century A.D. The virtuous man, he said, should first look for them by night, for then the leaves will shine in the dark, "but if anyone tries to pluck them, they rise in the air and fly away." They will, however, remain in the ground if a man strikes them with an iron rod. Josephus warned his readers that there was great danger attendant upon uprooting the plants, for when rudely and inconsiderately removed from the earth, they give a cry that can cause the hearer to go insane or to be driven mad. Shakespeare knew this, and told his audiences so in *Romeo and Juliet* (act 4, scene 3): "And shrieks like mandrakes torn from the earth"; and again in Act 8, scene 3: "What with loathsome smells and shrieks like mandrakes torn out of the earth, that living mortals, hearing them, run mad"; and yet again in Henry VI (II), Act 3, scene 2: "Would curses kill as doth the mandrake's groan." John Webster had Ferdinand in the *Dutchess of Malfi* (act 2, scene 4) say: "I have this night digg'd up a mandrake, and I am gone mad with't." The Arabian physician, Ibn Beither, said that the Europeans were mistaken, for it was not the shriek that killed, but that a root demon who lost his home would invade the digger. Ben Jonson knew of this horror, for in his *Masque of Queens* he had an old woman say, "I last night lay all alone on the ground to hear the mandrake's groan . . . and plucked him up." The hag was exceedingly brave; most people followed Josephus's advice and tied a dog to the crown of the plant so that when the cur responded to a whistle or to a bone, it would die as the plant came screaming out of the earth (Figure 4.37). It is unlikely that Josephus invented this method; a Persian name—derived from the Assyrian—is *sag-ken,* dug by a dog.

The most elaborate technique comes down to us from the Latin herbal of Apuleius Platonicus written in the fifth century A.D. Apuleius was probably a Greek who lived in Rome, wrote in Latin, and not only collected the wisdom of the evening fireplace stories, but also the aphorisms of philosopher-botanist-physicians. His *Herbarium* was translated into Anglo-Saxon in about the tenth century and profoundly affected the English view of the efficacy of many herbs, for echoes of Apuleius are found in the herbals of Parkinson, Gerarde, and others.

> This wort is mickle and illustrious of aspect and it is beneficial. Thou shalt in this manner take it. When thou comest to it, then thou under-

Figure 4.37
Dioscorides receiving the mandrake from a goddess of medicine. The dying dog is a representation of the legend of the method of digging the plant.

Figure 4.38
Uprooting the mandrake. Taken from a sixteenth-century herbal.

Figure 4.39
Male and female mandrakes. Taken from *Hortus Sanitatis* by Cuba published in Paris in 1489.

standeth it by this, that it shineth at night altogether like a lamp. When first thou seest its head, then inscribe it instantly with iron lest it fly from thee; its virtue is so mickle and so famous that it will immediately flee from an unclean man when he cometh to it. Hence, as we before said, do thou inscribe it with iron and so shalt thou delve about it as that thou thouch it not with the iron. But thou shall earnestly, with an ivory staff, delve the earth. And when thou seest its hands and feet, then thie thou it up, Then take the other end and tie it to a dog's neck so that the hound may be hungry. Then cast meat before him so that he may not reach it, except he jerk up the wort with him. Of this wort it is said that it hath so mickle might that what thing soever tuggeth it up, that it shall soon in the same manner be deceived. Therefore, as soon as thou seest it be jerked up and have possession of it, take it immediately in hand and twist it and wring the ooze out of its leaves into a glass ampulla.

Use of a dog as surrogate was repeated over and over again in herbals dating from the eleventh century (Figure 4.38). The *Beastiary* of Philip of Thawn written in 1122 repeated the dog story for the benefit of French collectors and Chai Mai, an illustrious Yüan dynasty scholar wrote in 1285 of this plant of the masters, *Yah-puk-lu*, that "grew thousands of li to the west and could be gathered only with the aid of a dog." Ben Jonson used the dog story in *The Masque of Queens*, and even earlier, Archbishop Aelfric solemnly recommended it, for the dog had no soul to lose.

There were different forms of mandrake. Dioscorides spoke of black and white mandragora, but it is likely that his black form was nightshade. Pliny said that the more robust plants were male, but Galen asserted that the male plant had a single tap root, while the female plant had a forked root with the two branches "spread wide apart." By the eleventh century, woodcuts in herbarium volumes included anthropomorphic representations of male and female mandrakes with primary and secondary sexual structures (Figure 4.39). The *Grete Herball* of 1526 said: "there be two maners, the male and the female; the female hath sharpe lines. Some say that it is better for medycyne that the male, but we use both. Some say that the male hath figure of shape of a man and the female of the woman, but that is fals. For natur never gave forme or shape of mankynde to an herbe." Nevertheless, John Parkinson's *Theatrum Botanicum* of 1640 still spoke of mandrakes and womandrakes.

The botany of the mandrake is fairly straightforward. The root is thick and fleshy, a taproot much like that of horseradish, but somewhat fatter; there may be a single tap root or more commonly two branches. Like horseradish or carrot, there is practically no stem, the leaves arising as a rosette directly from the root crown. The leaves may be very long, up to 30 cm or more, with broad blades which are hairy and have crimped or wavy margins. They tend to lie on the ground so that a single plant will cover a circular area of more than half a meter; there are few other plants that look anything like it. Flower stalks arise directly from the crown and bear solitary flowers with greenish, blue-violet, or purple petals which are unmistakably tomatolike although much larger than tomato or potato flowers. The fruit is a fleshy, many-seeded berry like the tomato and is yellow at maturity; it is sweet and edible when ripe, with an odor variously described as pleasant or mildly offensive. There are two botanical varieties, *Mandragora officinarum vernalis*, which flowers in the spring and *M. o. autumnalis* whose flowers develop in the early fall. Both are native to the Mediterranean basin and the Middle East and have been cultivated and naturalized throughout southern and central Europe. Both were introduced into Britain before the tenth century, possibly by the Romans, and were naturalized before William invaded in 1066.

Mandrake roots have been used medicinally for centuries; Homer reported that Petrocius treated the arrow wound of his friend Eurpylus by rubbing it "with a bitter root that took away pain." Hippocrates apparently had considerable respect for its ability to soothe: "The persons who are dull and sick and would commit suicide must take an infusion of mandrake root in the morning in a small dose, less than the amount which would cause delirium." In his *On Diseases*, II, 43, he recommended its use for malaria, and in his *On Diseases of Women*, II, 199, he said that for white or bloody flux, mandrake juice was to be mixed with sulfur, a swab dipped into the mixture and inserted into the vagina while the patient was enjoined to lie motionless and to sleep on her back. For females falling into fits (hysteria) and losing the power of speech, "a bolus . . . remedies the evil." Plato made mention in *The Republic* of its use to render insensible a sea captain who suffered from melancholia. Plato's pupil, Theophrastus, recommended dried mandrake mixed with meal as an invaluable poultice for wounds and suggested that scrapings of the root be mixed with vinegar and taken as a sleeping draught; he warned his readers that excessive use was dangerous.

Nevertheless, its soporific and analgesiac properties were blessed by physicians as they blessed opium, and they usually used both to alleviate pain. Theodorie Borgognoni, professor of medicine at Bologna during the early thirteenth century developed the *spongia somnifera:* a mixture of opium and mandrake juice was compounded in vinegar, taken up in sponges, and inserted into the nostrils of patients undergoing surgery. Contemporary clerics suggested that God used the sponges to put Adam to sleep before removing one of his ribs (Gen. 2:21). Roger Bacon approved the use of mandrake juice as a cure for insomnia and Shakespeare was aware of the sedative properties of mandrake. In *Antony and Cleopatra* (act I, scene 5), Cleopatra cries, "Give me to drink mandragora that I might sleep out this great gap of time my Antony is away." Iago tells Othello (act III, scene 8), "Not poppy nor mandragora . . . shall ever medicine thee to that sweet sleep. . . ." In *Cymbeline* (act V, scene 5) there is "a certain stuff which, ta'en would cease the present pow'r of life; but in a short time all offices of nature should again do their function" leading one to think that it was a draught of mandrake that Friar Laurence gave Juliet. Marlow confirmed the general belief that mandrake could cause a deathlike sleep. In *The Jew of Malta* (act V, scene 3)" Barabas says, "I drank poppy and cold mandrake juice and, being asleep belike they thought me dead and threw me over the walls." The Romans knew that this procedure was common, and there was a standing order that all men who were crucified were to be mutilated before their bodies were turned over to relatives. Is it possible that the sponge given Jesus was steeped in mandrake?

Images carved from large mandrake roots or from the counterfeit white bryony were in the form of angels (white magic) or devils (black magic) and were used appropriately. One of the charges laid against Joan of Arc was possession of a mandrake poppette; she denied it and denied even knowing what a mandrake was, but her denial was unacceptable to the court because it said that everybody knows about mandrake. Possession was punishable by death until the middle of the eighteenth century and with good reason: a well-developed root contained sufficient tropane to kill and the evil magic could "drain dry the cows grazing upon nine meadows, stop up the hens and rot the cabbages in the garden." The French knew the root as *maine de gliore* and the Germans spoke of *alraun* (Figure 4.40), both words referring to spirits inhabiting the earth. The root could be helpful to bring wealth and love or would glow when near buried treasure. Tavern keepers kept a root on hand knowing that its presence would draw customers who would drink until their pockets were empty. A man on his way to court was enjoined, "And when thou goest to law, put Erdman (alraun) under thy right arm and thou shalt succeed whether right or wrong." Alraun protected its owner in war; Rudolph II of Austria had it emblazoned on his war shield.

But, take care! Once you have acquired a mandrake or womandrake it will never leave you: it will return like the phoenix if you throw it into a fire; it will come dripping wet back to you if you try to throw it in a river or caked with mud if you try to bury it. If you die, it will give the owners of your house no rest as it searches nightly for you in the closets and under the stairs. Beware, oh beware!

Figure 4.40
Fifteenth-century drawing of the alraun in Germany.

ADDITIONAL READINGS

Barnard, C. "The Duboisias of Australia." *Economic Botany* 6(1952):3–17.

Harner, M. J. *Hallucinogens and Shamanism.* New York: Oxford University Press, 1973.

Heiser, C. B., Jr. *Nightshades, the Paradoxial Plants.* San Francisco: 1969.

McCue, G. A. "The History of the Cultivated Tomato." *Annals Missouri Botanical Garden.* 39(1952):349–353.

Parrinder, G. *Witchcraft. European and African.* New York: Barnes & Noble Books, 1958.

Ray, O. S. *Drugs, Society and Human Behavior.* St. Louis: C. V. Mosby Company, 1972.

Safford. W. "Daturas of the Old World and New; an Account of Their Narcotic Properties and Their Use in Oracular and Initiary Rites. *Annual Report,* Smithsonian Institute, 1920, pp. 537–567.

Thiselton-Dyer. T. F. *The Folklore of Plants.* London: Chatto & Windus, 1889.

Thompson, C. J. S. *The Mystic Mandrake.* New Hyde Park, N.Y.: University Books, 1968.

CHAPTER 5
🙠 Medicinal Plants

Herbs, Herbals, and Herbalists

Here begynneth a newe mater, the whiche sheweth and treateth of ye vertues and proprytes of herbs, the whiche is called an herball.
RYCHARD BANCKES, title page of his *Herball*, 1525

Medical archeology has shown that ills to which flesh is heir have been with us since we made the mistake of climbing down from the trees and ruining our backs by standing upright. Bacterial disease, injuries, and organic ailments have been diagnosed from bones, mummies, and from cave paintings. Some of the earliest written records have been of epidemics. Primitive man began to develop systems of medicine based on causation by supernatural forces and on treatment by a combination of religion and herbal medications. As settled communities developed, priests of the Great Mother received from Her knowledge of plants which could ease human misery. As written records supplemented and then replaced oral traditions, the healing arts developed.

For us in the Western world, medicine can be dated from ancient Greece. Homer sang of healing roots. The father of medicine, Hippocrates of Cos, organized a college of medicine in 490 B.C., the Temple of Aesculapius. Hippocrates taught medicine derived in part from the Code of Hammurabi of Babylon (2100 B.C.), the Elbers papyrus of Egypt (1500 B.C.), and records of herbs used by peoples of the Near East. The Temple of Aesculapius continued after the death of the founder, with its philosophy modified by the intellectual ferments provided by Socrates (470–399 B.C.), his pupil Plato (427–347 B.C.), and Plato's pupil Aristotle (384–322 B.C.). Aristotle was first and foremost a naturalist. As tutor to Alexander the Great, Aristotle traveled with military expeditions, gathering herbs, and learning of their properties from captured physicians. His star pupil, Theophrastus (372–287 B.C.), expanded the master's work in the first book devoted to useful plants. The *Enquiry into Plants* described the plants of the Mediterranean basin, classified them by use, as herbs, trees, and shrubs, and discussed the medicinal values of these plants.

As leadership passed from the Greeks to the Romans, medicine moved west. Superb engineers, the Romans drained swamps, built a water supply system that brought 300,000 gallons of pure water each day to the city, and instituted medical services for all citizens. The *Valetudinaria* followed the model of the Temple of Aesculapius, and physicians expanded their herbal healing arts to include plants of North Africa, Gaul, and Britain. In 77 A.D., Pliny published the first parts of his massive *Natural History* which eventually included 37 volumes, several of which were on plants and their uses. Pedanius Dioscorides (50–? A.D.) was a widely traveled army surgeon whose book, *De Materia Medica*, discussed over 600 medicinal herbs, their preparation, and their use.

This book formed, as we shall see, almost the sole basis for clinical medicine for 1500 years, supplemented only by the works of Claudius Galenius (129–199 A.D.). Galen was a Greek physician who served as court physician to Marcus Aurelius. He asserted that "disease is contrary to nature and can be overcome by that which is contrary to disease." His several volumes included botanical materia medica, pharmacy, and therapeutics.

As the Roman empire fragmented and the dark ages descended on Europe, observations of nature were replaced by the concept of authority. Dioscorides and Galen, although heathen, were accepted as authorities by the Roman Catholic Church. Thus, their writings formed a canon and a dogma, and no person who wished to remain a communicant could do other than follow the directions in these books. From about 200 to 1100 A.D., no books on herbal medicine, or on any science, were written in the Western world. Scribes copied Galen and Dioscorides, taking care to avoid introducing any new information.

MEDIEVAL MEDICINE

In searching for the basis of the disappearance of natural philosophy, historians have suggested several factors. A series of wars followed the dissolution of the Roman empire. The attendant failure of agriculture due to a scarcity of people to till the land led to recurrent famines, general squalor, and fostered the spread of epidemics that wiped out most residual intellectual values. Infant mortality exceeded 60 percent, and the average life span was less than 30 years. England experienced the Black Death in 664, 672, 679, 683, and 728, while equally frequent and equally deadly pandemics occurred on the continent. In today's terms, many people of Europe were insane—melancholia, depression, alcoholism, and suicide were the norm, and critical evaluation of natural phenomena was impossible. Europeans spent their short lives in abject fear and terror with demons, witches, and evil ones as their constant companions (Figure 5.1).

Disease was a manifestation of evil or of God's displeasure and could be cured by prayer, exorcism, visits to shrines, flagellation, and recourse to curse-lifters, who used herbs with virtues known to them through oral records or through the writings of Galen and Dioscorides. Intercession to heaven was formally ascribed to specific saints: leg diseases to St. John, cancer to St. Giles, toothaches to St. Appolonia, etc., and their assigned plants formed the medical treatments. The hierarchy of the Roman Catholic Church, composed of men subjected to the same stresses as were the rest of the population, echoed the same superstitions. Knowledge was useful insofar as it exemplified doctrine and dogma. Medicine, too, had to serve only God, and His priests assumed the duties of attending the sick. Herbs for healing were cultivated in monastery gardens, and hospitals were adjuncts of nunneries. Innovation or even observation of the natural world had no place in dogma,

Figure 5.1
Hound's tongue, *Cynoglossum officinale*. The rough leaves bear some resemblance to the rough tongue of a dog. Witches' brews containing "tongue of dog" and "adder's tongue" actually contained plant leaves.

and, besides, death was a release from misery and offered the possibility of salvation and eternal bliss.

As conditions worsened in the eighth and ninth centuries, people of Europe developed a folk medicine that was a complex of garbled Dioscorides, ill-translated Galen, ancient superstition, and religious fervor stirred together and imbibed with a prayer. Since few could read or write and those who could wisely refrained from challenging authority, herbal remedies were passed orally from generation to generation, becoming altered in the process.

DOMINANCE OF THE ARABS

The responsibility of preserving intellectual thought passed to the Saracens. Medical practice was the province of Arabian physicians such as Rhazas (965–919), Avicenna (981–1037), Avenzoar (1113–1162), and their Jewish colleagues like Moses Maimonides (1135–1204). Greek and Roman medical tomes were translated into Aramaic, Syriac, or Persian which permitted the introduction of some Indian herbal lore to enter the West. Commentaries on Galen and Dioscorides included drugs of North Africa, Byzantium, and other lands conquered by the followers of Mohammed. A medical college was established in Alexandria in the early eighth century and by the tenth century, medicine was centered in Salerno, a former Greek and then Roman health resort. Alexandria and Salerno attracted heretic Nestorian Christians and dispersed Jewish physicians who then traveled throughout Europe serving the medical needs of the noble courts. These physicians were known throughout Europe; Chaucer noted the debt owed them in the "Prologue" to the *Canterbury Tales*. Among the first herbals to appear was the hand-written *Paradise* [Garden] *of Wisdom* in 850 A.D. and the *Knowledge of Drugs* in 1025. An important work was the *Minhaj al-dukkan* (Handbook of Drugs) published by an Arabian Jew, Kòhên al-attar (Cohen the druggist), in 1259. Precise direction on collection and storage of drug plants, on methods for distilling, compounding and testing, on weights and measures, and on the values of each compound were included. Arabian medicine was deliberately going beyond the mere parroting of the wisdom of Dioscorides.

During this period, few European herbals appeared, most of them copies of Galen and Dioscorides, although some did "comment" by adding local lore. The *Leechbook of Bald* was an early Anglo-Saxon manuscript. *De Morbis Acertis* by Caelius Aurelianus and the *Herbarium Apuleii Platonici* were of immense value and utility (Figure 5.2). Albertus Magnus of Cologne (1193–1280) published *De Virtutibus Herbarium* (Healing Values of Herbs), and Hildegard von Bingen, a Benedictine abbess, wrote *Liber Simplicus Medicina*, one of the first books to discuss the role of the nursing sister. As priests and nuns became less worldly, medicine passed to uneducated laypersons and to the barbers who became surgeons. Gerald of Cremona (1114–1187) wrote *Canon Medicinae*

Figure 5.2
"Nymfea" the lotus, *Nymphaea*. Taken from a ninth-century manuscript copy of *Herbarius Apuleii Platonici*.

which translated into Latin the Saracen medical writings captured when Toledo was freed from Moorish domination. In 1317 John of Gaddesdon published *Rosa Medicinae* at Oxford which incorporated the Arabic information of Gerald into Dioscorides's treatments plus, for possibly the first time in Europe, his own clinical experiences: "For nothing is set down here but what has been proved by personal experience either of myself or of others and I, John of Gaddesdon, have compiled the whole in the seventh year of my lecture." The impact of this book on the medical profession was immediate; members noted that John was not subjected to the inquisition for his writings and concluded that they, too, could speak more freely—albeit circumspectly.

BOTANICAL RENAISSANCE

The invention of the printing press in Germany in 1448 served both God and man. The Gutenberg Bible permitted reading of the Scriptures by people other than the priests, and the printing of John of Gaddesdon's book ushered in the era of the herbal. Father Barthalomaeus Anglicus published *Liber de Proprietibus Regum* in 1470, and Cunrat published *Das Püch der Natur* in 1475, both volumes more clinical herbal medicine than Galen. The *Herbarius zu Teutsch* of 1485 was a mixed Germanic-Latin copy of the *Latin Herbal* of 1484 in which Galen and Dioscorides were accorded second place to the lore of contemporary physicians (Figure 5.3). The Latin *Ortus* [Hortus] *Sanitatis* of 1491 went even further than its predecessors in reducing the importance of the ancients. Within a few years, *Le Grand Herbier* was published in France, demonstrating that the vernacular—the language of the people—was replacing the Latin which the Church and its universities had insisted on for centuries.

The early sixteenth century saw the publication of two English books that specifically used the name herbal in their titles as an indication that the medicines were almost solely derived from plants. Rychard Banckes, apparently a printer of London and not a medical specialist, published *Banckes's Herball* in 1525. The book proved so popular that successive editions and translations appeared all over Europe. Its provenance is unknown. *The Grete Herball* was published the following year by Peter Treveris (Figure 5.4). Treveris prepared an excellent translation of *Le Grand Herbier* of France which was a version of the *Herbarius zu Teutsch*, but Treveris included a number of British remedies which are still in use: horehound for coughs, henbane as a narcotic, and rosewater and glycerin for soothing the skin. There were many remedies for melancholy which gives us some idea of the unhappy state of mind of many people in the early sixteenth century. The book is very medieval; remarkable powers were ascribed to magic stones, to the unicorn horn, to "powdre of perles," and to spells and incantations, but great weight was also placed on the efficacy of "wedys of ye feldye." Treveris was deeply concerned that the common people were at the

Figure 5.3
"Nenufar" the water lily, *Nuphar*. Taken from a revision of the *Latin Herbal* published in 1499.

Figure 5.4
"Nenufar" the water lily, *Nuphar*. Taken from *The Grete Herball* by Peter Treveris published in London in 1526.

mercy of doctors and apothecaries and the ignorance of "ye olde women," and he stated: "wherefore brotherly love compelleth me to wryte thrugh gyftes of ye Holy Gost shewynge and enformynge how man may be holpen with grene herbes. . . ."

Succeeding years saw publication of herbals in many languages. In order to transmit information about the plants useful to man, the herbals began to include plants of food or of horticultural value. Otto Brunfels published *Herbarium Vivae Icones* (Kinds of Living Plants) in 1536 with masterly woodcuts by an artist who worked from living material and did not conventionalize or stylize his drawings. With this as a model, Hieronymus Bock published the *Kreütterbock* in vernacular, even ribald German and included plants used by the people of the Palantine. Leonhard Fuchs (for which the genus *Fuchsia* is named) was a doctor of arts and professor of medicine at Tübingen. His *Historia Stirpium* of 1542 was in Latin, but he arranged the plants alphabetically by common name and discussed over 400 German and 100 foreign plants including, for the first time, New World plants like Indian corn and "the Grete pompion." Pierandrea Mattioli published *Commentarii in Sex Libros Pedacii Dioscorides* (A Six Volume Commentary on the Works of Dioscorides) with magnificent illustrations. The publication went through several editions, selling 30,000 copies at ten ducats each, due in large part to the fact that the commentary was the major part of the book and Dioscorides' writings were subtly down-graded. William Turner's *A New Herball* came out in 1551 with plant names in "Greke, Latin, Englishe, Duche and French wyth ye cummune names that herbaries and apothecaries use." Since England was almost completely removed from the authority of the Roman Catholic Church, Turner could afford deliberately to debunk Dioscorides and could also object to the superstitious witchcraft that had infiltrated clericalism.

The latter part of the sixteenth century was made botanically notable by the publication of several herbals which tremendously influenced books published in the seventeenth century. Mattias de l'Obel (the genus *Lobelia*) received his medical education at Montpellier, but settled in England where he was a go-between for the now-heretic English and the Catholic botanists of Europe. His *Stirpium Adversaria Nova* of 1570 is noted for excellent woodcuts. De l'Obel modeled his study on the 1560 edition of Konrad Gesner's *Historia Planatarum,* an encyclopedia of all known plants, much less medically oriented than any previous herbal. The Portuguese and Spanish herbals of Garcia de Orta, who wrote about the plants of India, and of Nicolas Monardes, whose 1569 book was translated into English as *Joyfull Newes Out of the Newe Founde World,* reflected the success of Latin countries in exploring in both east and west. Sunflowers, tobacco, chili peppers, and other plants were described and illustrated, and the European scientific community avidly bought and studied these books. One other sixteenth century herbal is a landmark in botany. Charles de Ecluse (Clusius) was a Montpellier-educated physician who served as court physician at Vienna and as

Figure 5.5
Egyptian lotus, *Nyphaea alba.* Taken from *Rariorum Plantarum Historia* by Carolus Clusius published in Antwerp in 1576.

professor of medicine at Leyden. He obtained New World plants from Sir Francis Drake and, among other books, published *Rariorum Plantarum Historia* in 1576 when he was 76 years old (Figure 5.5). Not only were many American plants illustrated, but Clusius also included the first comprehensive overview of the fungi.

During this period, herbals were the only permanent record that physicians and natural historians had of plants. Animals could be kept in royal zoological parks and their skins and skeletons stored for examination. Medicinal plants were grown in "physick gardens" (gardens of medicinal plants), but they seemed impossible to preserve in any condition so that the dried specimen could be identified or, even more importantly, could be used to determine whether another dried plant was the same. Herbaria, collections of dried plant specimens, did not exist until an Italian physician and professor of botanical medicine at Bologna, Luca Ghini, conceived the idea about 1550 that plants could be dried between blotters under mild pressure so that their form was retained. By the end of the sixteenth century, herbaria were developing in most of the universities of Europe, usually in the medical colleges where they were used to compare plants received as possible drug sources with plants known to contain effective medicines. As botanical science developed independently of medicine, herbaria were located in botanical gardens and in departments of natural science.

These early herbaria are exceedingly valuable, for they contain specimens which were used to name a plant for the first time—these are the "type specimens." World-famous herbaria, containing literally millions of herbarium sheets, are found at the Royal Botanical Garden at Kew in Surrey, the Gray Herbarium at Harvard University, the New York Botanical Garden, and in several other centers throughout the world. Each specimen is labelled with the name of the collector, the date collected, where collected, the taxonomic diagnosis (family, genus, species), and information on the ecology of the location, thus providing a wealth of information for all fields of botany.

THE AGE OF HERBALS

The early seventeenth century has been called the age of herbals because of the numbers of important publications and the impact of these books on both scientific thought and medicine. Certainly among the more valuable was *The Herball or Generall Historie of Plants* published in 1596 by John Gerarde. Gerarde (1545–1612) was superintendent of the estate gardens of Lord Burghley, Master in the Company of Barber Surgeons, and an apothecary with his own physick garden. The text and many of the illustrations were taken from the 1554 *Crüyderboeck* of Rembert Dodoens, but Gerarde so amended and commented on Dodoens that his herbal was essentially a new book.

Gerarde was a man of his times. He believed in trees which bore geese and others that had barnacles as fruit. Nevertheless, he took a

Figure 5.6
Title page of *The Herball or Generall Historie of Plantes.* This is the revision by Thomas Johnson of John Gerarde's herbal. Courtesy Library, New York Botanical Garden.

firm stand against witchcraft and magic in herbal medicine, intelligently discussed plants from the Americas, and attempted to standardize the prescriptions then in use. To a great extent, Gerarde's fame—and he was very famous—rested not on *his* Herball, but on the revised editions by Thomas Johnson published in 1633, 21 years after Gerarde's death (Figure 5.6). Johnson was an herbalist, a Free Brother of the Worshipful Society of Apothecaries, and a doctor of physick. The second edition had over 2500 woodblock pictures made in France, and these illustrations, the accuracy of the plant descriptions, and a clearly written text caused the book to be a financial success.

AN ESSAY ON BOTANICAL ART

The condensed, telegraphic, and ungrammatical word descriptions of plants in plant identification books such as *Gray's Manual of Botany* are nineteenth-century inventions and are very different from those appearing in the sixteenth- and seventeenth-century herbals. Thus, *Banckes's Herball* of 1525 describes the Venus'-hair fern (*Adiantum capillus-veneris*):

> This herbe is called maydenheere or waterworte. This herbe hath leves lyke to fernes but ye leves be smaller, and it groweth on walles and stones and in ye mudde. . . .

198 *Medicinal Plants*

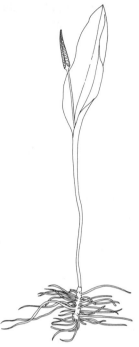

Figure 5.7
Adder's tongue fern, *Ophio-glossum vulgatum*. Taken from *De Historia Stirpium* by Leonhard Fuchs published in Basel in 1542.

Figure 5.8
Star of Bethlehem, *Ornithogalum umbellatum*. Taken from *Hortus Floridus* by Crispean van de Passe published in Arnheim in 1617.

The eighth edition of *Gray's Manual* published in 1950, describes the same plant:

> Fronds 1–5 cm long with a continuous main rachis, often pendulous from a slender horizontal chaffy rhizome, the polished dark stripes 0.3–3.0 cm long; the blade ovate-lanceolate, 2–3 pinnate at base. . . .

Obviously, Gray's description is accurate, each botanical term invented to define a structure with precision and with a minimum of words. Yet, herbalists were able to ensure that their readers could identify the plants through the use of botanical illustrations. In early herbals, these illustrations were so crude that identification was difficult; there was no artistic or esthetic standard by which to compare them. The concept of naturalism in botanical art of the sixteenth century was established by Flemish painters including Hans Memling, the brothers van Eyck, and especially by Hugo van der Goes, whose Portinari Altarpiece of ca. 1450 presented a masterly view of identifiable plants. Dürer's *Das Grosse Rasenstück* of 1503 shows a worms-eye view of a piece of turf with each leaf, each blade of grass, and each dandelion flower presented with accuracy and naturalism. Jan Brueghel (1568–1625) was one of the pioneers in developing the idea that accurate botanical detail could go hand-in-hand with beauty. It was, however, impossible to use directly the painter's techniques in printed books designed for mass distribution, and the woodblock print developed.

The woodblock was invented in China and Japan at least a thousand years before it was either reinvented or introduced into Europe. It allowed a flexibility and a precision that could not be achieved with the crude scratchings on soft copper plates that served to illustrate early herbals like the *Herbarius Apuleii Plantonici*. Woodblocks were made by artists, men like Hans Weiditz who did Brunfel's *Herbarium Vivae Eicones* of 1531 or the team of Albrecht Meyer (artist), Heinrich Füllmaurer (transcriber of drawing to block), and Veit Speckle (blockcutter) who did the illustrations for Leonhard Fuch's *Die Historia Stirpium* of 1542 (Figure 5.7).

By the end of the sixteenth century, the woodcut was supplanted by line engravings and etchings which were previously used by Dürer and Rembrandt. *Hortus Floridus* published in 1617 by Crispien van den Passe shows the degree to which accuracy and beauty could be combined (Figure 5.8). Work such as this established a tradition and a standard of excellence equalled, but not surpassed, by water colorists who dominated botanical art in the eighteenth and ninetenth centuries. Even color photography has rarely come up to these standards and, in most instances, is not as botanically useful as are well-executed drawings.

Two other herbalists are worthy of mention. John Parkinson wrote two books, the first *Paradisi in Sole: Paradisus Terrestris* appeared in 1629. It was a strange book for the time. Not only was at least half of it devoted to gardening, but the herbal portion contained a religious approach to even the names of plants. Parkinson assumed that Adam had received from God knowledge "of all naturall things" and that all plant names "descended to Noah afterwards." Parkinson's second book, *The Theatre of Plants or An Herball of Large Extent* published in 1640, described almost 40,000 plants. This was probably the first book that could properly be called a *flora*—a systematic description of plants. Only in the nineteenth century did floras become, out of necessity, restricted to descriptions of the plants of a circumscribed region, for example, flora of Saskatchewan or plants of Thailand.

Nicholas Culpepper's *The English Physician* of 1652 is notable for its brusque, dogmatic, and combative tone, the author's pen dripped in vitriol when discussing the medical profession and anyone else with whom he disagreed. Culpepper was interested in bringing medical treatments out of the hands of doctors and directly into the control of the poor. He believed, not incorrectly, that the physicians of the day were more interested in financial reward and social prestige than in healing. Unfortunately, Culpepper's education in medicine consisted of having been apprenticed to a London apothecary, and the book is a mishmash of herbal medicine, a reversion to the almost discredited doctrine of signatures, plus the addition of a large amount of astrological medical superstition derived apparently from Paracelcus. An example of Culpepper's medicine can be seen in his discussion of the burdock (*Arctium lappa*):

> Venus challengeth this herbe for her own, and by its leaf or seed you may draw the womb which way you please, either upwards by applying it to the crown of the head, in case it falls out; or downward in case of fits of the mother by applying it to the soles of the feet; or if you would stay it in its place, apply it to the navel; and that it is a good way to stay the child in it.

Independently of trying to figure out the various antecedents of "it," it seems that this plant, applied externally, can guard against miscarriage, can ensure abortion, and can encourage conception.

BEGINNINGS OF MODERN MEDICINE

The European herbals were almost passé by the middle of the seventeenth century. One of note was William Coles's *The Art of Simpling* published in 1657. This book has as its subtitle *An Introduction to the Knowledge and Gathering of Plants* and was the forerunner of the modern pharmacopoeia. In clear, unambiguous language, Coles told his readers what herbs were medicinal, how they were to be gathered, preserved, compounded, and administered. There was emphasis on doctrine of signature plants, but Coles firmly rejected Culpepper's astrological medicine. Johnson's revision of Gerarde, Culpepper, and Coles formed the basis for medical practice throughout Europe, and copies were taken by colonists to North America and Australia.

Although some of Europe's physicians were formally educated at schools of medicine, many obtained what little training they had through the apprentice system, and many more simply set themselves up as doctors. Henry VII signed a statute dealing with this matter that stated: "The moste persones of the said crafte of surgeons have small cunnynge," and the personal physician of Elizabeth I noted that "medicine was in the hands of tinkers, horse-gelders, rogues, rat-catchers, idiots, bawds, witches, sow-gelders and proctors of spittle houses." Britain organized the Royal College of Surgeons to supervise examination and licensing of physicians. Women were forbidden to become members because "they partly use sorcery and witchcraft." The hospitals, originally operated by religious orders, were secularized.

Although Theophrastus had a shop in which he sold dried herbs—apparently the first apothecary or drug store—subsequent herbalists and physicians either gathered their own plants or grew them in private physick gardens. In the late Middle Ages and even through the eighteenth century, a brisk trade in home-remedy plants was maintained by hawkers of herbs who set up boothes at markets and fairs, extolling the virtues of their "simples," dried animals, and magic stones to local people who could not afford the services of physicians. They would, for a small additional fee, give the purchaser the proper incantation to be used with the potion. James I, son of Mary Stuart of Scotland, ordered that the selling of herbs should be restricted to apothecaries because "grocers are merchants, but the apothecary's trade is a mystery [an art or craft]." The Worshipful Society of Apothecaries was formed in 1617 and, to protect its reputation and financial position, distinguished between herbalists, defined as gatherers of plants, and apothecaries, who compounded and sold herbal remedies. The society was faced with several serious problems. Medicinal plants were arriving in London from all over the world. Samuel Pepys could stroll to the docks and see sarsasparilla roots from Jamaica, sassafrass leaves from the Massachusetts Bay Colony, senna and myrrh shipped home by the East India Company, and rhubarb from China. There was, however, no good way to insure the purity or the potency of the herbs. As alchemy evolved into physics and chemistry, the apothecaries began to develop crude methods of analysis.

Apothecaries also recognized that the drugs from their plants varied from year to year and, in an effort to secure standardized material, they started to grow their own—a return to the physick garden concept developed in the Middle Ages when medicinal plants were grown on hospital land. Italy took the lead; the first secular physick garden was in Pisa in 1543, with plants arranged according to their assumed medicinal properties. Padua's medical school followed suit in 1545 and had the first professorship of materia medica and director of the physick garden. Paris, Rome, Heidelburg, Leyden, and Montpellier had physick gardens by 1590. In England, Oxford apothecaries purchased land across High Street from Magdalen College to develop a garden, while the Worshipful Society of Apothecaries organized the famous Chelsea

Garden along the Thames River in 1673. One of the first physick gardens in North America was that of Dr. John Bartram of Philadelphia whose garden was planted in 1782. The most famous American garden was established in 1801 by Dr. David Hosack of the Columbia University College of Physicians and Surgeons in New York. By 1890, the garden was completely surrounded by the growing city, and Columbia University rented the land for the eventual construction of Rockefeller Center. The garden was moved up to the wilds of the Bronx where it became the New York Botanical Garden.

Another problem recognized by seventeenth-century apothecaries was the great variation in the formulation of prescriptions. Each person learned a different recipe, and every home compounder used the one passed down from his or her ancestors. As early as 1542, the guild of barber surgeons in Nüremburg prepared a volume which it called a pharmacopoeia giving direction for mixing herbal medicines. Other such books appeared during the sixteenth century to supplement these herbals. It was at this time that the apothecaries reinstituted the practice of making each prescription with the sign of Jupiter, ♃, to indicate that—by this ancient oath—they had neither adulterated the medicine nor had they included any poisons. This has come down to us as ℞. England's apothecaries produced the London Pharmacopoeia in 1618 which, in addition to herbal remedies, included medical preparations of cockscombs, woodlice, and the hair balls from goat stomachs. Pharmacopoeias of many countries have gone through numerous editions, adding new remedies, removing others which proved ineffectual, and modifying directions for gathering, storing, and compounding. Chairs of pharmacognosy, materia medica, and medicinal chemistry were set up at medical schools. Boards to examine and license apothecaries were all part of the medical scene by the late eighteenth century. The era of modern pharmacy had dawned.

ADDITIONAL READINGS

Arber, A. *Herbals: Their Origin and Evolution,* New York: Macmillan (Hafner Press), 1970.
Trease, G. E. *Pharmacy in History.* London: Brilliere, Tindall and Cox, Ltd., 1964.
Wheelwright, E. G. *Medicinal Plants and Their History.* New York: Dover, 1974.

Nostrums and Cures

Purge me with hyssop and I shall be clean.
PS. 51:7.

When the settlers of North America reached Jamestown, Plymouth, or French Canada, they brought with them the herbal remedies common to their home countries. Some were effective and others were ineffective—the nostrums of modern times. They also brought their valued copies of Gerarde, Parkinson, Culpepper and Coles, but realized that plants listed in these books did not grow in the new country. Undaunted, they selected from the wilderness those plants which seemed to match the ones pictured in their herbals and chose others by the doctrine of signatures. They also learned about plants from the Indians. Although the plants were sometimes very different from those they knew, they took comfort from the fact that the concepts underlying Indian medicine were the same as in Europe.

AMERICAN INDIAN MEDICINE

Indian medicine was closely allied to religious belief, and the doctrine of signatures was as well developed in North America as it was in Europe, Asia, or Africa. Yellow plants were good for jaundice, red ones were used when blood was involved, and threadlike roots for intestinal worms. Belief that illness was caused by improper behavior, by failure to honor and placate a superior Being, or by inadequate attention to sacred objects or spirits led to psychosomatic guilts. These could be cured by exorcism, analysis of dreams, and by hypnotic dancing, drumming, and chanting. Tranquilizers, hallucinogenic substances, and foul-tasting or disgusting medicines to induce cleansing vomiting were used by most tribes, and white settlers were gratified to observe that Indians believed as firmly as they did in purges and enemas. Settlers remarked on how healthy the Indians were, not having the biological sophistication to realize that many tribes practiced euthanasia, and that it was a rare individual who could boast of having seen 700 moons. The full spectrum of disease and disability seen in any diverse population was found among the Indians, and lacking immunity to European communicable diseases, they were easy prey to smallpox, scarlet fever, tuberculosis, and venereal diseases introduced by the white man.

Nevertheless, Indians used many plants which were effective and which have been adopted into modern medical practice. Early Spanish explorers of the west coast found Indians using bark of a small tree (*Rhamnus purshiana*) as a laxative. Rediscovered in 1895 by Lewis and Clark, *Cascara sagrada* is still in the U.S. pharmacopoeia. Jalapa (*Ipomoea purga*) from Mexico, the common mayapple (*Podophyllum peltatum*), and senna (*Cassia spp.*) were effective cathartics. Salicylic acid was named for the willow (*Salix spp.*) from which it was first isolated, and Indians chewed willow bark to relieve headaches and ease sore muscles; aspirin is a derivative of salicylic acid.

COLONIAL MEDICINE

There were few physicians in the colonies, and most of them were barber surgeons skilled in bloodletting, bone setting, and the use of the silver tonguescraper. Herbal medicine was in the hands of goodwives, and cures were in the hands of God. An exception was Dr. John Josselyn who, armed with the 1633 edition of Gerarde, dosed the Massachusetts colonists and the farmers of the outlying districts. Josselyn wrote *New England's Rarities* in 1672 in which he compared English herbs with their North American counterparts. The Virginia and Charlestown colonies had herbalists who relied more on Culpepper than Gerarde, Parkinson, or Coles. Lawrence Hammond, for example, recommended for the *mergrums* (migraine headache):

> Mugwort and Sage a handfull each, Calomel and Gentian a good quantity. Boyle it in Honey and apply it behind and on both sides of ye head very warm and in 3 or 4 times it will take it quite away.

In the late seventeenth century, these remedies were supplemented with *The New Jewell of Health* by George Baker, "wherein is contayned the moste excellent secretes of Physick and Philosophy."

In the forefront of medical botanical work were a group of men, all physicians, who used gathered herbs and who started botanical physick gardens of their own. The lieutenant governor of New York, Dr. Calwalader Colden, introduced, through his daughter, the Linnaean system into the colonies. Dr. John Clayton (for whom the spring beauty, *Claytonia caroliniana* was named by his friend André Michaux), Dr. Alexander Garden (the genus *Gardenia*), and Dr. Michael Sarrazin (*Sarracenia purpurea*—the pitcher plant) were active physicians in the early eighteenth century and were equally active botanists who collected plants for European herbaria and physick gardens. Dr. Caspar Wistar (the genus *Wisteria*) was president of the American Philosophical Society and one of the founders of the medical college at Philadelphia. Dr. John Bartram of Philadelphia was appointed Plant Collector to His Majesty George III at £50 per annum to supply plants to the Royal Botanical Garden at Kew. Bartram "botanized" all along the eastern seaboard from the dangerous Mohawk country around Lake George in New York down through swamps of the Carolinas and exchanged pressed plants and fresh material with his friends and professional acquaintances in Europe. This tradition of the physician-botanist was maintained throughout the nineteenth century. John Torrey, whose flora of the eastern United States is classic, was professor of medical botany at the Columbia University College of Physicians and Surgeons.

His star pupil and later collaborator was Asa Gray who, after receiving the M.D., was called to Harvard University as professor of botany and director of the Harvard herbaria and the botanical gardens.

As Americans moved west after the Civil War, settlers were very attracted to Indian medicines. Sweatings to rid the body of "poisons," purgings, and the use of strong-tasting mixtures appealed to the people. The stronger smelling and the more vile tasting the concoction, the better—some medical historians have called the 1860–1890 period the age of heroic cures. Castor oil was one favorite. So were sulfur and molasses tonics or mustard plasters capable of severely burning the skin. Many nostrums were guaranteed to cure after one spoonful—no normal person could stand any more. Children went to school wearing around their necks cloth bags containing powerful-smelling herbs which, their mothers insisted, would keep away "germs." Most frontier doctors received their medical degrees by correspondence from diploma mills and some, who could not scrape up the $200–300 for a degree, called themselves naturopaths. Many credited their skills to secrets obtained while they were captives of the Apache or the Sioux. The settlers understood the medicine these men were using. The plants were familiar, and appeals to the wisdom of the noble savage, the "unspoiled creatures of nature" (who had been pushed onto reservations and were then imbued with a romantic aura), struck responsive chords in the minds of people who had moved west to escape the educated authority of the east. Samuel Thompson, a New Hampshire farmer, promoted the idea that Indian healers discovered herbs and natural remedies far superior than the unknown materials included in prescriptions written in Latin. Quacks and self-appointed saviors almost took over the health care of rural North America.

MEDICAL HUCKSTERS

Although originally based on herbal medicine derived from Gerarde et al., "Indian cures" became big business. Medicine shows, featuring "real" Apache, Sioux, or Blackfoot chiefs in full costume were among the few diversions in small towns. Literally drumming up trade, the doctor would come out and talk about health and then tell his audience that he had just the thing for what ailed them. Kickapoo Indian Sagwa Remedy, Ka-Ton-Ka, Dr. Morse's Indian Root Pills, and Dr. Lerox's Indian Worm Eradictor were sold to the accompaniment of drums, war whoops, and How!

Patent medicines were not the sole province of the traveling medicine man; newspaper advertisements featured Ayer's Sarsaparilla for complaints of liver and kidney complete with case histories:

> J.W. of Lowell, Massachusetts was troubled with want of appetite, oppressive weakness and severe pains in the small of the back; all indications of serious derangement of the kidneys and liver. Ayer's Sarsaparilla made him a well man.

Ayer's contained extracts of yellow dock, sassafrass, senna, poke-weed, wintergreen, cascara, quinine bark, prickly ash bark, glycerin, potassium iodide, and iron sulfate in alcohol. Lydia E. Pinkham's Vegetable Compound was the nation's favorite for "female weakness" as evidenced by "indisposition of the female reproductive apparatus amongst other female complaints." Containing a diuretic and 18 percent alcohol ("This is added solely as a solvent and preservative"), it was nipped and sipped by women all over North America. Paine's Celery Compound for nervous diseases had 21 percent alcohol and Hotsetter's Bitters was 41 percent alcohol—this is 82 proof, and was a blessing for the Americans of the bible belt who had taken the teetotaler's pledge.

> "Men!! Are You Slipping? Are you LESS ACTIVE than you used to be? Are you A GOOD HUSBAND? If you have doubts, don't delay!!!! Use . . ."

Not all patent medicines were innocuous. Some contained concentrations of senna, cascara, or jalap sufficient to cause diarrhea leading to dehydration, and others contained extracts of lobelia which caused a "cleansing vomiting" dangerous enough to injure the patient. Tuberculosis was called catarrh and, said one ad, "Childs' Catarrh Specific will effectually and permanently cure any case of catarrh, no matter how desperate." Other cough and catarrh medicines contained enough morphine or even heroin to cause addiction. Cancer, heart disease, diabetes, ulcers . . . you name it and the friendly grocer had a bottle for it on the shelf. There were, in fact, medicines to cure diseases which never existed except in the fertile brain of the inventor (Figure 5.9). With pressure from the American Medical Association, the U.S. Public Health Service, and drug companies that prepared ethical drugs (sold by prescription only), legislation in the form of the Pure Food and Drug Act of 1906 was passed to curb the more flagrant abuses, but loopholes in the laws still permit the sale of compounds of dubious value and preserve a multimillion dollar industry.

CAUSES AND CURES

If the herbal remedies given in any of the large number of recent books on "using nature's ways" are compared with those given in Gerarde, Parkinson, or even Dioscorides, one is struck by similarities that border on plagiarism. Even the terminology is the same: *pectorals* relieve chest congestion, *carminatives* calm the stomach, *discutients* dissolve tumors, and *vulnaries* heal internal wounds. These terms date back beyond the tenth century, and some were used by Hippocrates. People who didn't feel "quite right" took tonics or, in increasing order, *aperients, laxatives, cathartics,* or, if draconian measures were necessary, they *purged* themselves. *Astringents* tightened the skin, *emolients* soothed it, *demulcents* smoothed it, and *maturating agents* cleaned its pores of boils and pimples.

Based on the Greek four elements, fire, air, water and earth, Hippocrates derived the four cardinal humours which made up the body: blood, phlegm, white choler, and black choler. From these, Galen derived the four temperaments: sanguine, phlegmatic, choleric, and melancholic. Temperaments could be modified by being hot, cold, moist, or dry. Seizing on these dicta, for there was no other philosophy of human physiology, herbalists searched out plants to control the imbalances among humours that resulted in disease. *Alteratives* changed the character of the blood, *depuratives* purified the blood, and *rubifacients* increased the flow of blood. *Chlagues* increased the flow of bile, and *expectorants* aided in eliminating phlegm from the chest. Hot, cold, moist, or dry states could be modified by *diaphoretics, diuretics, emetics, febrifuges, refrigerants, sialogogues* (promoting salivation), *sudoriferics* (increasing perspiration), and *hepatics* (activating the liver). But from today's perspective, the entire medical system for which these herbal remedies was developed is difficult to rationalize with knowledge now available on the etiology of disease. A broken leg or a bleeding wound were the same problem to Dioscorides as to the medical faculty of the University of Chicago, but a *pectoral* will not cure tuberculosis. Essentially, herbalists of the fourteenth through eighteenth centuries—and those of today—were treating symptoms. Modern medicine is based on two fairly recent concepts, that of the controlled experiment developed in the eighteenth century and the discovery of bacteria in the nineteenth century.

ℰ AN ESSAY ON THE CONCEPT ℰ OF A CONTROLLED EXPERIMENT

There has been a great deal of nonsense written about the concept of the controlled experiment. Certainly the philosophers of the golden age of Greece didn't think of it; they believed that knowledge could be obtained by pure reason. The scholastics of the Middle Ages were so involved with the authority of the ancients that they, too, rejected direct experimentation; publication of Galileo's studies resulted in his excommunication. Only slowly did the idea develop that one could obtain direct information about how things worked only when the experiment had a built-in control or check. If I think that photosynthesis involves the production of oxygen, it is not sufficient for me to place a plant in the light and determine whether oxygen is liberated. At the very least, it is also necessary for me to put a precisely equivalent plant under conditions where photosynthesis does not occur, but where all other physiological and biochemical processes do occur, and then see whether I can detect oxygen formation. As my experiments become more sophisticated, the controls will correspondingly become more rigorous. If I believe that a particular bacterium causes sore throats, I can prove this only by otaining a pure culture of the suspected bacteria, infect *a large number* of test organisms with the bacteria, and have an equally large number of identical test organisms which have not been inoculated with this particular live bacterium. To complete the proof, I must honestly determine that only the infected test organisms get sore throats, and then I must reisolate the bacterium from the diseased animals and repeat the experiment again with the isolated bacteria. The requirement for a large sample size is even more recent than the idea of a control. In the bacterial experiment, it is always possible that some of the inoculated organisms may be resistant to the bacteria and, were I to use only one inoculated mouse, it might be immune, and I would conclude that this bacterium doesn't cause sore throats. To make the study even more rigid, I should use inbred strains of mice to provide genetic uniformity, and all the mice—experimental and control—should have been raised under precisely the same conditions.

The application of the concept of a controlled experiment is nowhere seen as clearly as in medicine. It is impossible to obtain a large population of genetically identical individual humans, and it is also impossible to maintain even a few people under absolutely identical conditions for even a few days. If the treatment is believed to be valuable, the experimenter is faced with the moral dilemma of deciding who will be treated and who will be a control. This is dramatically illustrated in *Arrowsmith* by Sinclair Lewis. Fortunately, plant scientists avoid these problems.

If I am run-down, think I have tired blood, take a herbal tonic, and feel better, there is no reason to assume that I had tired blood (whatever that might be), that the botanical cocktail cured my tired blood, or that this herbal remedy would work for your tired blood. There has been, you see, no control on the "experiment." To limit their liability, writers of modern herbals use the same weasel-words that Gerarde or Culpepper used in the seventeenth century: "This herb is reputed to . . .;" "teas made from the roots of this plant are said to . . .;" "it has been reported that a poultice of leaves of this plant relieve. . . ." A comparative reading of a dozen randomly selected modern herbals reveals that the same plant is supposed to be effective for a wide range of abnormal conditions. Garlic, a most excellent and necessary food seasoning has been recommended for falling hair and baldness (usually genetic in males), sore throats (caused by bacteria), sprains, unspecified aches, slowly healing cuts (possibly related to diabetes), peptic ulcers (caused by modern life?), the blahs, and over 20 other conditions. In two words: ImPossible!

MODERN PHARMACEUTICS

This polemic should not indicate that plants used in folk medicine cannot be effective. The pharmacist's shelves contain plant products derived from the herbal remedies of Europe and many other cultures. The Chinese invented vaccination 2000 years before Jenner learned about cowpox; they understood the circulation of the blood 1000 years before William Harvey; and they used advanced surgical procedures, including acupuncture, for centuries. Although the Chinese coated their herbal medicine with the same mystical-religious gloss as did the Europeans, rhubarb roots for laxatives, *Ephedra* as a source of ephedrine, and many other effective chemicals have been extracted or synthesized from herbs used in Chinese medicine.

So, too, have we benefited from the herbal medicine of another great Eastern culture, that of India. Precise reference in the Rg-Veda to a plant called *chandra* can be found. Chandra is a Sanscrit word for moon and an oblique indication that the plant was used in the moon's disease—lunacy. The plant, *Rauwolfia serpentina,* was known to Westerners as an effective reducer of fevers and a cure for dysentery and for many other apparently unrelated complaints. The common denominator in the uses was that the plant seemed to be an excellent sedative. This escaped the attention of Western physicians. Only in the 1940s was the fact that extracts of *Rauwolfia* were amazingly effective in reducing blood pressure brought to the attention of the West, although it was known by everybody in India down to the untouchables. Pills made from the dried roots were equally effective in reducing the violence of the insane to a degree and with a safety unmatched by barbiturate narcotics. An alkaloid, reserpine, was isolated and characterized in 1952–1953, allowing standardized dosages to be administered. Schizophrenics, who make up over 50 percent of the patients in our mental hospitals, could be treated with reserpine and could resume virtually normal lives in society. The concept that mental and nervous disorders can be successfully treated and controlled with chemicals has developed to the point that many synthetic chemicals have been made and tranquillizers are a way of life.

HUMAN FERTILITY

An area of intensive search and research is human fertility. Depending on the culture and its economic, political, social and religious milieu, fertility control has been of more than passing interest for thousands of years. Rome viewed sterility as an offense against the state. China, where sons were a religious necessity, mandated that a barren wife could not be permitted even to die in her own home. In modern South American countries, where *machismo* is a cultural factor, a man without children is "not quite a real man." In almost all societies, infertility is rarely considered the man's "fault." Herbs that could increase female fertility were prized and are described in medical writings throughout the world. Many of the recommended plants are examples of the doctrine of signature concept. The Chinese recommended the peach fruit because of its resemblance to the female genitalia; Greeks had their women eat nuts and pomegranates; Romans included breast-shaped fruits in the diets of their wives. The Mohawks, Cherokee, and Plains Indians fed phallus-shaped roots to brides, Near Eastern civilizations favored the mandrake (Chapter 4), and the Japanese and Chinese chose the morphologically similar ginseng. At this time, no plant product has been demonstrated to be effective in enhancing fertility or curing sterility.

Repression of fertility has long been a matter of great interest. A search through the ethnobotanical literature reveals that many plants have been used as contraceptives. Viewed from the perspective of modern pharmacology, most of these are nostrums, depending upon supernatural powers for effectiveness. Several do, however, show some promise. The Cherokee used the spotted cowbane (*Cicuta maculata*) whose tuberlike roots contain coniine and other alkaloids similar in structure and action to the ones in the hemlock (*Conium spp.*) that Socrates drank. Indian women of the western and southwestern plains drank an infusion of the roots of stoneseed (*Lithospermum ruderale*) which contains chemicals that abolish the normal menstrual cycle, decrease the size of ovaries, thymus, and pituitary glands, and interrupt the secretion of other endocrines. Similar chemicals are found in one of the sweet peas and in *Lycopus*, the bugleweeds of the mint family. Estrogen-like hormones are also found in a clover (*Trifolium subterraneum*), in stems of soybean plants, and in tulip bulbs, all of which have been used for contraception by various cultures.

The trend in modern medicine is to either extract the effective medicinal from the plants in which they are synthesized or to find out what the active compound is and then make it in the laboratory. Not only does the latter method eliminate the possibility that other constituents of the plant can cause undesirable side effects, but it permits precise standardization of dosages. In many instances it is still cheaper to have the plants gathered in the wild or grow them on herb farms, then isolate the effective compound, than it is to attempt synthesis of the molecule.

ADDITIONAL READINGS

Carson, G. *One for a Man, Two for a Horse.* Garden City, New York: Doubleday, 1961.

Haggard, H. *Devils, Drugs and Doctors.* New York: Harper & Row, 1929.

Kelly, H. A. *Some American Medical Botanists.* Troy, N.Y.: The Southworth Company, 1914.

Leighton, A. *Early American Gardens for Meats or Medicines.* Boston: Houghton Mifflin, 1970.

Mettler, W. *History of Medicine.* Toronto: Blakeston Company, 1947.

Rubin, A. A. *Search for New Drugs.* New York: Dekker, 1972.

Vogel, V. J. *American Indian Medicine,* University of Oklahoma Press, 1970.

Ergot and Ergotism

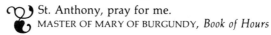 St. Anthony, pray for me.
MASTER OF MARY OF BURGUNDY, *Book of Hours*

The drug known as ergot is defined pharmacologically as the dried sclerotium of a fungus, *Claviceps purpurea,* developed on rye plants (Figure 5.10). This purplish-brown, somewhat hornshaped structure has over 150 names including spurred rye, horned seed, and *mutterkorn.* In the United States, the inclusion of ergot in materia medica dates back to a scientific paper presented before the Massachusetts Medical Society in 1813 when Dr. Oliver Prescott suggested that ergot could ease difficult births. Actually, Dr. John Stearns of New York demonstrated five years earlier that ground-up ergot could quicken childbirth. Although unknown to Prescott and Stearns, their reports only repeated observations made for centuries. In the *Kreuterbuch* of 1582, Adam Lonicer (for whom the genus *Lonicera*—honeysuckle—is named) noted that ergot was used by German midwives. The ancient Chinese probably knew of ergot's ability to cause uterine contractions, and Avicenna, a Moorish-Arabian physician of the tenth century, used the material for the same purposes.

Search for the active principal in the fungus has occupied the attention of many bio-organic chemists. The active alkaloids constitute less than 0.1 percent of the dry weight of the sclerotia. The isolation of

Figure 5.10
Head of rye bearing sclerotia of the ergot fungus.

the first (of many) alkaloids from ergot was made by a French pharmacist, C. Tanret, in the nineteenth century, but it wasn't until 1918 that A. Stoll, a Swiss, obtained the first chemically pure and biologically active alkaloid—ergotamine.

Most ergot alkaloid is isolated from sclerotia formed in the grain heads of rye plants. In order to obtain a large-scale infection of rye plants, a variety of techniques were developed in the 1930–1940 period. When the grain is just beginning to flower, large machines mechanically shake the plants, and, at the same time, a spore spray is directed at the flowers. When sclerotia are mature, which is simultaneous with maturation of the rye, the fungus is picked by hand. Yields depend on all the imponderables which attend the maturation of any crop, and deviations from yet-unknown environmental optima result in variable concentrations of the active alkaloids. Understandably, the variation in potency is of considerable concern and harvests are appropriately mixed to provide a product of uniform alkaloid content. Cultivation is further complicated by host-parasite interactions, the genetics of which are unknown; races of the fungus may develop from a single gene mutation at frequencies of about 1 per million. Since a field inoculation consists of trillions of spores, genetic uniformity is very unlikely. The ergot industry in the United States, Germany, Russia, and Spain produce huge quantities. In 1945 the United States imported 44 tons of ergot (each sclerotium weighing less than ¼ oz) and produced 100 tons in Minnesota, Nebraska, the Dakotas, and Illinois.

MEDICAL USES

Methods of isolation and purification of ergotamine and other active alkaloids were worked out shortly after the end of World War II, primarily in the laboratory of Albert Hofman in Basel. Prior to the use of sophisticated chemical techniques, crude biological tests for activity were used. For example, a sample was injected into a White Leghorn rooster, and the time required for the rooster's comb to turn blue was measured. Alternatively, frogs were injected and constrictions of blood vessels in the toe webs were observed.

The color change and vasoconstriction reactions supply clues for the medical use of ergot. Basically, it causes blood vessels to contract—thus reducing the danger of hemorrhaging during and immediately after birth. It also causes contraction of muscles, serving to assist in shoving the unwilling child into the world. Blood vessel constrictions and uterine contractions are accompanied by nerve blockage, alterations in motor functions, and some brain dysfunctions. When provided in appropriate concentrations, all of these ergotamine actions are useful in easing childbirth.

The knowledge derived from reactions to ergot alkaloids noted in childbirth was applied in the 1925–1930 period by internists treating migraine headaches. Afflicting a small but significant proportion of the population of every country, migraine sufferers experience periodic, persistent, and unilateral headaches, frequently accompanied by nausea and mild hallucinations. There is vasodilation of blood vessels in the brain, and hormonal balances are sharply altered from the normal pattern. Ergot, causing physicochemical changes in humans which are almost diametrically opposite the migraine syndrome, is widely used to control, but not to cure, the agonies of migraine. Unfortunately, ergot therapy is effective in only about 40 percent of migraine victims.

Whether used in childbirth or for migraine, ergot therapy is not without its deleterious side-effects, although fewer than might be expected of a drug with such potency. Obviously, there are wide variations in tolerance, and reports of toxicity can usually be traced to differences in patients rather than to variations in the composition of the drug. Especially sensitive newborn children whose mothers were treated during childbirth will occasionally show reddening of the skin or breathing difficulties, but these symptoms usually pass without permanent damage. This is fortunate, because there is no countertherapy for ergot poisoning.

ERGOTISM

For uncounted centuries, however, ergot has been one of the cursed scourges of mankind. It has plagued the body and mind ever since we began to use grasses for their edible seeds. In Europe it was called the holy fire (*ignis sacer*), St. Anthony's Fire, the ignius beatae Virginis invisibilis, or the infernalis.

Three major cereal grains have been used to make bread. The common bread was an unleavened product made from barley, but only wheat and rye flour make a raised or leavened bread. It is likely that Thrace and Macedonia, but not Greece, grew some rye, but the plant wasn't introduced into much of Europe until the Christian era. France began to grow rye about 300 A.D., and Britain obtained her starting seed when the Teutons invaded the island. Rye was not grown as a major cereal grain in Europe and European Russia until about the fifth century, and attempts to pinpoint just when historical records of ergotism began is difficult. Thus, an epidemic suspiciously like ergotism broke out among the Spartans in 430 B.C., and a plague of 857 A.D. in the Rhineland also matches the clinical symptoms. A disease "like fire" was reported in Paris in 943, from Aquitaine-Limousa in 994 with 4000 deaths, and from Rheims in 1041 with 2000 deaths. From that time on, instances of ergotism have been recorded in sufficient detail so that we can be sure of its cause.

Ergotism results from the ingestion of sclerotia of ergot ground up in rye flour. Two major types of ergotism are known, gangrenous and convulsive. In the former, severe constriction of the blood vessels results in swelling as blood accumulates in the hands or feet, with burning sensations alternating with intense cold. Numbness follows within a few days and this, in turn, is followed by blackening of the limb, horrible odors, and eventually merciful, but unbearably painful death.

Convulsive ergotism accurately describes the symptoms. Twitching of head, arms, and hands is followed by contractions of muscles throughout the whole body. The afflicted typically roll themselves into a ball and then stretch themselves out at full length, the actions accompanied by terrible pains. Vomiting, deafness, blindness, and hallucinations usually follow. Feats of superhuman strength, and the conviction that flying is possible have been noted. If the victim recovers, and about 30–40 percent do, hallucinations can continue aperiodically for up to a year. Domestic animals who eat ergot-contaminated grain or table scraps exhibit identical responses. It is said that dogs will tear bark from trees until their teeth fall out and that ducks will strut like roosters, attacking people and other animals. Depending upon the weather, the genetic constitution of the host and the fungus, the care taken to eliminate sclerotia before milling grain into flour, and the amount of bread eaten, devastating outbreaks of ergotism could occur. And they did occur in Europe on an average of once every five to ten years.

Innumerable people had ergotism before its etiology was recognized in 1673 by a Parisian lawyer-physician, Denis Dodart. Up to the middle of the eleventh century, over 20 massive epidemics were reported in France alone, and by the middle of the fourteenth century, over 50 epidemics had been reported from central Europe. Lacking any knowledge of its cause, it was reasonable to call upon the saints to intercede with heaven for succor. But which one? St. Anthony the Great was born in Egypt in the first century A.D. and established the idea of monastic life. Long the patron saint for erysipelis, a bacterial disease of the skin which causes swelling and burning, it seemed logical for him to become the intercessor for this disease as well (Figure 5.11). In 1039, a French nobleman, Gaston de la Vollaire, built a hospital in the Rhone Valley, obtained relics of St. Anthony, and asked monks to serve in the hospital. These men formed the Order of St. Anthony and dedicated themselves to nursing the survivors of ergotism. In the *Book of Hours* by the master of Mary of Burgundy (1480), St. Anthony is asked for protection against the disease. The Holy Fire disease was soon called St. Anthony's fire.

Although Dodart's identification of the cause of ergotism was known among the few educated physicians of the seventeenth century, the direct connection between *mutterkorn* and St. Anthony's fire didn't become general knowledge until the eighteenth century. The dark, heavy, sour, but very nourishing bread of central and eastern Europe contained so much ground-up weed seed that the dark sclerotia went unnoticed. Since bread was truly the staff of life, the persistence of the peasants in eating ergot-contaminated bread is not surprising. When a high probability of starvation had to be weighed against possible ergotism, people made the only logical choice.

Between 1580 and 1900 there were 65 major ergot epidemics in-

Figure 5.11
St. Anthony, patron saint of ergotism, praying for a victim of the affliction. The gangrenous type of disease is seen. Taken from a woodcut in *Fendtbuch der Wundartzney* by Gersdorff published in 1517.

cluding 29 in Germany, 11 in Russia, 10 in Sweden, and 5 in the United States. The introduction of the potato to Europe undoubtedly reduced the death toll by substituting for bread. In 1722, Peter the Great mounted an invasion of Turkey to obtain for Russia the still-coveted ice-free port to the seas. His cavalry ate ergotized black bread and 20,-000 men and horses were stricken; the invasion was called off. Between 1770 and 1780, epidemics raged through Germany and France with over 8000 documented deaths. In the winter of 1812–1813, Napoleon's troops and horses ate bread baked from rye commandeered from the Ukraine, and the resulting epidemic of ergotism contributed to his Russian defeat and turned the retreat from Moscow into a horror. In 1812, Austria passed a law stating that inadequately cleaned rye would be confiscated, and other European countries quickly passed similar legislation.

Although the gangrenous form of ergotism has almost disappeared—with no major outbreak since its last appearance in France in 1885—there is a growing suspicion that it is reappearing in Central Africa from sclerotia developing on cereal grains other than rye. The convulsive form, however, has increased in severity and is responsible for outbreaks of ergotism in the twentieth century. There was a severe outbreak in the Soviet Union in 1926 and a smaller one in England in 1928–1929 when the Jewish community imported rye from central Europe. During the well-studied Soviet Union epidemic in 1926, flour containing 2 percent ergot was found to be enough to cause convulsive ergotism and some samples of rye flour contained up to 7 percent sclerotia. It has been claimed that convulsive and hallucinogenic ergotism struck a town in Provence in 1951, but the French government denied this, stating that there was an inadvertent contamination of the flour by an insecticide; this bureaucratic explanation does not conform to the symptoms noted.

As home milling decreased during the late nineteenth century, purity control became easier, and large sclerotia, which are bigger than the grain, could be removed by sieving. Removal of smaller sclerotia and fragments is facilitated by the observation that when rye is mixed with salted water the sclerotia float and can be skimmed off. Nevertheless, rye grown for the organic foods market is not subject to federal control.

Although no records were kept prior to the nineteenth century on ergotism in animals, ergot alkaloids have long caused serious economic problems. Canada decreed in 1971 that no more than 0.1 percent by weight of ergot is permitted in feeds for poultry and other domestic animals, but even this level severely reduces the weight gain of cattle. Many other wild grasses are susceptible to infection by ergot, including those that form the food of migrating birds. The reports of abnormal behavior of wild plant-eating animals may be due to their feeding on ergotized grasses. It is, for example, not unusual to observe squirrels which seem to be "insane"—climbing a few feet up a tree and somersaulting back down again for hours on end. Cats, too, can show all the symptoms of a "real high." The parallels between these behavioral modes, some aspects of the convulsive type of human ergotism, and the "spaced" appearance of highly sensitive women in labor who have been given ergotamine suggest that the ergot alkaloids may, in fact, be potent hallucinogens.

LSD

As part of a long-term chemical study of ergot alkaloids, two Swiss chemists, Stoll and Hofmann, prepared a large number of compounds from ergot. Among these were a number of lysergic acid derivatives, the most familiar of these being the diamine of lysergic acid—LSD (Figure 5.12). Hofmann apparently was the first to experience the effects of LSD when he inadvertently touched his contaminated fingers to his lips and then attempted to bicycle home. The trip took several hours, rather than the usual half-hour. One early experimenter experienced not only initial hallucinogic effects of LSD, but reported that for several months thereafter he would see blood coming from the faucet whenever he opened the tap. The subculture use of LSD and the somewhat hysterical reports of uses of LSD in psychiatry have had at least one useful spinoff. They have directed the attention of botanical and medical investigators to the effects of hallucinogenic materials among the Indians of South America.

In a compendium on the plants, animals, and rocks of New Spain, Francisco Hernandez reported in 1580 that the Aztecs used the seeds of the snake plant, *coaxihuitl*, in religious ceremonies. Called *ololiuqui*, infusions of the seed deprived a man of his senses; those who used it "believed in the owl, sucked human blood and could communicate directly with their gods." To make themselves fearless, Aztec warriors mixed ashes of poisonous insects, tobacco, and ololiuqui and rubbed the mixture on their bodies. Not unexpectedly, Jesuit upholders of the only true faith railed against this diabolical seed and drove the cult into hiding. The users hid their knowledge of the plant for close to 400 years. In 1897, Dr. Manuel Urbina concluded that ololiuqui was a

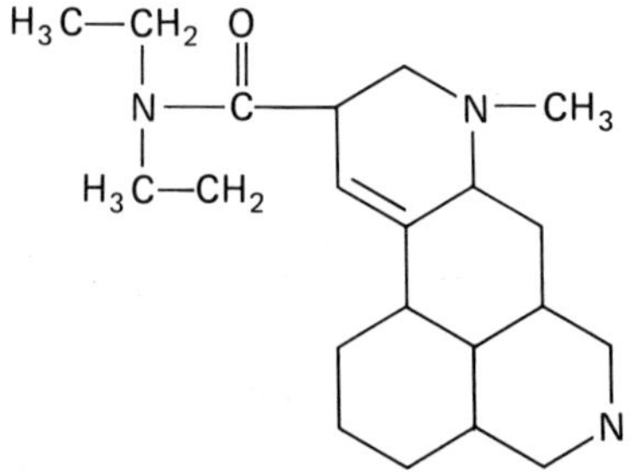

Figure 5.12
Chemical formula of LSD.

D-lysergic acid diethylamine

morning-glory. Most *norteamericano* botanists couldn't believe that the seeds of such a pretty plant could harbor hallucinogens, and the definitive authority of William Safford was invoked to discount both Urbina and an anthropologist, B. P. Reko, who in 1919 was so rash as to identify the plant as a morning-glory, *Rivea corymbosa*. The adequacy of identification by Urbina was confirmed in 1939 by a botanist, R. E. Schultes of Harvard. At about the same time, seeds of another member of the morning-glory family, *Ipomoea violacea* (a close relative of the true sweet potato) was found to be used by the Oaxacan Zapotec tribe for divinatory rites under the name *tlitliltzin* or *badoh negro*—the sacred black seeds. By 1955, it was clear that the hallucinogenic materials were alkaloids of the ergot type and by 1960 Albert Hofmann isolated several, including ergometrine, an alkaloid useful in childbirth. Other members of the same botanical family, the Convolvulaceae, also contain active substances.

Although Schultes and Hofmann published their studies in the *Botanical Museum Leaflets* of Harvard University, a periodical not noted for wide distribution outside plant taxonomic circles, the years 1964 and 1965 were marked by shortages of morning glory seeds in the United States. Fortunately, most people who took the seeds didn't prepare them properly and quit in disgust, but severe psychic effects and a few deaths were reported. The conviction that the user could truly fly out of a fifth story dormitory window became all too common.

ADDITIONAL READINGS

Barger, G. *Ergot and Ergotism*. London: Gurney & Jackson, 1931.
Christensen, C. M. *Molds, Mushrooms and Mycotoxins*. University of Minnesota Press, 1975.
Dowling, H. F. *Medicines for Man*. New York: Knopf, 1970.
Fuller, J. C. *The Day of St. Anthony's Fire*. New York: Macmillan, 1968.
Gray, W. D. *The Relation of Fungi to Human Affairs*. New York: Holt, Rinehart and Winston, 1959.
Gröger, D. "Ergot," chapter 12. In S. Kadis, A. Ciegler, and S. J. Ajl (eds.). *Microbial Toxins*, volume 8 (*Fungal Toxins*). New York: Academic Press, 1972.
Hofmann, A. "Ergot." In T. Swain (ed.). *Plants in the Development of Modern Medicine*. Cambridge, Mass.: Harvard University Press, 1972.
Schultes, R. E. "Hallucinogens of Plant Origin." *Science* 163 (1969):245–263.
Van Rensburg, S. J., and B. Altenkirk. *"Claviceps purpurea—Ergotism."* In I.F.H. Purchase (ed.). *Mycotoxins*. New York: Elsevier, 1974.
Youngken, H. W., Jr. "Ergot—A Blessing and a Scourge." *Economic Botany* 1(1947):372–390.

Jesuit Powder

That in marshes there are animals too small to be seen.
But which enter the mouth and nose and cause troublesome diseases.
VARID, gentleman farmer of Rome, 100 B.C.

Throughout man's troubled history, few diseases have played so tragic a role as malaria. It has killed or incapacitated more people than all plagues, wars, and automobiles. Over 10 percent of the U.S. overseas armies in 1943 had malaria. In 1938–1939, there were over 300,000 reported cases of malaria world-wide, including 20,000 deaths in the Brazilian province of Rio Grande de Norte. The Spanish-American War of 1898 saw four times as many troops inactivated by malaria as by wounds. The Army of the Potomac under General George McClellan lost at Chickahominy because it didn't have enough healthy troops to oppose General Lee. The Pilgrims' decision to settle in New England was made, according to William Bradford, because, "hott countries are subject to greevous diseases . . . and would not so well agree with our English Bodys." In 1596, the Earl of Cumberland captured Spanish Puerto Rico but couldn't hold it because his forces were decimated by malaria. The city of Florence was depopulated in the second century B.C. by malaria. Alexander the Great died of it in June, 323 B.C. Among the twelve labors of Hercules was the slaying of the nine-headed Hydra, the monster that brought misery and ruin to the people, and the elimination of the man-eating birds of the Stymphalian marshes—allusions to malarial infections. Untreated malaria may kill about one percent of those infected. The survivors, prone to relapse, may suffer from anemia, weakness, sexual impotence, chronic abortion, or secondary infections—all of which lower the value of the individual to self, family, and community.

ETIOLOGY OF MALARIA

The name malaria was coined in the seventeenth century by Dr. Francisco Torti by combining the Italian names for "bad" and "air," and it has been called the shakes, the fevers, the ague, and many other things, none affectionate. Hippocrates reported several clinical types of malaria. He believed that imbalances in the ratios of the four humors—blood, phlegm, black bile, and yellow bile—caused the disease. It was asserted for a thousand years that invisible worms were carried on the dank night air into the body and caused changes in the humoral ratio.

Yet it was also known that swamps and mosquitos were involved, for malaria was rarely found in dry and windy areas, and it disappeared during the winter. Empedocles of Acragas (Sicily) in the fifth century

Adapted from "The Fever Bark Tree," *Natural History* 85(1976):10–19. Copyright American Museum of Natural History.

B.C. had marshes drained to rid local towns of malaria. From India, still an endemic area, a lyric poem from the fifth century A.D. *Susrata* says:

> The green and stagnant waters lick his feet
> And from their filmy, iridescent scums
> Clouds of mosquitos, gauzy in the heat
> Rise with his gifts: Death and Delerium

Only during the middle of the eighteenth century was the relationship of mosquitos to the fevers accepted, and it was not until 1880 that a French physician, Charles L. A. Laveran, found microscopic parasites in red blood cells of human victims of malaria for which he received the Nobel Prize. In 1902, the British physician-bacteriologist Ronald Ross received the Nobel Prize for discovering another stage in the life cycle of the parasite in the gut of the *Anopheles* mosquitos. Giovanni Grassi proved that the *Anopheles* mosquito carried the parasites to the blood stream of humans and by 1899, the complete life cycle of the parasite, called *Plasmodium*, was known. There are four major species of malarial parasite, each causing different clinical types of the disease. The need for mosquito control became so obvious that marsh draining and other control measures were adopted throughout the world.

Since mosquito larvae develop in water and must come to the surface to breathe, draining was accompanied by pouring oil on water to alter surface tension and prevent the mosquito larvae from "holding" to the surface, but this procedure was ineffective. DDT came into general use as an insecticide in the 1940s, and in spite of almost criminal misuse, it saved millions of lives. But it is impossible to kill all female mosquitos. Even a few people with malarial parasites in their blood can serve as a reservoir for the insect and can become the base for a whole new epidemic. As far back as the fifteenth century physicians were dreaming of some medicine which could cure the disease, what we today would call a chemotherapeutic agent specific for the malarial parasite. One was found in a plant, and its history starts in Lima, Peru, the capitol of New Spain.

Figure 5.13
The "tree of the Jesuit's powder." It is doubtful whether the illustrator ever saw the tree; he drew from verbal descriptions. Taken from *Rariorum Plantarum Historia* by Carolus Clusius published in Antwerp in the edition of 1601.

THE QUINA TREE

After Lima was founded in the 1520s it became a proud and wealthy city with almost feverish financial activity. Riches were quickly amassed, and young Castilian gallants vied for permission to make an assured fortune from minerals so easily taken from the defeated Indians. The turnover of these hopeful merchant princes was high; malaria was endemic in Peru. Yet, as Church fathers noted, Indians in the bush were not excessively bothered by the disease. Through converts and slaves (usually one and the same) they heard that on the eastern slopes of the Andes Mountains was a tree whose bark, powdered and mixed with water, would cure the fevers. The natives called the tree *quina*—the fever bark tree (Figure 5.13). An Augustinian monk, Father

Calancha, published a book in 1639 about his experiences in New Spain in which he noted that the bark of the fever tree "cures the fevers and tertians." The Augustinians were a contemplative and less worldly order than the Jesuits, and it was the Society of Jesus that recognized the political potential inherent in this powder. Sending expeditions into the mountains, they soon had a monopoly on bark from Peru, Colombia, and Bolivia. By some means, they obtained an official monopoly by 1650. At first they trickled it back to Europe and then systematically built up their markets through discrete advertizing of the fact that they and they alone could cure the fevers then sweeping Europe. They had little competition, for standard therapy for malaria was still based on the Hippocratean dicta of humors. Treatment was a combination of bleeding the already debilitated patient, bedrest, and cooling cloths placed over the body.

RELIGIOUS MEDICINE

The Society of Jesus formally decided that this wondrous, God-given plant should be used wisely—some would say jesuitically—for "the greater Glory of God and for good and useful Christians." Rome, especially the Vatican, had first call on it for the need was especially great. Summer fevers were raging in Italy, and travel through Campania to Rome was an invitation to death. The princes of the Church remembered a previous conclave which resulted in the death by malaria of eight cardinals and 30 attendants, and postponed their conclave until late fall. Secular and religious princes, bishops, and noblemen—then down through the hierarchy—were ordered in priority lists. The Chinese emperor, the great K'ang Hsi, had several bad malaria attacks in 1693, and Jesuits in attendance at his court introduced the bark and saved his life; K'ang Hsi, was grateful but never became a convert.

It was obvious that regulation of the bark's use and study of the best method of treating malaria with it needed to be centralized. Control was vested in Juan de Lugo, S.J., native of Madrid, cardinal upon the insistence of Pope Urban VIII, eminent theologian, and sometime professor of philosophy at Gregorian University. De Lugo had a superb organizational mind, one capable of both administration and research; the society could not have chosen a better man. He devoted over ten years of a busy administrative life to evaluating dosages, attempting to extract the active principle, and in 1651 oversaw the publication of *Schedula Romana*, a book in which instructions on the use of the powder were detailed. With the imprimature of the Society and the tacit approval of the pope, no Catholic physician would dare openly to obstruct the use of the powder although many were, for a variety of reasons, opposed to it.

Meanwhile, Spanish galleons were lumbering back to Santiago and Cadiz with gold and silver bullion and with bales of bark bearing the mark of the Order. With malaria decimating French and English colonies on the eastern seaboard of North America, buccaneers who plundered the Spanish main were as welcome for their free-spending ways as for their loads of Jesuit bark. Some of the bales were trans-shipped in good English bottoms back to Britain where the London ague was endemic. The bark, commonly called "Jesuit's powder," was available by 1658 at John Crook's Bookshop, but because of fears of a popish plot or curses placed on it by "those Jesuit Devils," Oliver Cromwell died of malaria the same year without being treated.

Cardinal de Lugo died in 1660. His death, and changing religious politics, pushed the powder further into disrepute even in Catholic lands. Distribution and control of the powder by the Jesuits was waning, several prominent persons died after taking the powder, an unidentified fever in Rome in 1655 was not controllable by the medicine, and the power of the Roman Church to demand obedience and conformity had decreased. Physicians had never liked the deft hand of the Jesuits slipping into their pockets, and with the Jansenite apostacy challenging the authority of the Church, attacks on Jesuit powder intensified. Popular prejudices against the Society were fanned by reformists and physicians, supplies were short because of piracy and slave revolts in Peru, and people didn't like swallowing a bitter draught without knowing more about what was in it.

THE FEVEROLOGIST

Rumors were, however, flying in England that a former apothecary's assistant in the county of Essex could cure the ague. To quell fears, he announced that he was not a physician, but a *pyretiato*, a feverologist, who by long and arduous study, "by observation and experimentation" had "a certain method for the cure of this unruly distemper." He further announced, "Beware of all palliative cures and especially that known by the name of Jesuit's powder, for I have seen most dangerous effects following the taking of that medicine." The pyretiato moved to London in 1668 and set up a very lucrative practice under the horrified noses of the Royal College of Physicians. When Charles II contracted malaria, he called for this interesting man, Robert Talbor. Because of suspicions that Talbor was really using the hated popish remedy on the monarch, loyal Englishmen carried placards warning that the Jesuits were trying to poison the king. Happily, Charles recovered and Talbor was knighted. Charles also forced the College of Physicians to make Sir Robert a full member and warned this outraged body that "you should not give him any molestation or disturbance in his practice."

Through Robert Talbor, Charles saw a way to improve the always touchy relationships between England and France. The dauphin, son of Louis XIV, suffered from fevers, and Talbor was sent to Paris as "royal envoy and physician to the king of England." The entire royal family was cured, and Louis, in a royal gesture, sent Talbor to cure the queen of Spain. In spite of Spain's colonies being the source of the curative bark, there wasn't any in the mother country because the Jesuits had recently been expelled from Madrid and they had taken their powder with them. Upon his triumphal return to France, Talbor changed his name to Talbot—a distinguished French name—and became a celebrity in Paris. Madame de Sévigné referred to him as *un homme divin,* and his bedside manner was most favorably commented on by the ladies of the court. As the Chavalier Talbot, and with Paris rife with malaria, he amassed a fortune. In 1680 he decided to return to England with his secret, but Louis paid Sir Robert 3000 gold crowns and the promise of a life pension for a sealed envelope containing the remedy formula on the promise of the king that it would not be opened until after Talbot's death. Talbor, again English, returned home in glory, became a fellow of St. John's College, Cambridge, and composed his own epitaph before his death in 1681.

> The most Honorable Robert Talbor, Knight and singular physician. Unique in curing fevers of which he delivered Charles II of England, Louis XIV of France, the most Serene Dauphin, princes, many a duke and a large number of lesser personages.

In January 1682, Louis opened his expensive envelope and in a book entitled *The English Remedy or Talbot's Wonderful Secret for Curing Agues and Feavers* the secret was revealed. It was, of course, Jesuit powder mixed with wine to disguise the bitter taste. The discomfiture of the physicians who had denounced Talbor was great, but even knowing that the bark was effective didn't make the British medical profession comfortable using it. When Charles had another bout of malaria in 1682–1683, he literally had to beg his doctors for Talbot's remedy; they simply would not accept the fact that they had been outdone by a mere apothecary's apprentice!

PLANTATION CULTIVATION

By the end of the seventeenth century, quinine powder, no longer Jesuit powder, was the standard treatment for malaria. Spain still controlled trade through its exclusive mandates in Peru and Bolivia. As demand increased, it became obvious that there weren't enough trees available to assure supplies. Bark collectors had to go further and further into the mountainous bush to find the trees, getting lost and dying of dysentery or from the poisonous darts of the head-hunting Jivaro Indians. This danger was envisioned by the Jesuits 100 years previously when they tried, unsuccessfully, to require that a tree be planted for every one cut down and stripped of its precious bark. In the middle of the eighteenth century, a group of French botanists studied this problem and concluded that there were at least four species of the tree. This was confirmed in Sweden when botanical specimens were sent to Up-

Figure 5.14
Flowering branch of the quinine-bark tree, *Cinchona officinalis*.

sala for classification by Linnaeus. Linnaeus gave the genus name *Cinchona* to the trees (Figure 5.14) to honor the viceroy of Peru in 1628–1639, inadvertently dropping the "h" following the initial "C" in Count Chinchona's family name.

By the middle of the nineteenth century, the fundamental facts about the fever bark tree were well known. The trees are native to mountains of Peru, Bolivia, and Colombia, growing best in areas of abundant rainfall, at altitudes above 1000 m, and up to 8000 m. The genus is in the Rubiaceae, the family which contains the coffee tree. The compounds that destroy the malarial parasite are all closely related alkaloids which, when purified, form white, crystalline powders with a very bitter taste. Quinine alkaloids are found in the bark and may represent over 7 percent of the weight of the bark. Two French chemists isolated the quinine alkaloids in the early nineteenth century and by 1870, quinines could be isolated and purified and chemical determination of active materials was a routine laboratory technique. With this information, standardization of treatment for malaria was accomplished by basing dosage on the alkaloid content rather than on the amount of bark mixed with water. There appeared to be no barrier to plantation cultivation and freedom from malaria seemed merely a matter of time.

The South American monopoly was broken in 1865. Charles Ledger, an English bark trader, established his business in Puno, Peru, across Lake Titicaca from Bolivia. Knowing that the best bark was found only at high altitudes, he secretly sent his servant-translator, Manuel Incra Mamini, into the Bolivian Andes near the headwaters of the Rio Beni where *Cinchona calisaya* trees with high-yielding bark grew. Manuel left Puno in 1861 and returned in 1865 mumbling about bad growing years and leading a shy young Indian woman with two small children. Ledger paid Mamini £150 for the seed he had obtained, and Manuel foolishly went home to Bolivia where he was arrested and charged with the traitorous crime of smuggling cinchona seeds out of the country. In jail for less than a month, he was severely beaten and allowed to die. The precious seeds were, by this time, in London in the care of George Ledger, Charles's brother. The seeds were first offered to the British government who refused them. George went to Amsterdam and managed to sell some to the Netherlands' government for 100 gulden with a promise of an additional payment if the seeds were viable.

The seed packet arrived in Java in December 1865, was opened, and smelled so badly that it was assumed that they were rotted. Fortunately, they germinated well and the Ledger brothers received their promised 500 gulden. Over 10,000 trees were outplanted the following year, and by 1873, several plantations were established. The Dutch East India Company sent an alkaloid chemist to Java who methodically tested samples of the bark of each tree, marking those with quinine yields exceeding 10 percent. These trees were used as grafting scions for fast-growing root stocks. By 1874, the superintendent of the planta-

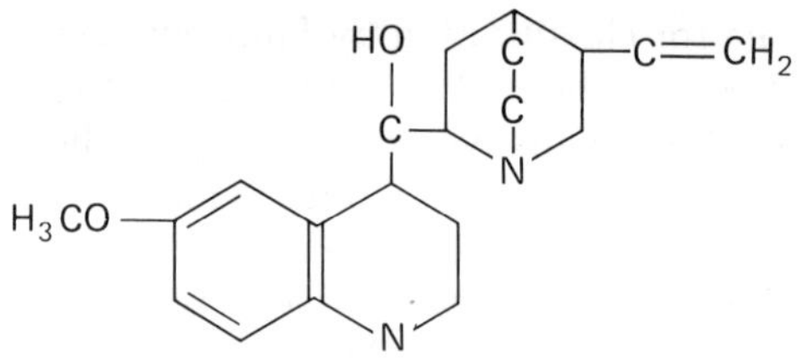

Figure 5.15
Chemical formula for
quinine.

Quinine

tions reported that within five years they expected to have over 2,000,-000 high-yielding trees . . . and they did. In 1881, South America had exported 9 million kg of bark, but by 1884 there were less than 2 million kg exported, and by 1890, the Dutch had a world monopoly on quinine. After some years of overproduction which led to reduced prices, controlled harvesting with ruthless stabilization of the world's supply was achieved by 1910, not to be altered until the Japanese overran the plantations in World War II.

With quinine under monopolistic control, attempts were made to grow high-yielding trees in other parts of the world, but none were economically viable operations. As World War II extended into the jungles of Asia, the Pacific islands, and into the parts of North Africa and southern Italy where malaria was very prevalent, supplies of quinine became worth more than their weight in gold. During the 1930s, the Winthrop Chemical Company in the United States had developed an antimalarial called atabrine used as a substitute for quinine. It was effective in wartime, but dosage was poorly understood, and it turned a patient's skin yellow. As the quinine shortage became a serious problem for warring armies, a crash program for the total synthesis of quinine started in the United States and England, and several effective antimalarial synthetics were developed by the mid-1940s, a bit late for the troops, but still a most worthwhile effort. Quinine isolated from bark is still cheaper than synthesized quinine (Figure 5.15) and is still being used to cure one major type of malaria that is resistant to synthetics. A small, but growing market for quinine has developed since the 1960s when tonic water with quinine was found to mix nicely with gin. Natural quinine, synthetic antimalarials and other mosquito-control measures continue to provide mankind with, if not complete freedom from malaria, at least reasonable control. And the people and their children whose lives have been saved are a major part of the world's population problem.

ADDITIONAL READINGS

Duran-Reynals, M. L. The Fever Bark Tree: The Pageant of Quinine. Garden City, N.Y.: Doubleday, 1946.
Haggis, A. W. "Fundamental Errors in the Early History of Cinchona." Bulletin of the History of Medicine 10(1941):417–459, 568–592.
Stanford, E. E. "The Story of Cinchona and the Mosquito." Nature Magazine 21(1933):65–68.

Taylor, N. "Quinine to You." *Fortune Magazine* (February 1934).

Taylor, N. *Cinchona in Java: The Story of Quinine.* New York: Greenburg Press, 1945.

Urdang, G. "The Legend of Cinchona." *Scientific Monthly* 61(1945):17–20.

The Purple Foxglove

The foxglove's leaves with caution giv'n
Another proof of favoring heav'n
The rapid pulse it can abate
And blest by Him whose will is fate.
WILLIAM WITHERING

Among the ills to which flesh is heir is cardiac insufficiency in which a weakened heart fails to pump enough blood through the body. Heartbeat is irregular and fluids collect in the arms, legs, and abdomen because the kidneys cannot perform their normal function. The swelling is known as dropsy or, more formally, as edema. This disease syndrome is not new. Ancient physicians knew of it, but lacking knowledge of the circulation of the blood discovered by William Harvey in 1628 and information on the function of the kidneys, treatment was limited to usually unsuccessful attempts to reduce edema with medicines which increased urine production (diuretic agents). Today, millions of people pop a small pill which regulates and strengthens the heartbeat and allows the kidneys to expell excess fluid quickly; cardiac insufficiency kills few people since the discovery of digitalis.

BOTANY AND MYTHOLOGY

Digitalis is a mixture of several naturally occurring cardiac glycosides synthesized by *Digitalis purpurea* and related species in the figwort (Scrophulariaceae) family (Figure 5.16). Native to Europe, western Asia, and central Asia, it is grown all over the world. The plant may be an annual—flowering the first year after planting—or a biennial—flowering the second year. The leaves are large, up to 30 cm long, covered with fine hairs (pubescence), and borne on a condensed stem producing a rosette rarely taller than 10 cm. When the plant flowers, it sends up tall stalks on which develop a raceme of thimble-shaped, white to pink to red flowers. The floral lip is structurally adapted as a landing platform for its pollinator, the bee, and colored spots on the corolla "light" the landing strip and direct the bee to the nectar-secreting cells at its base. Because foxglove is cross-pollinated, there is great genetic variability in the one million seeds formed in the flowers of a single plant—a fact of some significance in evaluating the quality and quantity of cardiac-stimulating chemicals formed.

Figure 5.16
Flowering heads of foxglove,
Digitalis purpurea. Courtesy
W. Attlee Burpee Co.

The Latin generic name, *Digitalis,* is something of a puzzle. It means "little finger" from the Latin *digitis* and is directly derived from the German name for the plant, *fingerhut.* The Latin term was applied by Hieronymus Tragus in 1539 and was repeated in Leonhard Fuchs's *De Historia Stirpium* in 1542. Several species are native to Europe, and all bear the common name foxglove or variants on the same theme: fox-bell, fairybell, fairyglove, fairyfox, fairythimble, or its equivalent in French, German, and other languages. According to one legend current during the reign of England's Edward III, a band of bad fairies gave *Digitalis* flowers to a fox so that he could put them on his toes to muffle the sound of his footsteps when he raided hen houses. Foxbell and fairybell are obviously derived from the shape of the flowers in their natural habitat of hollows and wooded glens favored by wee folk as well as by foxes. The common names, Our Lady's glove or Our Lady's thimble, are examples of the attempts by Church leaders to relate everyday sights to the holy family. Some have suggested that the common names are derived from old Middle English "Fuche's glew"—fox music—from a likeness of the flowers to a Saxon musical instrument which consists of a series of bells hung on an arched support.

Digitalis was a medicinal herb for centuries; Dioscorides praised it as a plant whose leaves, applied to the skin, could cure many diseases. Juice pressed from the leaves became an ingredient of salves applied to cuts, bruises, and the leg ulcers common in an era of inadequate diets and lack of soap. Rural people made hot water infusions of leaves and drank foxglove tea to experience an inexpensive but dangerous intoxication. One herbalist said that "it will scoure and clense the brest of the thicke toughness of grosse and slimie flegme and naughtie humours," and another reported that it would cure "pimply bodie, sore head, sore ears, heat of the maw and churnels"—whatever they were. John Gerarde wrote that "foxgloves are bitter, hot and dry with a certain kind of clensing qualitie joined therewith; yet they are of no use, neither have they any place amongst medicines," but John Parkinson disagreed, believing that digitalis ointments would cleanse out sores and ulcers and could cure epilepsy. Nicholas Culpepper reported that an ointment containing the juice of foxglove leaves would cure the king's evil (tuberculosis of the lymph glands) and that "it is one of the best remedies for a scabby head that is." In the New London dispensatory of 1687, William Salmon wrote:

> Fox-glove expectorates thick phlegm, if drunk with thick mead [it] clears away obstructions of the liver and spleen, is an extraordinary good wound herb, prevalent against King's Evil and may be used instead of gentian. Two handfuls of the herb taken with polypody helps the epilepsy.

ENTER WILLIAM WITHERING

The modern history of digitalis in cardiac insufficiency is completely tied up with the life of a single person. William Withering was born in 1741, the son of a Wellington, England apothecary. A bright, middle-class boy, he played the flute, bagpipes, and harpsicord, was a skilled archer, a good golfer, and participated in local dramatics. In 1766 at the age of 25, he received his medical degree from the University of Edinburgh, distinguishing himself in several courses, but, as he wrote his parents, despising the required botany course: "The offer of a gold medal by the professor would hardly have charm enough to banish the disagreeable ideas I have formed of the study of Botany." He established a general practice in the town of Stafford in Shropshire where one of his first patients was the gentle and delicate Helen Cookes. William courted her by collecting flowers for her to paint. Through this influence of a good woman, he became devoted to the "lovable science," married Helen in 1772, and, in 1776, published *A Botanical Arrangement of All the Vegetables Naturally Growing in Great Britain According to the System of the Celebrated Linnaeus, With an Introduction to the Study of Botany."* The book so interested the distinguished Erasmus Darwin that Withering was invited to share the lucrative practice of Charles's grandfather in Birmingham. Withering maintained a clinic in Stafford, making the 60-mile trip each week. On one trip, as he stopped to change horses, he was asked to visit an old woman suffering from the dropsy. Dr. Withering concluded that she could not survive and was astounded when, a few weeks later, he found his patient alive and apparently healthy. In his own words:

> In the year 1775, my opinion was asked concerning a family receipt for the cure of the dropsy. I was told that it had long been kept a secret by an old woman in Shropshire, who had sometimes made cures after the more regular practitioners had failed. I was informed also that the effects produced were violent vomiting and purging; for the diuretic effects seemed

to have been overlooked. This medication was composed of twenty or more different herbs; but it was not very difficult for one conversant in these subjects to perceive that the active herb could be no other than Foxglove.

The real story is slightly different. Mrs. Hutton, locally known as a witch, grew herbs in her garden and also collected plants from which she brewed various concoctions and sold them to the local people as remedies. Villagers believed that she conspired with the devil to make people sick so that she could cure them. She would wave her arms over the person's head, mumble incantations, stare intently at the person, and then run out into the garden, pull up plants, and brew tea which she sold to the now thoroughly frightened patient. The "family receipt" was purchased by Dr. Withering for "several gold sovereigns," including the critical information that only the foxglove had any effect on dropsy and that the other plants made the patient throw up to prove how powerful the medicine was.

Still maintaining a successful general medical practice in Birmingham and his clinic in Stafford—plus the responsibilities of a growing family and the public service that is the lot of a prominent member of a respected profession—Withering spent close to ten years studying digitalis therapy in dropsy. Chemistry was not sufficiently advanced to permit the isolation of the active ingredients: biology in general and human physiology in particular were just in their infancy. Questions included which part of the plant was most active (the leaves); could the leaves be dried (they could); what was the best solvent for the active material (alcohol was good, but the extract had undesirable side effects—cold water was best); should one pick leaves in early or late summer (mature leaves from two-year old plants were the most active); and, most importantly, what was the optimum dose and how frequently should it be administered. In 1785, he published *An Account of the Foxglove and Some of Its Medicinal Uses*, a clinical study so detailed and so accurate that if one had no other information than is contained in the Account, one could use digitalis effectively and safely:

> Let the medicine therefore be given in the doses and at the intervals mentioned above; let it be continued until it either acts on the kidneys, the stomach, the pulse or the bowels; let it be stopped upon the first appearance of any of these effects, and I will maintain that the patient will not suffer from its exhibition, nor the practicioner disappointed in any reasonable expectation.

The report was something of a bombshell in England where the standard treatment for dropsy was to puncture the water-logged tissues with a scalpel (unsterilized) and stretch the patient over bedsprings to allow the fluid to drip into buckets. Withering was elected to the Linnaean Society and the Royal Society of London, received honorary degrees in Germany and France, and had a genus of tropical American plants, *Witheringia*, named in his honor. His medical practice grew apace, but he found time to botanize and even discovered a new crystalline form of barium carbonate which was named witherite.

Nevertheless, foxglove treatment of dropsy was still viewed with suspicion since it had a history of association with old herb women, and many doctors were, following the fifteenth-century herbalists, using it for king's evil and epilepsy. If, they said, it could cure dropsy, it should certainly cure these other diseases, and when it couldn't, it was clearly useless. Furthermore, if a little was good, a lot is better, and when some patients died of overdose, the fault clearly lay with Withering. Yet in a time when concepts of witchcraft and of divine punishment in the form of disease were still extant, when the doctrine of signatures was still believed, when the relationship between cause and effect was unclear, and when the idea of a controlled experiment was scarcely understood, a paragraph in Withering's *Account* is worth reading:

> As the more obvious and sensible properties of plants such as colour and taste and smell have but little connexion with the diseases they are adopted to cure; so their peculiar qualities have no certain dependence upon external configuration. Their virtues must be learnt either from observing their effects upon insects and quadrupeds; from analogy deduced from the already known powers of some of their congeners, or from the empirical usages and experiences of the populace. This last lies within the reach of every one who is open to information, regardless of the source from whence it springs. It was a circumstance of this kind which first fixed my attention on the foxglove.

William Withering died of tuberculosis in 1799 and was buried in a vault on which a *Digitalis* plant was engraved.

Although he noted that digitalis "has a power over the motion of the heart, to a degree yet unobserved in any other medicine, and that this power may be converted to salutary ends," there is no evidence that he related dropsy to cardiac insufficiency. With the discovery of the stethoscope by René Laennec in 1820 and the development of medical pharmacology in the middle of the nineteenth century, the effects of digitalis on the heart became clearer, although much still remains to be studied. Indeed, we now know that digitalis is not a single compound, but a mixture of four or five cardiac glucosides (Figure 5.17) with inter-

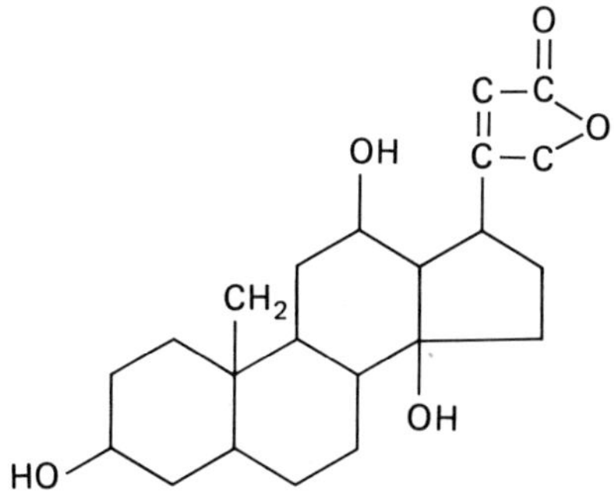

Figure 5.17
Chemical formula for digoxigenin.

Digoxigenin

acting effects on the heart. Recent clinical studies have shown its effectiveness in treating glaucoma (increased fluid in the eyeball), in neuralgia, asthma, and several other physiological diseases.

Digitalis purpurea arrived in North America after the Revolutionary War when Withering sent seeds to Hall Jackson of Portsmouth, New Hampshire, in 1787 to be cultivated in Jackson's medical garden. Escapes from this initial planting and subsequent introductions of ornamental plants have distributed the plant throughout New England, Oregon, Washington, Pennsylvania, and New York; a few hundred acres supply the medical needs of the world.

ADDITIONAL READINGS

Dowling, H. F. *Medicines for Man.* New York: Knopf, 1970.
Estes, J. W., and P. D. White. "William Withering and the Purple Foxglove." *Scientific American* 212(6)(1965):110–119.
Fisch, C., and B. Surawicz. *Digitalis.* New York: Grune & Stratton, 1969.
Gordon, B. L. *The Romance of Medicine.* Philadelphia: Davis, 1944.
Kranz, J. C., Jr. *Historical Medical Classics Involving New Drugs.* Baltimore: Williams & Wilkins, 1974.
Marks, G., and W. K. Beatty. *The Medical Garden.* New York: Scribner, 1971.
Trease, G. *Pharmacy in History.* London: Balliere, Tindal & Cox, 1964.

CHAPTER 6
Economics and the Politics of Food

Supermarket Botany

Mouth watering Garden Delights to brighten summer meals!
AD IN *Burlington (Vt.) Free Press,* 1977

Just as surgeons block out the operating room when carving a roast, so a botanist with a shopping list pays little attention to plant science when pushing a cart in a supermarket. Yet, items on the shelves can form the basis for an entire botanical education. Spice racks contain leaves, stems, flowers, fruit, and seeds of many species. Beverage sections contain coffee and cacoa seeds, tea leaves, and extractives from all parts of plants. Toothpaste contains skeletons of diatomaceous algae, and ice cream is made with emulsifiers from marine algae. Many plastic household items contain plants, and, when one has gone through the checkout counter, the groceries are packed in paper bags.

Our primary food plants are cereal grains, all but one members of the grass (Gramineae) family. Wheat, corn, rice, oats, and rye are eaten in many forms including breakfast foods, baked goods, as starch additives, and as food for the domestic animals that we consume; a hamburger is mostly plant products. All cereal grains are single-seeded fruits called caryopses, characterized by starchy endosperm tissues which serve to nourish the young embryo and seedling. Barley, rye, wheat, corn, and rice endosperm are fermented into beer and spirits. Several grasses are sugar sources; cane sugar from *Saccharum officinarum* stems, molasses from sorghum (*Sorghum vulgare*) stems, and liquid sugars from hydrolysates of the starch of corn. The only basic "cereal" grain which is not a grass is the one-seeded fruit, an achene, of buckwheat (*Fagopyrum esculentum*), a member of the same family (Polygonaceae) as rhubarb. Most buckwheat in North America is used as cattle feed, but some finds its way into pancakes; the groats and kasha of eastern Europe are milled buckwheat.

LEGUMES

Seeds are important items in our diet. Sunflower, sesame, and cotton seeds are used for their edible oils and in cooking as well as snacking. The world's most important seeds are from the pulse or legume (Leguminosae) family, a huge taxon with close to 500 genera and 11,000 species. Broad beans (*Vicia*), cowpeas (*Vigna*), chickpeas or garbanzos (*Cicer*), lentils (*Lens*), garden peas (*Pisum*), soybeans (*Glycine*), and several other genera are all Old World plants, originating in Asia, Africa, or Europe. Mid-America's gifts included the beans (*Phaseolus*) including lima, scarlet runner, black, brown, speckled, black-eyed, and kidney beans. Long before Western man invaded this hemisphere, cultivars of *Phaseolus* had been spread by Indians throughout the hemisphere; beans, corn, and squash were the dietary staples. As food, legumes are excellent sources of carbohydrate and protein found in cells of the two massive cotyledons which enclose the embryo. Their strong, impermeable seed coats allow legumes to store well; fully edible (although not viable) beans have been collected at archeological sites dated before the tenth century. In most legumes, total protein levels approach those of meat, but their amino acid balance is less adequate. Edible oils of peanuts and soybeans exceed 40 percent by weight of the seeds, making these plants important industrial sources as well as nutritious food. Carob beans (a chocolate substitute), tonka beans (a vanilla substitute), and licorice root are food flavorings. A few legumes—jequerity bean, several lupines, and senna—are found in patent medicines sold in drug sections of supermarkets. Many supermarkets now cater to the foreign-food gourmet; mung bean sprouts, soybean sauce, miso for Japanese soups, and Chinese bean pastes can be found next to the chow mein noodles.

The peanut, too, is a mid-American plant (Figure 6.1). It is unique in the family since after fertilization the flower stalk elongates, bends down, and pushes the ovary into the ground; maturation of the fruit and contained seeds occur only underground—hence the European name, groundnut. It is not a nut, but a true pod composed of the dry pericarp of the ovary. The brown-to-red seed coats are the testa. Most legume fruits are dry at maturity, but immature pods are edible, for example, Chinese snow peas, string and snap beans, and the yam bean (*Pachyrhizus erosus*). Although a Brazilian native, peanuts were introduced to Portugal in the sixteenth century and did not reach North America until the seventeenth century when African slaves brought

Figure 6.1
Peanuts, *Arachnis hypogaea.*
Courtesy U.S. Department of
Agriculture.

seeds to Georgia and the Carolinas. Some of the crop is smeared on bread, fed to circus elephants, and nibbled with beer, but even more is used for industrial and salad oils and for animal food.

ℰ AN ESSAY ON SELECTION

During the millennia since human beings took to farming, they have domesticated for their food fewer than 200 of the several hundred thousand species of flowering plants. Of these, some have been made so specialized that they cannot perpetuate themselves in the wild. Corn (Chapter 2) is certainly one of these, and sugar cane may be another. Most food plants can, however, be traced back to wild species which are still extant, and the domesticated plants can be crossed with their wild ancestors. Taking literally the biblical injunction that man is to have dominion over the life of the earth, people have grown wild plants and have selected those individuals possessing physical and chemical characteristics that they wanted. The plants selected were generally the result of chance alterations—mutations—which affected one or more visually apparent (phenotypic) characters. Color, taste, growth rates, size, and development of specific tissues or organs have each been, consciously or unconsciously, used as criteria for these selections. We have done the same thing with domestic animals. The breeds of dogs, derived from a few species of wild canines, now present an almost bewildering array of size, shape, and attributes which, if breeding is not left to casual encounter, can be perpetuated.

Perpetuity of uniformity is frequently easier to achieve in plants than in animals. Many plants are self-fertile. Others are cross-pollinated, but are typically grown in monocultural associations in large areas so that fertilization is with male generative nuclei the genetic potential of which is almost identical to that in the female nuclei. Still others are replicated by vegetative propagation which ensures genetic uniformity. We use "seed potatoes," stem sections of sugar cane, and graftings to start food plants. Our cereals and vegetables are grown from seed the production of which is carefully controlled to ensure genetic and phenotypic uniformity.

Obtaining desired variations in plants is, however, less easily achieved. Deliberate or inadvertent cross-breeding has resulted in desirable new cultivars. Hybrid corn and Green Revolution cereals (Chapter 2) are outstanding examples of this, but they required considerable sophisticated skills and many years of hard work. Occasional chance mutations have been exploited. Differences in the patterns of spots on bean seed are not related to nutritional value, but Indian tribes in Meso-America selected beans with culturally desirable patterns and perpetuated them to the exclusion of other patterns. Ethnobotanists have been able to trace the social interactions among tribes by examining their beans. The chance mutation of one branch of one orange tree was recognized and perpetuated by grafting buds of that branch onto other trees; the mutant produces oranges without seeds—our navel orange. The red grapefruit was a chance mutation in Texas. The thousands of cultivars of apple, pear, and other fruit trees were obtained in the same way. By selection of mutants and their propagation, we have obtained new plants. They are not new species, for they can be crossed with the parent type and differ from the parent in relatively few genetic characters; most breeds of dogs are cross-fertile—just look around the neighborhood. By selection and reselection, it is possible to obtain plants which do not look much like their ancestral wild types. A doubled and redoubled American Beauty rose looks very little like *Rosa damascena* and a pomeranian lapdog doesn't look or act like a wolf. Yet, the Pom can mate with a wolf, and the multipetaled rose can cross with the wild species.

If we use one dictionary definition of a monster as an organism which differs markedly from the normal, and we define "normal" as the usual or the expected, man has deliberately created monstrous plants to serve their needs. We have bred and selected for giant seeds, fruits, flowers, and vegetative organs. We have selected plants for altered or unusual metabolic pathways favoring the accumulation of chemicals we want. We have selected and perpetuated cultivars of banana, grape, sugar cane, and other plants that do not produce viable seeds or never flower.

Nowhere is this selection and perpetuation of the monstrous more apparent than in the genus *Brassica* of the mustard (Cruciferae) family. The genus contains perennial, biennial, and annual herbs with close to 100 wild species found in north temperate parts of the world. The family is easily recognized by the characteristic flower, having four petals which form a cross (hence the family name) and six stamens of which two are short, and producing a special fruit, the silique. Although it is likely that the cultivated brassicas are derived from several wild species, a reasonable ancestor is the colewort (*B. oleracea*) a scrubby, perennial native to Europe and Asia (Figure 6.2). From colewort, we have selected and developed a wide array of common vegetables. One line resulted in kale (var. *gemmifera*), cauliflower, and broccoli (var. *botrytis*). A second line in the genus was selected for storage organs, giving us the kohlrabi (*B. caulorapa*), whose shortened stem enlarges into an aboveground tuberous vegetable, and the rutabaga (*B. napobrassica*), whose tuberous storage stem develops below ground. The turnip's (*B. rapa*) storage organ is also a tuber with the true root extending down from the swollen stem. It is questionable whether all the leafy vegetable members of the genus are derived from colewort, but there is little doubt that they are all closely related. Several species of mustard are found wild and are exploited for their edible leaves (*B. juncea*) and particularly for their seeds (*B. nigra and B. alba*) used as condiments. Pak-choi (*B. chinensis*) is an important leafy vegetable in the Orient as are the closely

Figure 6.2
Possible evolutionary relationships among species and
varieties in the genus
Brassica.

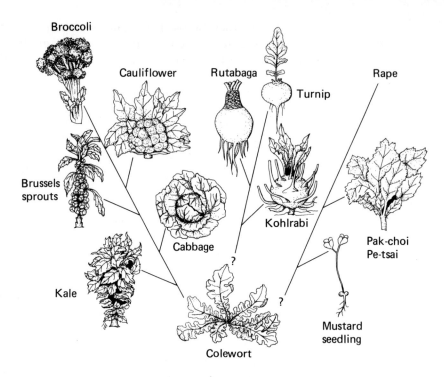

related petsai (*B. pekinensis*) and false pak-choi (*B. parachinensis*). Rape or colza (*B. napus*) is grown primarily for its seeds which yield an edible oil.

FRUITS AND VEGETABLES

When asked to contrast fruits and vegetables, botanists hem and haw, knowing that scientific definitions bear little relation to common usage of words (Figure 6.3). For botanists, a fruit is a ripened ovary, its contained seeds, and structures which mature with the ovary, but there is no definition for a vegetable. We think of vegetables as plant parts eaten during a meal and fruits as parts eaten for dessert. Since people have no trouble with these words when used in ordinary speech, it is botanically useful to group plants by the organ or structure eaten.

ROOTS

Roots anchor plants into the ground and serve to absorb and conduct water and minerals from the soil up to the stem of the plant. Many roots are also storage organs, maintaining supplies of photosynthate—starch and sugar—transferred down via the phloem (Chapter 1) to cells where they can be again mobilized for upward transmission via the xylem to the above-ground portion of the plant. Thus, roots may be

Figure 6.3
Bountiful harvest of vegetables. Courtesy U.S. Department of Agriculture.

good sources of carbohydrates, but only poor-to-fair sources of proteins, fats, and vitamins. With few exceptions, roots used for food are taproots; fibrous roots are rarely eaten. Familiar tap roots are those of carrots (*Daucus carota*) and parsnips (*Pastinaca sativa*) and celeriac (*Apium graveolens*) in the carrot (Umbelliferae) family, horseradish (*Armoracia rusticana*) and radish (*Raphanus sativus*) in the mustard (Cruciferae) family, and beets (*Beta vulgaris*) in the Chenopodiaceae. All of these plants are biennials, accumulating starch and sugar in their roots during the first year's grown and mobilizing the carbohydrates for energy needed the second year for flower and fruit production.

The sweet potato (*Ipomoea batatas*) in the morning-glory (Convolulaceae) family is frequently called a yam, but the true yam (*Dioscorea batatas*) is a member of the Dioscoraceae, a monocot family allied to lilies and amaryllis. The sweet potato is a New World plant, probably native to South America, while the yam originated in southeast Asia. Both are grossly enlarged roots of vines and are more nutritious than any of the other food roots, containing up to 5 percent protein and unusually high levels of iron, calcium, and minerals. The yellow carotenoids and vitamins A and D are fat-soluble substances which can be used by man. In tropical areas, there has been considerable selection resulting in dry, mealy, and gelatinous cultivars with white-to-almost-red tissues.

In the American tropics, and now in other tropical areas where the white potato cannot grow, the major noncereal carbohydrate is manioc or cassava (*Manihot esculenta*) of the poinsettia (Euphorbiaceae) family. Also called yuca, sagu or—more familiar to North Americans—tapioca, this Brazilian native is an easily cultivated root crop. Manioc roots look something like sweet potatoes, but grow to larger sizes. Cultivars called bitter manioc contain toxic concentrations of prussic acid (hydrocyanic acid) which must be squeezed from the shredded pulp before cooking. Others, called sweet manioc, are directly boiled or baked, much as we use sweet potatoes or white potatoes. Tapioca consists of dried manioc starch.

A few other roots may be found in the market. Chervil (*Chaerophyllum cerefolium*) and parsley (*Petroselinum crispum*) of the Umbelliferae family are used in Europe in soups, sugar beets supply sucrose, and root beer is flavored with the roots of *Smilax spp.* Sassafras teas consist of the bark of young roots of *Sassafras albidium*, a plant in the bay leaf (Lauraceae) family.

TUBERS, CORMS, AND RHIZOMES

One sure way to fail a botany examination is to classify the white potato as a root. There are three kinds of underground stems. A rhizome is a creeping, horizontal stem, growing parallel to the soil surface, usually a few centimeters below the ground. We eat a few rhizomes, notably the Jerusalem artichoke. Not from the Middle East but from North America and not an artichoke, the English name is a corruption of the Italian *girasole articiocco* or edible sunflower, an accurate name because the potatolike rhizome is from a sunflower (*Helianthus tuberosus*) in the aster (Compositae) family. The stored carbohydrate is not starch, but inulin, a polymer of the hexose sugar fructose. Although fructose, one of the two components of sucrose, is easily metabolized, humans lack enzymes capable of hydrolyzing inulin, and the Jerusalem artichoke is a poor food source. Its starchy taste is, however, appreciated by people on carbohydrate-restricted diets. The rhizomes of arrowroot (*Maranta arundinacea*) are ground into a flour to make biscuits consumed with apparent delight by young children. Two members of the ginger (Zingiberaceae) family are important rhizome crops. True ginger (*Zingiber officinalis*) is used in oriental cooking (Figure 6.4) and is dried

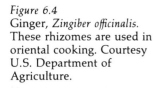
Figure 6.4
Ginger, *Zingiber officinalis.* These rhizomes are used in oriental cooking. Courtesy U.S. Department of Agriculture.

and ground into the major spice in gingerbread. Turmeric is the major yellow component in curry powders and chutneys and was an important natural dyestuff for cloth.

When gardeners speak of planting crocus bulbs, they are really planting corms, the enlarged, fleshy, bulblike stem bases of plants. Few corms are seen in North American markets. Water chestnuts (*Elochoris dulcis*) in the sedge (Cyperaceae) family are rarely sold in their fresh— and infinitely more tasty—form and the canned product is used. The famous poi of Hawaii is a paste made from the corms of the taro or dasheen (*Colocasia esculenta*) plant. Some corms, those of arum (*Peltandra virginiana*), Jack-in-the-pulpit (*Arisaema triphyllum*), and skunk cabbage (*Symplocarpus foetidus*), all in the arum (Araceae) family, are gathered by "stalkers of the wild asparagus."

Tubers are defined as thickened, subterranean stems possessing a number of undeveloped buds or eyes. The white potato is the primary example. Tubers develop at the ends of long, thin rhizomes which branch from the main stem of the plant. A close look will show that the potato tuber is distinctly stemlike. The eyes are in a spiral as are the buds of the above-ground stem. The internal arrangement of tissue systems is that of a stem and not that of a root.

AERIAL STEMS

Few above-ground stems are used as food. Three are usually found in supermarkets. Asparagus *(Asparagus officinalis)* in the lily (Liliaceae) family originated in the Mediterranean basin and is now grown wherever gourmets are to be found (Figure 6.5). The plant is a perennial with the stems arising from rhizomes. The scales on the stalks are true

Figure 6.5
Asparagus served with hollandaise sauce. Courtesy U.S. Department of Agriculture.

leaves, but only when too old and tough to eat does the plant produce
the fernlike cladophylls—actually stem branches—which are used dec-
oratively. The special taste of asparagus is due to asparagine, a nitrogen-containing compound related to amino acids. Canned white asparagus is produced by mounding earth around the emerging spears to
blanch and etiolate them, and, as is usual with etiolated tissues, they are
very tender and succulent—as well as expensive and insipid. A stop at
the gourmet section will show cans of bamboo shoots (Figure 6.6). Depending on the species, this grass may have a woody stem which may
grow to more than 30 m. Young shoots arising from rhizomes of several
genera (*Bambusa, Phyllostachys, Guadua*) are collected, the outer scales
removed, and the inner tissues used for Chinese stir-fry dishes, for
boiling, and for baking. Bamboo shoots are not particularly nutritious,
but they enhance the character of other plants and meats in oriental
dishes. Several stem tissues are gathered for salads including twigs of
Portulaca, chervil (*Anthriscus cerefolium*), pokeweed (*Phytolacca americana*)
and roselle (*Hibiscus saddariffa*). Other stems are used as herbs and
spices: cinnamon bark, angostura (*Cusparia febrifuga*) bitters, and dill-
weed (*Anethum graveolens*) will be found in the spice section down the
next aisle.

PETIOLES

The structure connecting the leaf blade to the stem is called the
petiole. It contains vascular tissues for transmission of materials to and
from the leaf, and it also serves to maintain the position of the leaf relative to the sun. Although many petioles are eaten as part of leaves, only
three have been developed for food. Two of these, celery (*Apium graveolens*) in the Umbelliferae and rhubarb (*Rheum rhaponticum*) in the
buckwheat (Polygonaceae) family, are common foods. Celery is a

Mediterranean plant, but it may also have evolved independently in the Himalaya Mountains. Until the fourteenth century, it was grown solely for its seeds which were used in medicine and as a food spice. Selection and breeding has been directed toward production of a long, tender, juicy petiole which has culminated in the Pascal cultivar which dominates the market. Before Pascal, celery was bleached and etiolated to give the white-to-pale yellow-hearts of celery. The ribs and strings on the convex side of the stalk are not the vascular bundles (these are small dots inside the stalk), but are bundles of strengthening cells called collenchyma which have thickenings of cellulose in each corner of the cell to aid in supporting the elongated petiole. In celery, the base of the petiole is enlarged where it joins the short, condensed stem. When this anatomical feature is carried to extremes, the enlarged clasping petiolar bases form a bulblike structure, as seen in fennel (*Foeniculum vulgare*), a member of a closely related genus.

Fennel seeds, like those of celery, cumin (*Cuminum cymunum*), coriander (*Coriandrum sativum*), dill (*Anethum graveolens*), and chervil (*Chaerophyllum bulbosum*) are umbelliferous culinary herbs, but one fennel cultivar, known as Florence fennel or *finocchio*, has been selected for large petiolar bases and is eaten raw in salads or boiled.

Rhubarb originated in Asia, and one species was grown in China for its rhizome which contains a gentle laxative. Dried rhubarb rhizomes were as valued in Renaissance Europe as were the fabled spices of the Orient. Only during the seventeenth century was the petiole found to be edible and, at the same time, the leaves shown to be poisonous. The toxic materials include oxalic acid, plus other poorly characterized compounds. The astringent taste of rhubarb sauce or rhubarb pie is due to the high acid content of the stalks, and a pinch of baking soda (Sodium bicarbonate) added to the cooking water will reduce this tart taste without destroying the flavor of the rhubarb.

Although formerly restricted to the northeast part of North America, one can now purchase canned fiddlehead fern, young fronds of the ostrich fern (*Pteretis pensylvanica*) (Figure 6.7). Tasting vaguely like asparagus, these young petioles plus immature leaves are boiled and served with butter and salt.

Figure 6.7
Young fiddlehead fronds of ostrich fern, *Pteretis pensylvanica*. Courtesy U.S. Department of Agriculture.

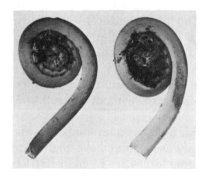

APICAL MERISTEMS

As discussed earlier (Chapter 1), the stem tip is composed of an apical meristem with its young leaves. The most obvious food apical meristem is the cabbage head, whose anatomy is a perfect example of an apical meristem. Brussel sprouts are undeveloped axillary buds, but their anatomy is also that of an apical meristem. Head lettuce, the most popular salad plant in North America, is *Lactuca sativa*, a member of the aster (Compositae) family. The plant is a biennial, native to Eurasia. It is depicted in Egyptian tombs dated to 4500 B.C., was grown in China by 800 B.C., and was eaten in Japan by 700 B.C., although it was not cultivated in Europe until the Middle Ages. The early cultivars were the leaf-types, with crisphead cultivars selected out by the seventeenth century. It is the heading types which are apical meristems surrounded by enlarged, overlapping leaves.

LEAVES

Many leaves are used as herbs, and these represent a number of botanical families. The mint (Labiatae) family is especially well represented by oregano, marjoram, basil, thyme, and the savorys. Water cress, curly cress, and winter cress are all in the mustard family; as are tarragon and wormwood, used as herbs in wines and vinegars. Bay leaves (*Laurel nobilis*), the universal garnish parsley, and several herb teas are made from leaves. Vegetables include members of the cabbage group, spinach (*Spinacia oleracea*) in the goosefoot (Chenopodiaceae) family, and the endive (*Cichorium intybus*) in the aster family whose roots, as chicory, are used to flavor Louisiana-style coffee.

Members of the lily (Liliaceae) family are among our most common vegetables. Onions, leeks, shallots, and chives (*Allium spp.*) are all true bulbs, defined as shortened, usually underground stems with fleshy overlapping leaves (Figure 6.8). The relation of a bulb to an apical meristem is seen upon dissection; the younger leaves are higher up on the compressed stem than the older leaves. In leeks, chives, and green onions (scallions), the edible portion is the fleshy bases of the strap-shaped leaves which form a cylinder. Garlic, cloves and shallots consist of closely attached small bulbs or bulbils, each composed of two leaves. The thin cylindrical leaf is surrounded and protected by a surrounding, larger, and fleshier leaf, the whole surrounded by a papery epidermal layer. The characteristic odor of *Allium* bulbs is due to an organic sulfide, allicin, which is moved in the blood stream to the lungs where it is exhaled—to the annoyance of one's neighbors—but how anyone can cook without the full array of *Allium* species is beyond comprehension.

With high concentrations of water-soluble B vitamins and minerals, leafy vegetables have high nutritional value and outstanding tastes. Some leaves are eaten in restricted parts of North America and

Figure 6.8
Onions, scallions, garlic, and shallots. Courtesy U.S. Department of Agriculture.

are either ignored or made fun of in other regions. People in the southern United States esteem collard, turnip, beet (Swiss chard), and mustard greens, usually boiled with bits of bacon or fatback, and the "potlikker" drunk as a soup. Young dandelion (*Taraxacum officinale*) in the aster family is esteemed in the Northeast; nettle (*Urtica spp.*), amaranth or Chinese spinach (*Amaranthus spp.*), and seakale (*Crambe spp.*) were introduced by immigrants from eastern Europe to the prairie states and provinces.

FLOWERS AND THEIR PARTS

Entire flowers are used as delicate flavorings; jasmine tea does contain jasmine blossoms, and the Chinese flavored orange pekoe tea with orange blossoms. The unusual crystallized violets made in France are whole flowers soaked in a sugar solution which is then allowed to dry and crystallize. Some flowers are used as spices; the clove is a flower bud as is the caper (*Capparus spinosa*) of the aster family. The most expensive flavoring is saffron, the stamens of *Crocus sativus*. Flowers of a chrysanthemum, *Pyrethrum*, are collected in Japan and East Africa for their ovaries which contain an effective insecticide. Cauliflowers (Figure 6.9) and broccoli are the entire inflorescence of the plant, and one eats both the flowers and their fleshy peduncles or flower stalks. The globe artichoke is the floral bud of *Cynara scolymus*, a thistle in the aster family (Figure 6.10). The edible portion is the bracts surrounding the floral receptacle, plus the receptacle itself. Definitely an acquired taste, the bracts are dipped into melted butter or mayon-

Figure 6.9
Cauliflower, *Brassica oleracea*
var. *botryis,* is a cabbage with
style. Courtesy U.S. Department of Agriculture.

Figure 6.10
Globe artichoke, *Cynara scolymus.* Courtesy U.S. Department of Agriculture.

naise and their fleshy bases eaten. The inedible choke consists of the
true floral parts and is anatomically similar to a sunflower head, but the
fleshy floral receptacle is considered to be a delicacy, frequently served
as artichoke "hearts."

FRUITS

The simple definition of a fruit given at the beginning of this section did not attempt to cover the structural variety of fruits that are
found in the produce section (Figure 6.11). A fruit, botanically, is
usually the ovary wall and its contents. The wall is the pericarp which
has three cell layers—the outer exocarp, a middle mesocarp, and an
inner endocarp. The base of the flower, the receptacle or torus, frequently contributes to the fruit and its contents. Seeds consist of an
embryo, food storage tissues, and the seed coats, the testa. For convenience and to define the developmental processes by which fruits assume their mature forms, botanists have attempted to classify the
almost infinitely broad range of fruiting structures. For our purposes,
the excellent classification scheme developed by Dr. Arthur Cronquist
will be used.

Figure 6.11
Bountiful harvest of fruits.
Courtesy U.S. Department of
Agriculture.

Banana

Grape

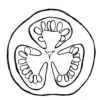

Tomato

Figure 6.12
Berries are fruits enclosing
one or more seeds. The epicarp layer forms the fruit
"skin," and both mesocarp
and endocarp layers develop
into the edible flesh.

Figure 6.13
Drupes are fruits with a
stone. The epicarp forms the
fruit "skin" with the edible
flesh developed from the mesocarp. The endocarp forms
the woody case surrounding
the seed. The coconut is a
drupe with woody carpic
tissues.

I. *Simple Fruits:* fruit derived from a single pistil or ovary
  A. Fleshy Fruits: fruits in which the pericarp cells remain alive
    1. Berries: pericarp fleshy, occasionally leathery, one or
    more carpels and seeds
      a. Berry: exocarp forms skin (banana, members of to-
      mato family, date, grape) (Figure 6.12)
      b. Pepo: rind from receptacle (all members of cucurbit
      family)
      c. Hesperidium: rind is exocarp + mesocarp, juicy tis-
      sue is endocarp (citrus fruits)
    2. Drupe: fleshy fruit with stone; epicarp forms skin,
    mesocarp forms flesh, and endocarp forms woody stone
    surrounding seed (peach, apricots, olive, coconut) (Fig-
    ure 6.13)

Coconut

Olive

Peach

Figure 6.14
A buckwheat, *Fagopyrum esculentum*, achene. The fused carpic layers form the seed case enclosing the curved embryo surrounded by starchy endosperm.

3. Pome: edible flesh from enlarged receptacle, core of pericarp (apples, pears)
B. Dry Fruits: pericarp cells dead at maturity
  1. Dehiscent fruits which open at seams at maturity
    a. Legume: fruit opens along 2 seams (peas, beans)
    b. Follicle: fruit opens along 1 seam (star anise, milkweed, kolanut)
    c. Capsule: several carpels, several seams (okra, vanilla bean)
    d. Silique: long, narrow capsule bearing seeds on margins (mustard seed)
  2. Indehiscent fruits which remain closed at maturity
    a. Achene: one-seeded fruit, pericarp dry and free from seed (buckwheat) (Figure 6.14)
    b. Caryopsis: one-seeded fruit, pericarp dry and adhering to seed (cereal grains)
    c. Nut: like achene with woody pericarp (hazelnut, acorn)
  II. *Aggregate Fruits:* fruit derived from fused ovaries and receptacles
A. Etaerio of drupes (blackberry, raspberry)
B. Etaerio of achenes: enlarged, fleshy receptacle bearing achenes (strawberry) (Figure 6.15)
C. Sorosis: composite fleshy fruit composed of spike (core) + floral parts (pineapple)
D. Syconium: fleshy fruit composed of a swollen, hollow receptacle containing many flowers (figs).

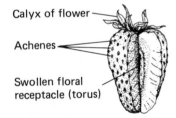

Calyx of flower

Achenes

Swollen floral receptacle (torus)

Figure 6.15
Strawberry, *Fragraria spp.* The edible "berry" is a false, accessory fruit composed of the enlarged floral receptacle; the "seeds" are achenes.

Some fruits can be included in this classification with difficulty. The pomegranate is a berry with a leathery pericarp; the edible pulp surrounding each seed is a fleshy, juicy testa or seed coat. Rose hips, used for jam and an excellent source of vitamin C, are etaerios of achenes enclosed within a fleshy receptacle.

    Considerable confusion exists among those fruits called nuts (Figure 6.16). The peanut seed is, of course, a bean. Several so-called nuts are botanical drupes. Almonds, cashews, coconuts, and macadamia nuts are the stones of the drupe usually sold after the dry pericarps

Figure 6.16
A selection of "nuts," most
of which are not true nuts.
Courtesy of U.S. Department
of Agriculture.

have been removed. The Brazil nut fruit is a capsule containing seeds whose testa forms the difficult-to-remove coat. Chestnuts, hazelnuts, and the oriental Litchinut are true nuts, but in litchi, the delicious flesh is an aril, a special layer of cells surrounding the seed and is botanically comparable to the aril of nutmeg which we use as mace or the edible flesh surrounding a pomegranate seed.

ADDITIONAL READINGS

Bailey, L. H. *Manual of Cultivated Plants.* New York: Macmillan, 1944
Brouk, B. *Plants Consumed by Man.* New York: Academic Press, 1975.
Cronquist, A. *Introductory Botany,* 2nd ed. New York: Harper & Row., 1971.
Lovelock, Y. *The Vegetable Book: An Unnatural History.* New York: St. Martin's Press, 1972.
Robbins, W. W. *The Botany of Crop Plants.* Philadelphia: P. Blakiston & Company, 1917.
Sargent, F. L. *Plants and Their Uses.* New York: Holt, Rinehart and Winston, 1913.
Schery, R. W. *Plants for Man,* 2nd ed. Englewood Cliffs, N.J.: Prentice-Hall, 1972.
Vaughan, J. G. A multidisciplinary study of the taxonomy and origin of *Brassica* crops. *BioScience.* 27(1977):35–40.
Verrill, A. H. *Foods America Gave the World.* New York: Farrar, Strauss & Giroux, 1937.

Sugar

Sugar and spice and everything nice,
That's what little girls are made of.
NURSERY RHYME

Although people talk about a "sweet tooth," it is small enervated buds on the tongue that perceive the five basic tastes: sweet, salty, acid, bitter, and metallic. A child wisely licks a peppermint stick, knowing that the sweet-taste buds are at the tip of the tongue. Sugars are not the only compounds that activate these buds, but sugars are the chemicals that give sweetness to most foods. All sugars are composed of carbon atoms plus hydrogen and oxygen atoms in the two-to-one ratio found in water, hence the generic name carbohydrates. Familiar carbohydrates are sugars with four and more carbon atoms; the most abundant are two monosaccharide sugars with six carbons each (glucose and fructose), the 12-carbon disaccharide of glucose and fructose (sucrose or table sugar), and two macromolecular polymers of glucose (cellulose and starch). Since there are four Calories per gram of sucrose (18 Calories per standard teaspoon), and its chemical energy is immediately available for body processes, sugars are excellent sources of energy; indeed, they are the primary energy sources for all living cells. The starches, long chains of glucose molecules containing several hundred or even thousands of hexose molecules, must be converted by starch-hydrolyzing enzymes into the monosaccharide glucose or its disaccharide, maltose, before they can be metabolized. If sugars are eaten in excess of daily energy requirements, the animal body converts them into storage fats, but sugars per se are not any more fattening than are other foodstuffs eaten in excess.

HONEY

Until the eleventh century, the sweet tooth of most of the world's people was satisfied with honey, supplemented in North America by maple syrup. Honey is a bee-concentrated nectar of flowers—Aristotle called it dew distilled from the stars and rainbows—containing a mixture of mono- and polysaccharide sugars plus nutritionally insignificant amounts of amino acids, minerals, and vitamins plus small amounts of ill-characterized compounds that distinguish sage or apple honey from that of thyme or clover honey. As evidenced by "a land flowing with milk and honey" from Exod. 12:8, it was highly prized and was a symbol of a peaceful, plentiful agricultural community. Bees were kept in Europe, the Americas, Africa, and Asia and the honey used directly as sweetener or fermented into sweet-tasting mead or *methgelyn* beers. Today, honey production is a moderately sized industry with 300,000 tons produced each year.

MAPLE SYRUP

No book by a Vermont resident could omit mention of maple syrup and maple sugar; one could be exiled for this sin of omission (Figure 6.17). In early spring, on a bright, sunny day after a cold night, the sun warms the dark bark and young wood of the sugar maple tree (*Acer saccharum*), permitting amylases to convert starches in ray cells into soluble sugars which move up the tree, providing the energy source for the eventual bursting of winter buds. The American Indian learned to subvert sap movement by gashing the trunk and collecting

Figure 6.17
Collecting sap from sugar
maple trees, *Acer saccharum.*
Courtesy U.S. Department of
Agriculture.

the clear sap which contains up to eight percent sucrose. By dropping
hot rocks into a birchbark container of sap, the sugar was concentrated
and carmelized to produce a thick brown syrup with close to 68 percent
sugar content. Additional concentration resulted in the formation of
coarse brown crystals of maple sugar. *Sisibaskwat*—"drawn from
wood"—was eaten directly and used to season meats and vegetables.
Some was traded to less fortunate tribes and, eventually, to white settlers. Colonists improved on the Indian's methods by boring holes in
the trees, inserting spouts of elderberry twigs, and boiling down the sap
in kettles. As a rule of thumb, it takes close to 40 gallons of sap to produce one gallon of syrup. Typical maple taste and odor develops only
when the sap is concentrated by heat. The mystique of wooden sap
buckets, ox-drawn sledges moving sap down a hill, and steamy sugar
houses warm from rock maple fires under boiling sap caldrons has
been lost. Today, electric pumps pull the sap from the trees through
blue plastic tubing directly to stainless steel, oil-fired evaporating pans
from which syrup flows into plastic bottles whose labels bear pictures
of nineteenth-century sugaring operations.

The grade, and hence the price, is determined by color, with light
amber syrups with delicate maple flavors commanding more interest
than do darker, more caramelized syrups with more pronounced flavors. One tradition, still observed, is the sugar-on-snow party at the
end of the season, when hot maple syrup is poured on fresh snow to be
sopped up with raised, unglazed doughnuts; the cloyingly sweet taste is
cut by bites of dill pickle. Most commercial maple syrups are com-

pounded of cane or corn syrup to which a small volume of real maple syrup is added for flavoring, but even this token can be eliminated by adding an extract of funegreek seeds (*Trigonella foenumgraecum*), a leguminous plant whose essential oils closely mimic the taste of maple syrup.

SUGAR CANE

The bulk of the world's sugar is obtained from two plants, the sugar beet (*Beta vulgaris*) and the sugar cane (*Saccharum officinarum*). Of the two, sugar cane accounts for about 60 percent of the total supply. *S. officinarum* is probably a man-made hybrid of *S. spontaneum, S. robustum, S. sinense,* and *S. barbari.* All have been in cultivation for so long that we can say only that sugar cane originated somewhere in the South Pacific, possibly in New Guinea, and that it was carried to the Asiatic mainland by humans. No mention of cane appears in the *Rg-Veda* of 1000 B.C., but it was noted in other Indian writings of 900 B.C. Sugar cane is a tall member of the grass family (Gramineae) capable of reaching 8 m in height with a stalk diameter of 10 cm. The stalk has a pithy center in which the sugary cell sap accumulates in cell vacuoles. At each stalk joint or node, broad leaves form. Seeds are produced in tassels and are sparingly fertile, but the plant is so genetically mixed (heterozygous) that stand uniformity can be assured only by vegetative propagation from sections of the stalk which, when planted, sprout from buds at each joint (axillary buds). Modern cultivars, so-called "noble canes," have been genetically engineered for optimum yields under varied conditions of cultivation, although a major restriction is a requirement for tropical or subtropical climates with abundant rainfall or modern irrigation. The plant matures in 12–15 months, but since new stalks, *ratoons*, develop from the rootstock, one planting may produce crops for a number of years. At harvest time, pith cells will contain close to 15 percent sucrose, all of it formed through the biochemical process of photosynthesis.

ℰ AN ESSAY ON PHOTOSYNTHESIS

Photosynthesis is the most efficient energy-converting process on earth; the transformation of the physical energy of light into the chemical energy bound into a sugar molecule can approach 30 percent. The over-all process can conveniently be divided into two phases, the first requiring light—the light reactions—and the second phase independent of light—the dark reactions which occur either in light or in darkness (Figure 6.18). Light, primarily red and blue wavelengths, is captured by the green chlorophyll molecules contained in the membrane system of the chloroplast (which is why leaves look green; the red and blue wave-

Figure 6.18
Photosynthesis as a chemical factory. In chloroplasts (the rectangular box), light energy is used by chlorophyll to split water to form molecular oxygen and hydrogen ions (H*). In a second set of light reactions, chlorophyll is excited and the excitation energy is used to link phosphate ions to form the high-energy compound ATP. Both the H* and the ATP are then used in the dark reactions to convert atmospheric carbon dioxide into carbohydrate. Reproduced from *The Garden Journal* of the New York Botanical Garden of October 1974 with permission of the author.

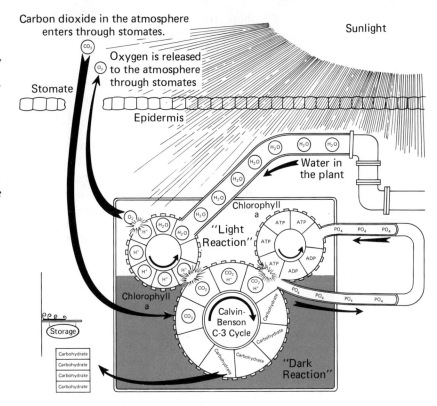

lengths of sunlight have been absorbed by chlorophylls, and the green wavelengths in white light are either transmitted through the leaf or reflected from its surface). Light absorption can also occur in other pigments in the chloroplasts, especially the orange carotenes and yellow xanthophylls, and the energy is transmitted to the chlorophylls. When radiant energy is absorbed by chlorophylls, electrons in the molecules are activated and shift to a higher energy state, roughly analogous to lifting a rock from the ground into the air. The electrons are held in the "excited state" for a fraction of a second and, like the rock suddenly released, can fall back down into the chlorophyll molecule with the excitation energy released. If, however, the electrons trickle through a series of compounds, much as water cascades through the paddles on a water wheel, the potential energy can be captured into chemical bonds of a compound, abbreviated as ATP (adenosine triphosphate), required for the photosynthetic dark reactions. In another set of light-driven reactions, excited electrons activate another compound, PN (pyridine nucleotide), which can hold the hydrogen ions also needed for the dark reactions. These hydrogen ions are obtained by a chlorophyll-mediated splitting of water with the oxygen from the water released from the plant as the molecular oxygen we and most other organisms must have—have you thanked a green plant today? The speed with which these reactions occur is measured in milliseconds, and, when each complete cycle is finished, the active chlorophyll molecules are again ready to absorb light and repeat the process. Since there are thousands of active, light-reaction centers in each granum of a chloroplast, thousands of grana in a chloroplast, thousands of chloroplasts in a leaf, and several dozen leaves on a sugar cane plant, the number of ATP and PN molecules produced per minute is numbered in the tens of millions per plant.

The raw material for formation of sugar through the dark reaction is the gaseous compound carbon dioxide. Composed of two oxygen atoms connected to a single carbon atom, CO_2 is normally present in air at the low concentration of 0.03 percent. During the day, when the "little mouths," the stomata, are open, CO_2 diffuses into the leaf cells and moves to chloroplasts. There it is attached to a carrier molecule, a 5-carbon compound called RuDP (ribulose diphosphate). The carbon atom in CO_2, being attached to two oxygen atoms, is in a fully oxidized state in which it cannot link up with other carbons to form the six-carbon chain of a simple hexose sugar such as glucose. To permit chain formation, not only must the oxidation level of carbon be reduced by shifting carbon-oxygen bonds, but hydrogen atoms must also be added. Both of these reactions are accomplished by the actions of the two compounds produced in the photosynthetic light reactions; the hydrogen atoms carried by PN are added to the carbon with required energy supplied by the ATP molecules. This "reductive fixation" of CO_2 into a carbohydrate is a cyclic process with CO_2 entering the chloroplastid, linking to the RuDP carrier, being reduced with energy from ATP, and then being released in solution into the cell with the carrier freed to accept another CO_2. The analogy with a factory taking in raw material, putting it on an assembly line, fabricating the end product, and shipping it to the consumer is quite apt.

There are, however, built-in inefficiencies in the photosynthesis factory. One is the low concentration of CO_2 in air. The efficiency of the light reactions is so great that there are excesses of both ATP and hydrogen-carrying PN formed relative to the CO_2 available. In some plants like the jade plant (*Crassula*) and the stonecrops (*Sedum*), stomata remain open at night with the incoming CO_2 held as an acid which can release the CO_2 again when the light reactions begin in the morning. In other plants, including corn, sorghum, bamboos, and sugar cane, the limitation of low CO_2 is overcome by a biochemical mechanism in which CO_2 is taken up by special cells and held until it can be used in the light reactions. Another saving is achieved by the repression of photorespiration, a process in which sugars are metabolized rapidly in light, but not in darkness. By increasing the amount of CO_2 available for photosynthetic sugar formation and by preventing the loss of sugar by photorespiration, more sugar is made. Some of this sugar is used as the energy source for the rapid growth of the sugar cane plant, some is used as units which are polymerized into cellulosic plant cell walls, some is used to synthesize amino acids and other compounds that comprise a plant cell, but so much sugar is formed that it accumulates in the vacuoles of the pith cells, available for human exploitation.

The exploitation of sugar cane started early in the history of agricultural development. By 850 B.C. it was a staple crop in India; the Manu code provided that any traveler could take two stalks from a cane field to gain strength for the journey. Joints of cane were chewed to get the sweet sap, and, by about 800 B.C., crude presses squeezed out the sap to be boiled down, producing a lumpy, dark-brown mass, much like dark grades of brown sugar. Small lumps were called *sakkara* (gravel) and larger lumps were called *khanda* (pebbles) from which we derived the word candy. Sherbets of dissolved sakkara and rose water were frozen for the rajah's table with snow carried hundreds of miles from the mountains; it was believed that this was excellent medicine to increase production of semen. China and Japan obtained cane by the third century A.D., and the Tang dynasty Emperor, Tsai Tsung, sent agents to India to learn the secrets of making sakkara. Nearchus, general to Alexander the Great, reported that on an expedition to the Indus Valley in 325 B.C. he saw "the sweet juice of a reed which is called sakkar." Dioscorides wrote in the first century A.D. of "a sort of hard honey which is called sakkarum found in canes or reed in India, which is grainy like salt and brittle between the teeth, but of sweet taste." Nevertheless, sugar cane did not reach the Middle East until 550 A.D. when Persians grew it in irrigated fields. When Persia was conquered by followers of Mohammed in the early seventh century, cane was introduced into Syria, the Mediterranean islands of Sicily and Cyprus and into Egypt. Dark-brown crystalline sugar was made in Egypt about 750 A.D., and small amounts found its way into Europe where it was a delightful but very expensive spice and medicine for coughs and stomach complaints. The crusaders saw, on the fields of Laocidia (western Syria), "certain reed-like plants . . . on account of its sweet juice we chewed against thirst," and when Jerusalem was captured and sacked in 1187, Arabian sugar was a valuable portion of the booty brought back to Europe.

By the middle of the twelfth century, there was sufficient demand for *sucra* to warrant the attempts of Venetian merchants to establish a trade center for the movement of sugar from North Africa into Europe and when, in about 1410, a group of merchants purchased a method for refining crude, brown sugar into a sparkling white product that could be molded into cones or loaves (Figure 6.19), a Venetian monopoly was established. In 1413, Venice sent 100,000 pounds of sugar to England in exchange for British wool. Prices were so high (about $3.50 an ounce) that few people could afford it; the common people ate honey and tasted sugar only when it was prescribed to disguise bitter herb medicines (Figure 6.20).

Figure 6.19
Molding sugar loaves. Taken from a fifteenth-century woodcut.

Figure 6.20
A sugar merchant of the fifteenth century. Taken from a contemporary woodcut.

Although cane was grown in Spain, the Moors so completely controlled its supply that Spanish sugar went first to Venice before being distributed to the rest of Europe. In 1420, Henry the Navigator, ruler of Portugal, obtained planting stock from Sicily and established a sugar plantation on the island of Madeira, on the Canary and Cape Verde Islands, and on St. Thomas Island off the African coast. Columbus brought cane to Hispaniola on his second voyage in 1493, and the first crop was harvested in 1509. The downfall of the Venetian monopoly was foreshadowed when the Portuguese established plantations in Brazil in the early sixteenth century. The conquest of tropical America was immediately followed by conversion of lush forests into sugar plantations—all of the West Indies became sugar-growing land. Impetus for expanding the sugar trade was the introduction of Arabian coffee, Chinese tea, and Aztec cocoa, all of which were more palatable with sugar. Catherine de Medici, queen of Henri II of France, introduced Florentine pastry-making and Venetian "sweetmeats" to the French court, and her contemporary, Elizabeth I of England, was so fond of sugar that her court is credited with inventing sweetened fruit pies. The addition of sugar to chocolate and the invention of marzipan, glazed fruits, and even crystallized violets hastened the development of the then crude art of dentistry.

SUGAR SLAVES

With a firm market, enterprising businessmen began to increase the volume of refined sugar produced each year. Spanish and Portuguese planters took seriously the admonition of Pope Nicholas V to expand their operations in order to "reduce to perpetual slavery the Saracens, pagans and other enemies of Christ." They grew enough sugar to eliminate the Arabian market. But in order to do so, they needed cheap labor. Sugar did not create slavery; it is sanctioned in the Testaments (Lev. 25:44–46) and the golden age of Greece was built on slave labor. African slaves were brought to Brazil by the Portuguese in 1519 because the native Indians were unwilling workers and often died when put to the lash. Spain followed in a few years, and Britain imported slaves into Barbados in 1630. By the end of the seventeenth century, good Catholics were wondering aloud why Dominican, Augustinian, and particularly the Jesuit orders dedicated to poverty should be accumulating wealth and power from the labor of slaves. True, the blacks were baptized, thus saving their souls, but their bodies were worn out within a few years. The Jesuits were expelled from France in 1767, partly because of their use of slave labor, while the Compagnie des Indies was shipping 300 vessels packed with slaves each year to plantations in the Antilles and to Jesuit-founded plantations in Louisiana. Because of black slaves, Spanish and French sugar outsold English sugar until almost the end of the eighteenth century. The tremendous traffic in slaves was justified according to most planters; a cost-accounting in 1700 demonstrated that self-perpetuation of slave populations was not economical—infant mortality was well over 50 percent and a new-born slave had to be fed and clothed and occupied the attention of its mother until the child was at least seven years old and capable of working in the cane fields. Slaves were brought to the sugar plantations until the early nineteenth century when, in revulsion, the world demanded that the slave trade be abolished.

With the demise of the slave trade, some indigenous slavery continued in the sugar islands, but the demand for cheap labor was unabated. In an effort to obtain needed field hands, Spain, Britain, and the United States engaged in what we now realize was the greatest unforced mass migration of peoples the world has ever seen. By act of Parliament in 1837, people from India were brought as indentured laborers to the West Indies. Close to 1.5 million people immigrated to the British sugar colonies. A quarter of a million Chinese were also brought in by English planters. The Spanish in Cuba introduced Chinese workers; the only long-term benefit was a cuisine that included Chinese-Cuban dishes. The Dutch imported Indians in 1800, and some Javanese and Indonesians in 1890.

HAWAIIAN SUGAR

Among the more remarkable mass-migration schemes was the one in Hawaii. Sugar cane existed on the islands for centuries; Captain James Cook ate cane in 1799 before he was killed. The first attempt to grow cane commercially was in 1803, but no market existed and the venture failed. At about the same time, a group of Calvinist missionaries left Boston to escape the "progressiveness" of the Congregational Church and settled in Hawaii to convert the heathen and build a new life. After successfully converting the rulers of the islands, they introduced heavy clothing (with drastic increases in tuberculosis) and Yankee concepts of land tenure. The bulk of the arable land was soon owned by missionary families. The Rev. Joseph Goodrich activated the sugar industry in 1829, and by 1840, the Rev. John Emerson, the Rev. Artemas Bishop, the Rev. Hiram Bingham, and several medical missionaries were in the sugar business. Abhoring slavery, the missionary monarchy decided to import labor (Figure 6.21), since "the native Hawaiian won't work and they can't be trusted." Chinese were imported in 1852 in large numbers, followed by Japanese, Portuguese, Koreans, and Philippinos. Caucasians were brought in as foremen and engineers, initiating the multicultural, polyglot society that we think of as typically Hawaiian. These successive waves of immigrations were, however, carefully orchestrated by the planter aristocracy to prevent any group from dominating the work force and demanding higher wages and

Figure 6.21
Major racial groups in Hawaii and the dates when immigration started.

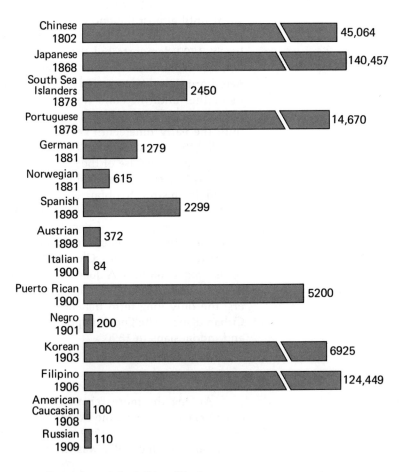

benefits. Small acreages are still leased on a year-to-year basis by the Bishop estate mostly to immigrant families, while the plight of the natives continues to worsen as they are excluded from virtually all aspects of the economic and political life of what was once their land.

SUGAR BEETS

Even before sugar cane was grown commercially in Hawaii or the Philippines, a new development in sugar occurred. In 1605, a Frenchman, Oliver de Serres, wrote: "The beetroot, when boiled, yields a juice similar to syrup of sugar," but, it was not until 1747 that the German chemist Andreas Margraff proved that the beet juice was chemically identical to sugar from the cane (Figure 6.22). Margraff's student, Franz Karl Archard, planted a field of beets under the patronage of Frederick the Great and, after a successful harvest, calculated that a good grade of sugar could be obtained for less than $0.06 per pound. Authorities in France and Germany were very interested. They were paying that much and sometimes more for cane sugar produced by their economic and sometimes political rivals, the English. British sugar merchants also became interested and tried, unsuccessfully, to buy Achard off with the equivalent of $120,000. King Frederick William III supported construction of the first sugar-beet factory in Silesia in 1800–1801. When news of this success reached France, Napleon ordered beet cultivation, establishment of schools to train sugar technicians, and large-scale production of sugar from beets to break England's tight blockade. Following the battle of Waterloo and the lifting of the blockade, the infant industry collapsed not only because cheap cane sugar again became available to France, but also because beet sugar was then a low-grade product with an unpleasant smell and French chefs com-

Figure 6.22
Mature sugar beet. Courtesy Holly Sugar Corp.

plained that it didn't "handle" the same way as cane sugar did. Germany persisted and, with its splendid agricultural schools and well-organized chemical industry, succeeded in preparing sucrose from beets with reasonable yields, low costs, and a quality only slightly below that of cane sugar. From 1870 until the start of World War II, Europe obtained a third of its sugar from Germany.

Unknown to Europeans, wild beets in California had, for centuries, supplied, in the words of Pedro Fages, a Spanish explorer of the 1760s, "a quantity of molasses, candy and sugar that is not unworthy. . . ." With the success of Germany and France in beet sugar production, it was only a matter of time until it was tried in the United States. Attempts were made in Massachusetts in 1838 to develop economically viable enterprises, but lack of technical skills and a series of untimely mischances caused these to fail. The first successful operation was in Alvarado, California, in 1869. By 1890, extensive plantings and efficient factories were in operation in California and in the intermountain areas of Colorado, Utah, and Nevada under the financial control of the Church of Latter Day Saints. Congress aided this infant industry by authorizing a two-cent-a-pound bounty on all sugar produced within the continental United States and by mobilizing the resources of the U.S. Department of Agriculture to develop beets with higher sugar content and resistance to diseases.

REFINING SUGAR

The technology of refining sugar has come a long way from the primitive boiling down of sap beaten out of lengths of cane. Under modern, intensive plantation practices, the mature sugar cane plants are mechanically topped, leaves removed mechanically or by setting the field on fire, and stalks harvested by machine or by the hot, dirty and hard job of slashing with machetes. Since sugar content decreases rapidly after cutting, harvested cane is rushed to mills where it is shredded and fed into serrated rollers capable of squeezing the juice out of 3000 tons of cane a day. The spent stalks, called *bagasse*, are used as a fuel for the mills, for paper, cattle food, and in the manufacture of plastics. Raw juice contains, in addition to sucrose, protein, fat, and other chemical constituents of plant cells, all of which must be removed. Lime and mild heat clarify the juice by precipitating these impurities. The clear juice is passed through filters under pressure, and concentrated by evaporation under heat and reduced pressure to produce a thick syrup containing sugar crystals and molasses, a dark brown mixture of caramelized sugars, and other substances. Massive centrifuges spin off the molasses (which is reworked to extract more sugar), and sucrose crystals are redissolved and recrystallized several times to produce refined sugar. Residual molasses can be used as cattle feed supplements, but much of it is used to produce rum.

Extraction of sucrose from sugar beets differs from that used for cane only in the procedure used to obtain the crude juice. Cells and tissues of the beet are tough and cannot easily be broken by crushing, so that it is necessary to shred the beets. The shreds are placed in tanks of hot water, allowing the sweet sap to diffuse from the cells. Clarification, crystallization, and refining steps are essentially identical to those used in the sugar cane industry. Beet molasses and spent beetroot pulp are used as cattle fodder. Modern refining of beet sucrose has resulted in a product chemically, physically, and nutritionally indistinguishable from cane sucrose.

Table sugar is the purest bulk chemical produced, and its grading is based solely on the size of the sugar crystals which ranges from large, rock-candy lumps to confectioners' XXXX or icing sugars which are as fine as flour. Brown sugars range from the almost white #1 grade through grade #15, a deep, brown-black product. The golden through medium browns, used in cooking, have increasing intensities of molasses flavor, but are not nutritionally superior to white sugars—they are sometimes made by adding molasses to white sugar. Liquid sugar syrups have high acceptance in baking and soft drink industries since the sugar is already in solution and also because they contain small amounts of glucose and fructose which repress sucrose crystallization. In jelly and jam processing, boiling acid fruits with sugar also hydrolyzes some sucrose into its component monosaccharides, serving the same purpose.

SUGAR LEGISLATION

Sugar is very big business. Modern cane mills which produce raw, 96 percent pure sugar cost about $75 million, and an equivalent sugar beet factory costs over $50 million; plants that refine sugar cost about the same. The refining business is capital-intensive, requiring highly skilled workers and management in addition to financing large inventories of both raw and refined sugars. Current world consumption is 80 million metric tons per year with per capita use in the United States and Canada—nations with the biggest sweet tooth—being close to 40 kg per person per year. With so many countries deeply committed to sugar production, sometimes to the virtual exclusion of other crops, regulation of supply and demand to stabilize prices is of concern to producers and consumers. The desire for sugar has also provided a base for direct and indirect taxation. In the United States, the first sugar duty was imposed in 1789, levied at three cents per pound to provide income for a young nation desperate for money to pay war debts. The duty and the tax reached a high point in 1815 when it was 12 cents per pound. Domestic supplies became available with Thomas Jefferson's Louisiana Purchase in 1803, Hawaiian sugar reached the mainland in 1835, and Philippine and Puerto Rican sugar was considered to be a domestic supply at the end of the Spanish-American War in 1898. In 1894, the United States introduced a sugar tariff to benefit and strengthen domestic suppliers. One factor in Cuba's successful struggle for freedom from Spain in 1896–1898 was this high tariff in which a 40 percent *ad valorum* duty was imposed on Cuban sugar. This caused the planters to seek to tie their fortunes to the United States; after the revolution, Cuban sugar was accorded "preferred status."

As new strains of cane and beets were developed, as technology became more sophisticated, and as acreages devoted to sugar increased, sugar production expanded rapidly after World War I. Prices plummetted to the point that producing countries could not sell enough sugar to obtain currency for the purchase of goods and services which, in turn, affected the industrialized nations. Retrenchment during the 1920 decade resulted in a shortage of sugar. The Jones-Costigan Act of 1934 and the 1937 American Sugar Act were passed to eliminate drastic price fluctuations, to ensure a stable ratio between domestic production and the balance of payments among trading nations, and to "reward friendly foreign nations." These acts, and subsequent amendments, provided that the U.S. secretary of agriculture each year must determine the domestic consumption of sugar and set quotas on U.S. production and imports from foreign suppliers. Although the percentages varied from year to year, roughly half of the total quota was allotted to domestic producers with Cuba, the Philippines, and Puerto Rico given special status. The cost of the program was paid by a half-cent-a-pound tax on all sugar; this was so efficiently managed that between 1934 and 1974, the U.S. Treasury was enriched by $600 million above administrative costs.

With an assured market—Cuba's quota was essentially its entire output—Cuba became a one-crop economy with all its eggs in a basket made of sugar cane (Figure 6.23). Eighty percent of Cuban exports in the 1950s was raw sugar to be refined in the United States, and 40 percent of the total profit went to U.S. companies. Political stability was guaranteed by financial and military support from the United States, the country remained undeveloped since reciprocal trade agreements mandated purchase of American goods, and American capital dominated the Cuban economy. The land was cultivated to cane to the point that food was imported into this fertile tropical island. A repressive military dictatorship, national pride, and the realization that labor-intensive practices on the sugar plantations and mills served only the interests of a tiny oligarchy, resulted in a revolution in 1959. Political and economic power was transferred to leaders whose political philosophy was anathema to the United States. The United States severed relations with Cuba in January 1960 and lifted the sugar quota in July of the same year. The Cuban sugar quota, which constituted up to one-third of U.S. consumption, was allocated to other, "more friendly nations," especially to the Philippines and the Dominican Republic, which by coincidence had been formed from Santo Domingo in 1960. Cuba, without its

special status, was forced to market the product of a national monoculture on the world market with economic consequences which had political consequences.

The American sugar acts were paralleled by those of other nations; over 90 percent of all sugar produced was under some sort of
market control. Quota systems worked well during World War II and
during the postwar recovery period, when sugar was in relatively short
supply, but strains and cracks developed during the late 1960s and
early 1970s as more and more sugar was produced and surpluses accumulated. This sugar glut was made more severe by capture of part of
the sugar market by liquid, high-fructose sweeteners made from corn
that undersold cane and beet sugar. The American Sugar Act was not
renewed in 1974, and a price war in sugar caused additional financial
strains in both producing and consuming countries accompanied by
wide fluctuations in prices and considerable speculation on futures
markets. The Philippines and the Dominican Republic were especially
hard hit since they had increased production since 1960 to meet their
enlarged quotas. Reports of near-starvation among the peasants served
only to depress prices further; there is always a profit to be made from
human misery.

It is apparent that monoagricultural economies are dangerous.
Crop failure or overproduction cannot be counterbalanced by other
crops or by an undeveloped industrial sector. Dependence upon foreign capital and imported necessities bind a monoculture nation to ex-
ternal control and possible exploitation. Economic stability seems to be
a prerequisite for social stability and political independence, and the
"sugar republics" are now beginning the agricultural and manufactur-

Figure 6.24
Horse-powered cane-grinding mill. Courtesy U.S. Department of Agriculture.

ing diversification needed to achieve these desired ends (Figure 6.24). It is unlikely that the future will see nations whose survival depends entirely on satisfying the sweet tooth of the world.

ADDITIONAL READINGS

Aykroyd, W. R. *The Story of Sugar.* New York: Quadrangle, 1967.

Deere N. *The History of Sugar,* 2 vols. London: Chapman & Hall, 1949–1950.

Edson, H. *Sugar: From Scarcity to Surplus.* New York: Chemical Publishing Company, 1958.

Nearing, S., and H. Nearing. *The Maple Sugar Book.* New York: Schocken Books, 1970.

Schalit, M. *Guide to the Literature of the Sugar Industry.* New York: Elsevier, 1970.

Taylor, F. G. *A Saga of Sugar.* Salt Lake City, Utah: Deseret News Company, 1944.

Thomas, H. *Cuba—The Pursuit of Freedom.* New York: Harper & Row, 1971.

Vandercook, J. W. *King Cane: The Story of Sugar in Hawaii.* New York: Harper & Row, 1939.

Van Hook, A. *Sugar: Its Production, Technology and Uses.* New York: Ronald Press, 1949.

The Potato

Our white potato is one of only six of the 2000 species in the genus *Solanum* which forms tubers. Wild potatoes grow in the southwestern United States, Mexico, Central America, and western South America from Venezuela south through Argentina, mainly at higher altitudes. Potato's affinities to tomato, bell-pepper, and eggplant are clear; the leaves, flowers, and fruit are similar. In fact, the tomato and potato were originally included by Linnaeus in the same genus, with later separation into different genera a matter of taxonomic convenience rather than necessity. The tuber is an underground storage stem formed at the ends of secondary roots which have developed from thickened underground runners, the stolons. The basic stemlike character of the tuber can be seen externally as the whorl of shoot buds—the eyes—and is obvious when the internal arrangement of the tissues is compared with that of the aerial stem.

POTATOES IN PERU

Potatoes are still the basic carbohydrate food of Indians of the Peruvian and Bolivian highlands who live at altitudes too high and too cold to support the growth of corn. They have been cultivated since at least 500 B.C. by peoples who selected and reselected wild plants to emphasize rapid maturation, increased tuber size, and to eliminate plants containing poisonous steroidal alkaloids. The result was a plant whose tubers were walnut-sized with red, yellow, blue, or brown skins and flesh that could be any of these colors. The potato is basically a source of starch with very little fat and minimal protein (about 2 percent by weight). Vitamins, especially water-soluble B vitamins and vitamin C are found just beneath the corky *periderm* or skin.

In all cultures production of primary food crops is accompanied by rituals designed to ensure a successful harvest, and the Indians of Peru were as aware of this need as were wheat, maize, or rice farmers. Their pottery includes jars in the form of potato tubers (Figure 6.25) or anthropomorphized potatoes dedicated to *Axo-mama*, the female generative spirit of the potato. Human sacrifices to Her were made, and seed potatoes were dipped in blood—preferably human, but usually animal—before planting. Planting, cultivation, and harvesting were done with the *choqui-taclla*, the footplow (Figure 6.26), and plants were

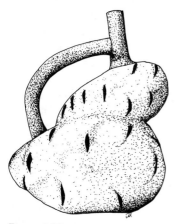

Figure 6.25
Pre-Columbian pottery jar from Peru in the form of a potato.

Figure 6.26
Peruvian Indians planting potatoes with a foot plow. This efficient tool is still used in the Andes Mountains. Taken from a mid-sixteenth-century woodcut.

tended by women who conducted "female" rituals over the developing plants. During harvest, any woman who found a blood-red potato had the right to run over to the nearest man and hit him in the face with it; this is probably of deep psychosexual significance and is not recommended for anyone without Inca genes. Harvested tubers were prepared for storage by converting them to *chuños*. Tubers were spread on the ground, allowed to freeze, and then stomped with bare feet to press out the moisture, a process repeated several times until the potatoes were completely dehydrated. Fresh potatoes were also used to make *chicha* beer. Supplemented with seeds and beans and occasional meat, the potato sustained people living in one of the world's most rugged environments.

By the tenth century A.D., the highland peoples had been subjugated by Inca tribes from Lake Titicaca in Bolivia. The Incas built an imperial federation led by a priesthood and an hereditary nobility whose high civilization was based on beans, maize, and the potato. Spanish conquistadores first saw the potato in 1532, and the 1533 report of Pedro Cieze de León said: "Of provisions . . . there are two other products which form the principal food of these Indians. One is a kind of earth nut which, after it has been boiled is as tender as a cooked chestnut, but it has no more skin than a truffle and it grows under the earth in the same way." Gonzalo Jimenez de Quesado called the tubers "earth truffles," reported the manufacture of chuños, and noted that although the product was bitter and tough, it was life-sustaining and

Figure 6.27
"Potatoes from Virginia."
Taken from *The Herball* by
John Gerarde published in
London in 1597.

constituted over a third of the Indian's diet. The natives called the plant *papas*, carefully distinguishing it from the sweet potato, *batattes*, which grew in warmer climates, but the Spanish didn't make this differentiation so that our word, potato, is a combination of Indian words. Chuños, tubers, and seeds were sent back to Spain by 1565–1570 as curiosities. Philip II sent planting material to the pope who distributed it to many papal ligations for growing in monastery gardens. The exact source of the potatoes that reached England by 1580 is still a minor mystery. John Gerarde described the plant in his *Herball* of 1597 (Chapter 5) and called it the Virginia potato (Figure 6.27) because many plants from the New World were "from Virginia" and also because no loyal Englishman would accept any Spanish name—the memory of the Armada was still fresh in their minds. Charles d'Ecluse (Carolus Clusius) obtained plants in 1587, and in his massive *Rariorum Plantarum Historia* of 1601 published illustrations of the "Peruvian papas." Gaspar Bauhin named it *Solanum tuberosum esculentum* in 1590, a name later accepted by Linnaeus, who dropped the last word.

INTRODUCTION TO EUROPE

With plants available in many gardens and the knowledge that the Indians lived on them, it was natural that they would be tried by Europeans. A hospital in Seville added them to soup as early as 1573, Spanish ships carried chuños as emergency rations, James I of England ate them publically in 1612, and in 1619 a Swiss gourmet described roasting and peeling them and the proper oil gravy to dress them with. Italians were using potatoes as cattle fodder in 1588, but most people assiduously avoided even touching them. They were not mentioned in the Bible and hence were not a food designed by God for man. Even in the nineteenth century, John Ruskin spoke of the "scarcely innocent underground stem of one of a tribe set aside for evil" because of the obvious familial resemblance to nightshade, datura, and mandragora, all plants of witchcraft, evil, and death (Chapter 4), and Clusius earlier noted that these solanaceous plants, like others of their ilk, might contain poisons or "could be used for exciting Venus." Because of their rough and scabby skin, it was asserted that potatoes could cause leprosy or scrofula; recognizing this danger, Burgundy passed a law in 1630: "In view of the fact that the potato is a pernicious substance whose use can cause leprosy, it is hereby forbidden, under pain of a fine, to cultivate it in the territory." As late as 1710, William Salmon, an English physician, said that "the English or Irish potato increaseth thy seed and provoketh lust in both sexes." This canard was still extant in the nineteenth century. Many Italians, seeing their cattle thriving, began adding them to stews in place of turnips, and a few other brave souls were spreading the word that boiled potatoes tasted fine so long as they were well salted. Nevertheless, potatoes were not a regular part of the European diet until the first quarter of the seventeenth century.

POTATOES ARE FOR PEASANTS

The failure of the general population to take up potato eating was of concern to the upper classes and the nobility. The grandduke of Tuscany recognized that if his peasants would eat potatoes instead of bread and pasta, the more valuable wheat could be reserved for his personal profit. After all, the entire sequence of potato cultivation could be accomplished with no tools other than a spade or hoe, a crop would mature in less than four months in virtually any soil, production was tremendous (an entire family plus their livestock could be fed on tubers produced on an acre of land), and their preparation for the table involved simple boiling. The potato could be the economic salvation of the landlord; efforts were made to promote their use.

Methods of weaning peasants away from expensive grain took different forms in different countries. Frederick Wilhelm, elector of Prussia, ordered potato cultivation in 1651 as "a national duty," but without much response from his otherwise dutiful subjects. His grandson, Frederick William I, passed a law stating that anyone who refused to plant and eat potatoes would, forthwith, have his nose and ears cut off, and many Germans decided that *kartoffels* were the best food available. His son, Frederick the Great, provided free seed potatoes and even distributed a recipe book, although the people generally just boiled them. Catherine the Great obtained potatoes from Prussia and mandated their cultivation for local consumption in European Russia; Ukrainian grain was reserved for use in the colder parts of Russia and for export. The French were initiated when prisoners of war returned home from Germany at the end of the Seven Year War with Prussia (Frederick having fed his prisoners on potatoes), bringing tubers back for planting in Normandy and the central provinces. In 1771, the Academy of Science of Besaçon offered a prize for the discovery of a new food which could take the place of cereal grains in the event of famine. Antoine-Augustine Parmentier won the award for recommending that the *cartoufle* (earth-nut)—later to be dubbed *pomme de terre* (earth-apple)—was the obvious choice, citing the reasons used by the grandduke of Tuscany and his own personal experience as a prisoner of war. In contrast to standing fields of grain, potatoes could not be set afire by invading enemy troops and the tubers could be stored underground—hidden from maurading bands. Parmentier had a flair for publicity; he presented King Louis XVI with a bouquet of potato flowers on His Majesty's birthday. In return, he was given a plot of royal land on which were grown potatoes served at a dinner attended by celebrities including the good Dr. Benjamin Franklin. Recipes, including the following, were distributed:

> To make a potato pudding, first boil and blanch potatoes and add the marrow of two bones. Add a pint of cream, ten egg yolks, five beaten egg whites, sugar, rose water, one half pound of bread, cinnamon, ginger, nutmeg, minced citron and orange, candied lemon and some sack [a light

wine from the Canary Islands]. Bake in a dish with puff paste and covered with a sauce made with thick butter, sack and sugar.

The peasant boiled potatoes and ate them with salt and oil. The King told Parmentier, "France will thank you hence because you have found bread for the poor."

In 1662, the Royal Society of London met to consider the possibility of planting potatoes throughout England. John Forster, Gentleman, published these deliberations in a book entitled:

England's Happiness Increased or a Sure and Easy Remedy Against All Succeeding Dear Years by a Plantation of the Roots called Potatoes: Whereby (with the Addition of Wheatflower) Excellent Good and Wholesome Bread may be Made Every Year 8 or 9 Months together, for Half the Charge as Formerly; Also by the Planting of These Roots Ten Thousand Men in England and Wales Who Know Not How to Live, or What to Do to Get a Maintenance for their Families, may on one Acre of Ground make 30 pounds per Annum. Invented and Published for the Good of the Poorer Sort . . .

It was suggested to Charles II that a tax on potatoes be levied, the proceeds to go directly to the royal treasury: "There is no reason why his gracious Majesty should not benefit by this annual license money of 50,000 pounds, inasmuch as his loyal subjects would thus benefit by the cultivation of the potato."

Actually, there had been some cultivation of the potato in the British Isles. Thomas Hariot, estate manager of Sir Walter Raleigh, planted them in 1588 at Youghal, Ireland. They were grown in the physick (medicinal) garden in Lambeth, London, in 1656, in Oliver Cromwell's wife's garden in 1664, and Robert Morrison, botany professor at Oxford, wrote in 1680 that it was a familiar English garden plant. With royal assent, planting stock was obtained from Raleigh's estate, fixing for all times the name "Irish" potato.

THE IRISH SPUD OR LUMPER

Three areas in Britain were most in need of cheap food: Wales, Scotland, and Ireland, with the most desperate need being in Ireland. From the time of Henry VIII, when the break with the Roman Catholic Church was complete, the subjugation of the Catholic Irish became deliberate royal policy; the Irish were to be obedient to the crown, and the land was to be "planted" with reliable, that is, Protestant, owners. In spite of a papal bull designed to prevent Elizabeth I from ruling, the subjugation went forward; the Desmond Revolt of 1568–1582 was ruthlessly crushed; and the Cromwellian upheaval of 1650 uprooted the peasants, leaving them without homes, crops, or cattle. The Penal Laws of 1690 completed the process by depriving the Irish of civil rights, ownership of a horse, the right to speak Gaelic, to get an education, to attend college, to marry a non-Catholic, or to hire a Protestant. The Irish became beggars in their own country and, each year during the seventeenth and eighteenth centuries, close to 40 percent of the people were on semistarvation diets, sustained by their religious faith and a burning desire for freedom from England's impossibly heavy yoke.

Growing conditions in Ireland were ideal for the potato—a cool, humid climate, deep and friable soil overlaying chalk and a long growing season. The prevailing moist westerlies that blew from the Atlantic checked the spread of insects carrying diseases known to reduce yields. The land tenure situation also contributed to the spread of potato cultivation. Ireland had been parcelled out to absentee landlords, all Anglicans, who used their estates for pleasure and who lived in England on profits accruing from lush grain fields, well-pastured cattle, and sheep whose mutton and wool nourished and clothed the middle and upper classes. To relieve themselves of any responsibility to the peasants, landlords granted long-term leases to middlemen who, in turn, further leased to subcontractors and sub-subcontractors with corresponding increases in the rents charged for the hovels occupied by the peasants. Occupying and tilling little more than an acre per family, the Irish paid their exorbitant rents, taxes, and tithes in cash (selling the pig) and in labor, including the domestic service of the women—the "Irish washerwoman" was not merely a carefree dance—and they were forever in debt to loan sharks, the *Gombeenmen* (boogymen). Prolific—to the delight of the Church; ignorant—to the delight of the middlemen; rebellious—to the delight of the red-coated military, the people were sustained by the knobbly spud or lumper.

The numbers of Irish people increased, virtually doubling between 1700 and 1800. Indeed, Europe's population which had been stable for centuries, jumped between 1725 and 1825 from 140 million to 266 million. The technology of the industrial revolution, advances in livestock breeding, the introduction of crop rotation, and systematic manuring contributed to the availability of the food which had been a limiting factor in population growth. But most demographic experts now believe that a major factor in population growth was the availability of this wonderful new food supply. By 1780 Ireland's food "is potatoes and milk for ten months of the year, potatoes and salt the remaining two." The Irish male consumed 12 pounds of boiled potatoes a day, every day. His wife ate about eight pounds, and his children no less than five pounds. Fishermen sold their catch and ate potatoes. Sheep and cattle were tended by peasants who tasted meat about three times a year, bread was over half potato starch, and drink was potato beer or *poteen*, distilled potato beer. Bloated stomachs were the least serious evidence of malnutrition; potatoes lack fat-soluble vitamins A and D, so that, even with milk in the diet, eye dysfunctions and rickets were common. Adam Smith, in his *Wealth of Nations* of 1776 warned that "should these roots ever become . . . the favourite vegetable food of the people. . . . populations would rise and rents would rise. . . ." He noted that "it is not uncommon . . . for a mother who has borne twenty children not to have two alive." Daniel Langhaus observed the same high birth and high death rate in potato-eating cantons in Switzerland. Dr. Thomas Beddoes in his *Essay on Consumption* of 1800 said: "this root . . . has probably contributed to the degredation of the human species." Yet, there were few options for the peasant; immigration was a fearful choice, for the Church warned that those who went to a pagan, that is, non-Catholic, country were in danger of losing their immortal souls. Besides, there was no money to pay for a passage to Australia (a hated "English" land), Canada (English or what was worse, French), or America.

THE POTATO MURRAIN

Crop failure was always on people's minds. Seed potatoes were blessed and holy water sprinkled on the plants on Ascension Day. Candles were lit to Mary to intercede for good crops. In 1809 a virus disease reduced yields by 30 percent and another virus infestation in 1833 was sufficiently serious that hardier and more adventurous people worked passage to eastern Canada and to the United States. On August 23, 1845, the prestigious *Gardener's Chronicle* headlined: "A fatal malady has broken out amongst the potato crops." The potato murrain had struck the plants all across Europe and was most severe in Ireland where the tenant farmers' patches were so close together that any murrain could sweep through the whole country. Green tops blackened and shriveled within a day (Figure 6.28) and young tubers had black patches on their still-thin skins. On September 13, 1845, the *Chronicle stated,* "We stop the press with very great regret to announce that the potato murrain has unequivocally declared itself in Ireland. Where will Ireland be, in the event of a universal potato rot?" The answer was obvious—starvation. Landlords were of two minds about this. The potential loss of rent was annoying, but on the other hand, this would allow them summarily to dispossess tenants, tear down their miserable shacks, and turn potato patches into productive grazing land. A British commission suggested that things weren't too bad; even if the potatoes were rotten, the peasants could wash the starch out of them and use this to bake bread. They knew, however, that starch alone could not sustain life without vitamins, minerals, and protein.

English newspapers were filled with suggestions on how blackened, shriveled, slimy, and stinking potatoes could be eaten (by the Irish) and theories on the cause of the murrain. The respected editor of *Gardener's Chronicle* asserted that it was because wet, cool weather caused the potatoes to take up too much water. One correspondent suggested that it was due to the electricity generated in the air by the tremendous and dangerous speeds of steam locomotives. Another insisted that it was just punishment from God visited on Catholics, and some Catholics said that it was caused by not being Catholic enough. Remember that this was 1845, 20 years before Robert Koch and Louis Pasteur proved the causal connection between microorganisms and disease. Reverend M. J. Berkeley, curate of Northamptonshire, wrote

Figure 6.28
Early symptoms of potato blight on the leaves.

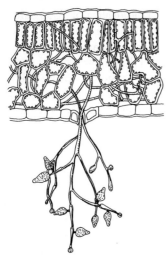

Figure 6.29
Spore-bearing hyphae of the blight fungus protruding from a stomata on the lower surface of a potato leaf.

the *Chronicle* that whenever murrain struck, he was able to observe a fungus on the blighted parts of the leaves and tubers. Although others had seen the fungus, most believed that it was a consequence of decay and not the cause. Controversy raged about this matter until the fungus, named *Phytophthora infestans*, proved to be the cause, its airborne spores being carried on the lightest breezes from plant to plant (Figure 6.29).

As with most plant diseases, the severity of the blight depended on the inherent genetic resistance of the plant and on weather conditions. In Peru, it is too cold for much development of the fungus, and there is scarcely enough rain to keep potato leaves wet long enough to allow spores to germinate and invade the leaves through the stomatal pores. In 1845 the weather was just cool enough and just rainy enough, the cultivars growing in Ireland were just susceptible enough, and the fungus was just virulent enough so that about 75 percent of the total crop was a complete loss.

FAMINE

Although the summer of 1847 started well, rain, gloomy weather, and cool winds developed during July, and between July 27 and August 3 the potato blight killed close to 80 percent of the plants. Food relief was instituted on a half-hearted basis, but the people were by then so weak that only massive relief with high protein meat supplies would have stemmed the growing tide of starvation and death. Soup kitchens were reluctantly set up, but a once-a-day meal of a pint of soup made with a pound of gristle, a peck of peas, a peck of barley, and five heads of cabbage in 150 gallons of hot water or a cup of maizemeal boiled in water had little positive effect. The government congratulated itself by noting that it not only provided food, but even cooked it.

The consequences were inevitable: three million people starved to death. Corpses could not be buried before they began to rot in the streets; typhus fever and cholera became epidemic; women aborted; and child mortality approached 65 percent. Only in 1974 was it discovered that blighted potatoes contain an alkaloid causing abortion and birth defects. Those that survived would never attain full vigor and would spend their lives with the reduced intelligence that comes from inadequate protein during the first years of life. With enlightened self-interest, some landlords "forgave" the rents and others hired leaky ships to take people to Canada or to the United States (Figure 6.30). Between 1846 and 1854, close to 2 million people died, and the famine set into motion a wave of immigration that reduced Ireland's population from over 8 million in 1846 to 4 million in 1900. The Irish constituted 35 percent of the immigrants to the United States during this period and became 15–18 percent of the population.

The new immigrants caused a huge, almost insolvable problem for the United States. In spite of being rural, they had no farm skills, since their only real experience was spading potato patches and hoeing

Figure 6.30
Irish emigrants leaving for
New York as depicted in
Harper's Weekly Magazine.

cabbages; and machinery was replacing scything. The immigrants had
no money to move out of the port cities. The near-starvation conditions
under which they had lived left them physically weak and prey to the
diseases of malnutrition. They arrived in a country suspicious of Catholics and flooded a labor market unable to employ even the people who
were in the cities before the Irish arrived. Ditch-digging and odd jobs
were the lot of the men and the women worked in factories for low
wages and formed the servant class for affluent Americans. Having no
knowledge of city life, they were cheated at every hand and, in self defense, formed enclaves within New York and Boston which served not
only to protect them, but also to mark them as a people apart. Anxious
to work, they established their base in the construction trades where, in
time, they branched out, eventually to become active in all aspects of
American life up to and including the presidency. But the road was un-
necessarily hard and bitter, and although these injuries are not so deep
as those received from the British over a 300-year period, they still
exist.

MODERN CULTIVATION

Close to 200 million metric tons of potatoes are grown throughout
the world with 80 percent of this amount produced in Europe and Russia. Only wheat is produced in larger tonnages, but the land area devoted to potatoes is only 11 percent of that devoted to wheat. Based
solely on total caloric value, the potato provides more carbohydrate
than do other major food crops (Figure 6.31), a fact that was recognized
when the rulers of Europe forced potatoes on their people in the seven-

Figure 6.31
Comparison of production,
areas cultivated, and nutritional yields of the four
major carbohydrate crops.
Data from 1970 harvest
season.

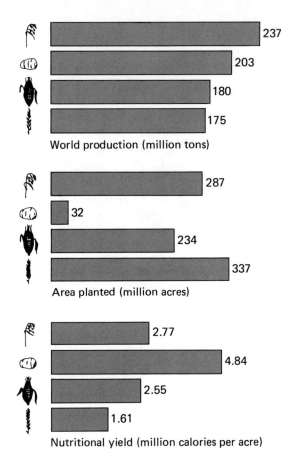

World production (million tons)

Area planted (million acres)

Nutritional yield (million calories per acre)

teenth century. Unless very large volumes are consumed with a supply
of vitamins A and D (milk, eggs, butter), a nutritionally adequate diet
cannot be obtained from potatoes.

Potatoes entered North America when English settlers brought
them to Bermuda as a food for slaves. The plant reached the mainland
in 1719, when a colony of Scotch-Irish settled in Londonderry, New
Hampshire, and grew the plant primarily as animal feed. There were
only a few kinds then available, a red-skinned type, a brown-skinned
type, and a dark-brown type, but, by 1760, there were many types with
names like 'Cluster,' 'Irish Apple,' 'Cups,' etc. Each of these types, now
called cultivars (cultivated varieties), were fairly uniform since they
were propagated vegetatively from pieces of the previous year's tubers
saved as "seed potatoes."

At least theoretically, the number of cultivars may be infinitely
large, although some are genotypically so similar as to be essentially
identical. To develop a new cultivar one would take seeds, plant them,
and select from the mature plants to obtain phenotypes that are desirable—earliness, size, skin and flesh color, taste, or disease resistance.

Prior to the Great Famine, disease resistance was not considered, for the concept of a pathogenic (disease-producing) organism was not known, but by 1880, gardeners were attempting to develop cultivars resistant to virus diseases, to wart, scab, rot, and blight. Selection for desirable growth and form was well under way by the middle of the nineteenth century. The Reverend Chauncey Goodrich of New York obtained seeds from South America in 1851, grew them in his garden, and from thousands of plants selected one named 'Garnet.' Garnet and selections from it became standard cultivars in the Northeastern United States and Canada. By 1870, when the concept of sexual reproduction was established, a science of plant breeding began.

There were over 400 cultivars by 1890. The difficulty with this early breeding work was that the breeders were merely reassorting genetic characters within a relatively small genetic pool, and the statistical probability of obtaining a genotype which included strikingly new phenotypic characters such as disease resistance was low. People got the idea, in about 1880, that they should obtain wild plants whose characteristics had not been genetically selected. Nations sent plant hunters back to the mountains of Peru, Bolivia, and Mexico to collect wild potatoes for cross-breeding. The United States and other potato-growing countries established experimental potato plots to grow wild potatoes for seed to be distributed to breeders. A truly blight-resistant potato has yet to be developed, but cultivars are now available which have moderate resistance to this and other diseases.

In addition to being plagued by viruses, fungi, and bacteria, the potato is food for a host of insects, among which is the voracious Colorado potato beetle (*Leptinotarsa decemlineata*), a small insect with characteristic yellow and black stripes on its wing covers (Figure 6.32). It was first described in 1824 in Colorado where it was fairly rare, living on leaves of wild members of the Solanaceae. As American farmers moved west before and after the Civil War, those that settled the prairies planted potatoes. The beetle, finding the potato much to its liking, began to move east. The beetle crossed the Mississippi River in 1865,

Figure 6.32
Colorado potato beetle feeding on potato leaves. Courtesy U.S. Department of Agriculture.

reached Ohio by 1869, and Maine by 1872. In spite of rigorous quarantine against potato importation into Europe, the beetle was in England by 1875 where it was ruthlessly exterminated. Beetles were seen in Germany in 1914, and because of the war, they were not hunted down and are now munching happily away on potato leaves throughout Europe. Toxic chemical sprays, especially chlorinated hydrocarbons, keep them from destroying the crops, but the attendant environmental damage is becoming evident throughout the world.

ADDITIONAL READINGS

Correll, D. S. *The Potato and Its Wild Relatives.* Austin: Texas Research Foundation, 1962.
Dodds, K. S. "Evolution of the Cultivated Potato." *Endeavour* 25(1966):83–88.
Dodge, B. S. *Potatoes and People.* Boston: Little, Brown, 1970.
Hawkes, J. G. "The History of the Potato." *Journal of the Royal Horticultural Society.* London: 92(1967):207–224, 249–262, 288–302.
Large, E. C. *The Advance of the Fungi.* New York: Holt, Rinehart and Winston, 1940.
Niederhauser, J. S. "The Blight, the Blighter and the Blighted." *Transactions of the New York Academy of Sciences,* series II, 19(1956):55–63.
Salaman, R. N. *The History and Social Influence of the Potato.* Cambridge University Press, 1949.
Woodham-Smith, C. *The Great Hunger: Ireland 1845–1849.* New York: Harper & Row, 1962.

Figure 6.33
Flowering branch of tea, *Thea sinensis.*

A Cup of Tea

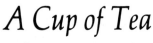 Tea is naught but this:
First you make the water boil
Then infuse the tea
Then you drink it properly
That is all you need to know,
SENNO-RIKYŪ (1585)

Tea is, next to water, the most popular drink in the world. In Great Britain, a "cuppa" is akin to chicken soup in Brooklyn, a panacea for everything from chilblain to broken marriages. Linnaeus originally named this plant *Camellia japonica* because his herbarium specimen came from Japan, but the shrub is native to north India and China and we now classify the plant as *Thea sinensis* (Figure 6.33).

In modern Western society, the word alkaloid conjurs up visions of drug addiction, degradation, and death. Yet one group of alkaloids, the caffeine alkaloids, is nonaddictive and is important socially, eco-

Adapted from "The Tea Mystique," *Natural History* 84(1975):12–29. Copyright American Museum of Natural History.

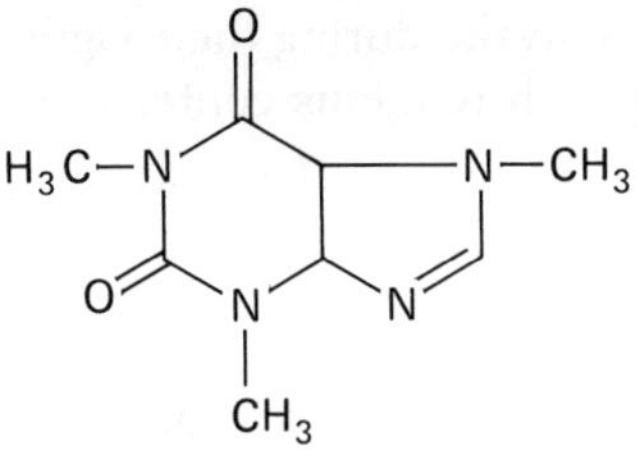

Caffeine

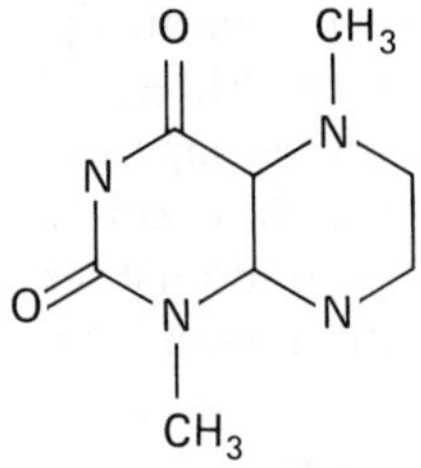

Theobromine

Figure 6.34
The nonaddictive caffeine alkaloids.

nomically, and medically (Figure 6.34). Chemically, caffeine is 1:3:7 trimethylxanthine, a compound made by plants but not animals as a branch of the same biosynthetic pathway in which nucleic acids are synthesized. It was isolated in 1820, and its chemical structure was determined by 1825. Although caffeine can be synthesized in the laboratory, isolation is cheaper than synthesis. Most supplies are obtained during the decaffination of coffee. Caffeine affects the central nervous system of man as a mild stimulant and also serves to increase the heart rate. In some people the antisoporific action of caffeine can lead to insomnia . . . the "no coffee tonight or I won't sleep a wink" syndrome.

ALL THE TEA IN CHINA

During the Han dynasty (206 B.C.–220 A.D.) a medical book reported that *ch'a* (Chinese for tea) could cure tumors, abscesses, and ailments of the bladder. It "lessened the desire for sleep" and "gladdens and cheers the heart." Infusions of young tea leaves in hot water gave a pleasant flavor to the flat, insipid taste of water boiled to prevent the spread of water-borne diseases. When Buddhism came to China, tea kept monks awake during all-night meditation and eased pangs of hunger during prolonged fasts. By the fourth century A.D., reliable records on cultivation, processing, and brewing were maintained. The preferred method of processing was the tea cake technique in which young leaves were mixed with rice, baked, the cake pounded into small pieces, and the fragments boiled with onion slices, ginger root, and orange to "render one sober or to keep one awake." Peasants had to be content with old leaves boiled in salted water. By the early Tang dynasty (608–907 A.D.), tea was an export product, taxes were assessed on it, and a three-volume, ten-part *Tea Classic* was written extolling the pleasure of tea drinking, giving directions for growing, storing, and brewing.

One example of the aesthetic importance of tea can be seen in a letter of 730 A.D. from a Chinese gentleman to a friend:

> I am sending you some leaves of tea that come from a tree belonging to the monastery of Ou I Mountain. Take a blue urn of Ni Hung and fill it with water which has been melted from snow gathered at sunrise upon the western slopes of SouChan Mountain. Place the urn over a fire of maple twigs that have just been collected from among very old moss and leave it there until the water begins to laugh. Then pour it into a cup of Huen Tscha in which you have placed some leaves of this tea. Cover the cup with a bit of white silk woven at Houa Chan and wait until your room is filled with a perfume like the garden of Fouen Lo. Lift the cup to your lips, then close your eyes. You will be in Paradise.

Baked tea cake or the infusion of leaves was supplemented during the Sung dynasty (960–1120 A.D.) by a technique in which quickly dried leaves were finely powdered and whipped with hot water into a thick green froth. During this period, a branch of Chinese Buddhism, the Ch'an sect, incorporated tea into its rituals. Subscribing to the Tao ideals of simplicity with contemplation as the way to enlightenment,

Ch'an Buddhists used tea to keep themselves awake during their vigils and to induce that state of calm harmony which religious contemplation requires.

THE TEA CEREMONY

For centuries, Japan was dependent upon China for protection and looked to the "Middle Kingdom" for enlightenment and culture. The Emperor Shōmu obtained tea in 729 A.D. from the Tang emperor. He invited a hundred Japanese Buddhist priests to join him in sipping this august beverage, but the Japanese did not like the bitter, almost black and cloudy liquid. When Ch'an Buddhism entered Japan in the middle of the thirteenth century as Zen Buddhism, the ritual ceremony of tea came with it. The founder of the Tenryu-ji Zen Buddhist temple in Kyoto modified the ceremony to be more compatible with Japanese concepts and gave the Japanese version its name, *Cha-no-yu*, the ceremony of hot-water tea. In 1477, the Ashikaga shogun-military ruler of Japan elevated Shoku, a monk of the Shomyu-jo temple of Kyoto, to the position of first high priest of Cha-no-yu. Shoku reconstructed from oral tradition the tea ceremony of the Chinese Ch'an Buddhists in which monks successively drank tea from the same bowl while contemplating a statute or a drawing of the saint Bodhidharma, known in Japan as Daruma—patron of Zen and of tea (Figure 6.35).

The formalization of the tea ceremony had profound repercussions in Japan. Zen, initially a small sect, struck a responsive chord in the minds and hearts of the military governors of Japan and their Samurai warriors. Soldiers all, they were trained to respect simplicity, austerity, and ritual. Those that survived the bloody fighting retired to a life of contemplation instead of running for high political office. Zen offered them concepts they understood, and the symbiosis between soldier and priest resulted in the secularization of Cha-no-yu. Toshimasa, eighth Ashikaga shogun, built the famous Silver Pavilion and, within its walls, his tea priest built the first tea hut. The tea hut itself was copied from Chinese paintings of the Sung dynasty. It was essentially a "rude" hut, straw-thatched, with a small door through which one entered on hands and knees. To enhance the feeling of austere harmony, the land surrounding the tea hut became a garden designed as an aid to Zen communication with and absorption into nature. The *uchi-roji* garden, a miniaturized landscape, was placed so that participants in the tea ceremony could view it through the small windows of the hut. The "nature as purified" roji developed into the Japanese garden as we know it today.

The hut itself had design features we now associate with Japanese life. Simplicity, elimination of luxury and ostentation, and the harmony necessary for quiet were stressed. Details of construction were codified: internal walls were of rough, gray plaster, window openings were of bamboo, beams should retain the marks of the adze, and the number of

Figure 6.35
Bodhidharma, Buddhist patron of the tea plant. Taken from a fifteenth-century Chinese woodblock print.

Figure 6.36
Flower arranging as part of
the Japanese tea ceremony.
Courtesy Japan National
Tourist Organization.

mats were specified. Although the *tokonoma* was not original with the tea
hut, it reached its present perfection as it came to symbolize the Zen's
alcove for a statue or painting of a saint. Precious scrolls, a lovely stone
on a teakwood stand, or some other object of contemplation could be
displayed in this alcove. It was the responsibility of the tea master to
place within the tokonoma objects of such profound and quiet beauty
that the observer's spirit of harmony with nature was enhanced. Although originally parts of altar decorations, flower arrangements came
to be placed within the tea hut alcove and this led to *ikibano*—flower arranging (Figure 6.36). In fact, the arrangers of flowers were the tea masters. Soami was tea master and flower arranger in the early sixteenth
century and developed the *rikka* style; Senno-Rikyū, his successor, developed the simple, austere, and delicate *nagiere* style, and his succes-
sors perpetuated and modified these arrangements into the schools of
flower arranging that we know today.

The tea ceremony, its hut, garden, and plant arrangements em-
body two deeply rooted Japanese ideals. The first of these is *furyu-no-
asobi*, approximately translated as an elegant amusement allowing one
to lose one's self in the joy of delicate peace and tranquility. The second
is the concept of *wabi*, or mellifluous simplicity in which every aspect of
an act is carefully orchestrated and integrated. The tea ceremony is not
simply the drinking of frothy, grass-green tea, but rather having the

Figure 6.37
Students learning the Japanese tea ceremony. Courtesy Japan National Tourist Organization.

entire ceremony exemplify that combination of serenity, grace, and beauty which is the spirit of Zen, participating in a ritual of perfect harmony and tranquility. A family may spend huge sums on a tea hut and ancient tea utensils because, today, the tea ceremony has become overlaid with self-serving, conspicuous consumption. Although still a Zen ritual, the ceremony is also a part of the proper mode of conduct of the upper classes in Japan. Young women take courses to allow them to lead tea ceremonies, and they learn Cha-no-yu much as our children from equivalent social strata learn to ride an English saddle or serve in a receiving line (Figure 6.37).

TEA IN EUROPE AND NORTH AMERICA

Tea entered the consciousness of Europe in the middle of the fifteenth century through narratives of travelers to the mysterious East. The Portuguese, who had established themselves on Macao off China's coast, sent back a few chests of tea leaves, hoping to develop a market. Little came of this venture, and traders confined themselves to silk, rhubarb for the constipated, and art curios. The Dutch tasted tea, found it pleasing, and after their conquest of Batavia, moved into the Chinese tea trade. Commercial interactions between Europeans and the tottering Ming dynasty were poor, and the ascendancy of the suspicious

Manchus in 1644 did not help; traders were confined to Canton in the south where ch'a was called *t'e*. The Dutch forced the Portuguese out and obtained a monopoly on the tea trade. Russia got its tea through a trading town, Kiakhta, on China's northern frontier, and, being isolated from much of Europe, developed a unique tea-drinking style. The samovar, glasses of tea sweetened with a spoonful of jam, and the tricky technique of sipping boiling tea through a lump of compressed sugar held between the teeth are still confined to Russia. They did, however, export the use of a lemon slice.

As tea trickled down socially from European nobility to the less exalted, it came under the scrutiny of the medical professions. Dr. Cornelius Decker, seventeenth-century publicist for the Dutch East India Company, recommended that "an obstinate fever could be cured by drinking every day forty to fifty cups of tea." French medical authorities were generally antitea, and the entire Collegè de Médicine denounced "this dubious drug." Its few advocates made such extravagant claims for its medicinal value that the French gave a Gallic shrug and went right on drinking wine. Germany preferred beer, and the Latin countries were priced out of the market by the Dutch whose hatred of Spain and Italy had an unhappy historical base.

In 1660, a London shopkeeper, Thomas Garroway, advertised the stimulating properties of tea in glowing terms: "it maketh the Body active and lusty, eliminates gripping of the Guts and pains in the Bowels." The British were, however, still coffee drinkers and were not about to pay the equivalent of $50 per pound demanded by the Dutch. Tea was obviously an article of snob appeal. Sam Pepys tried it that year and didn't like it. The wife of Charles II, who was Portuguese, served it after her wedding, and, since the nobility drank it, Pepys tried it again in 1667 and, following the example of his king, liked it. Charles II gave an absolute monopoly of importation to the British East India Company, freezing out the Dutch who were at war with France, then an ally of England. With an assured market, the Company established a regular tea run from Canton and Amoy to England, and the place of coffee in Britain's social life was threatened. From 100 pounds in 1680, Britain imported a million pounds of tea by 1700. There were several hundred shops in London serving tea, but none of them served women. Thomas Twining built an annex to his coffeehouse, The Golden Lion, that catered solely to women who made Mr. Twining a wealthy man. Tea was at first sipped from handleless porcelain cups as used by the Chinese (Figure 6.38). To protect delicate ladies' fingers, the British pottery firms of Spode and Wedgwood added handles to cups, and with the rediscovery in Dresden in the middle of the eighteenth century of the ancient Chinese art of porcelain making, elaborate tea services became appropriate wedding gifts. Translation of these designs into metal by Georgian silversmiths followed.

By 1780, coffeehouses had become tea houses, gin mills of the slums became tea bars, and imports rose to 14 million pounds a year.

Figure 6.38
A family tea party in England. This portrait was made before the British added handles to the Chinese cups.

Samuel Johnson called himself "a hardened and shameless tea-drinker, who has for many years diluted his meals with only the infusion of this fascinating plant" and confessed that he would drink up to 12 cups of tea at a sitting. John Wesley, who founded Methodism, first denounced weak-willed people who substituted one evil, tea, for another, liquor, but he finally succumbed and had a special pot made for him by Josiah Wedgwood. Tea became the approved drink for the antiliquor forces, and the temperance hotels that still exist in provincial towns of England have excellent and inexpensive tea bars.

But the East India Company was in trouble. Anticipating huge sales to the American colonies, they had millions of pounds of tea in storage which the colonists wouldn't buy because of the tea tax. The colonies were drinking tea smuggled into eastern ports by the Dutch and through Canada by enterprising French-Canadians who had little enough love for the English. As textbooks say, the rest of the American tea saga is history.

Europeans developed their own tea ceremonies. Afternoon tea originated in about 1750 with Anna, wife of the seventh duke of Bedford, who complained of a "sinking feeling" in the late afternoon and decided to have a snack between the noon meal and late evening supper. George III regularly had tea and cake at court, thus sanctifying "the tea." Afternoon "low tea" was a light meal of watercress sandwiches, scones, and bread and butter with jam. The British developed precise codes of behavior governing the way a cup was to be held, proper pouring technique, and whether the milk should be added before or after the tea. One gentleman, Henry Sayville, complained bitterly about the feminization of the men who participated in these "absurd social activities" instead of rallying round the bar of the nearest pub for a pint of ale and a clay pipe. High Tea became a full meal, replacing the former family supper. Mrs. Beeton, the Julia Child of the Victorian era, described a high tea as a "moveable feast which included hot dishes, fowl, hams, salad, cakes, tarts, trifles and fresh fruit."

Iced tea was invented in the late nineteenth century by Richard Blechynden, an English public relations man representing a Ceylonese tea company at a St. Louis trade fair. Arriving on a beastly hot day with a group of Singhalese waiters in full native dress, he was unable to find anyone who wanted to drink hot tea. With the genius of a good PR man, he poured the hot tea over ice, added sugar and lemon, and created a sensation. The tea bag was the inadvertent invention of Thomas Sullivan, a New York importer and wholesaler. In 1908 he decided to send out samples in small silk bags instead of the usual tin caddy. Some unknown person dropped a bag into hot water and an entire new industry was born.

TYPES OF TEA

There are many kinds of tea. An ancient Chinese scholar said, "The essence of the enjoyment of tea lies in appreciating its color, fragrance and flavor," and these characteristics are determined by many factors. There are two major species, *Thea sinensis* from southwest China, northeast India, and Cambodia, and the less valuable *T. assamica* from southeast Asia. The shrub can grow to about 6–8 m, but under cultivation it is kept to about 1 m. Ideal growing conditions include well-drained, slightly acidic soils, ample rainfall, and especially elevated altitudes. The best teas are grown in the mountains of Assam, Formosa, and Japan, with most of the North American tea grown in India. Attempts to grow tea in the Western Hemisphere have not been successful, although plantations were established in the U.S. South during the early part of the twentieth century. Parts of East and South Africa and South America show excellent potential for tea production. The fine teas of Ceylon and of India are all mountain-grown, usually at elevations of over 1500 m.

Young bushes, started from the seed of selected plants or from rooted cuttings, are grown until they become large shrubs and are then severely pruned back to encourage formation of many twigs. At intervals of a week to 10 days, the plant undergoes a flush of tip growth producing two very small, tender young leaves and an unopened leaf bud. The leaves and bud are picked by hand; as a consequence, cheap labor is necessary to keep production competitive. Experienced pluckers can gather 15–20 kg per day, enough for about 4–5 kg of tea. Older, coarse leaves bring lower prices and produce a beverage with stronger and less fragrant odors and flavors.

Independent of the cultivar grown, three major tea types can be prepared. Black tea, the most common type, is produced in a series of steps. The picked leaves are spread on racks to wither for a day in the sun. This reduces water content and initiates enzymatic processes that release aroma. From the withering racks, leaves are put through machinery which ruptures leaf cells, releasing enzymes that cause chemical oxidation of tannins and phenolic compounds. The enzymatic processes, called fermentations although they are actually oxidations, are continued in a cool, damp atmosphere until the leaves become a bright, shiny copper color, much like a new penny. Finally, the tea is dried—fired—to stop fermentation and to remove excess water. It is during the firing that tea leaves acquire the dark color we see in black tea. The withering process is not carried out for production of the green tea types which retain a slightly "grassy" taste. The rarely seen Oolong teas are produced by a very short withering process. Green and Oolong teas are found primarily in Oriental establishments in North America and tend to have delicate flavors and aromas and very light colors.

There are close to 3000 different kinds of tea, taking their names from the districts where they are grown, from the processing techniques used, from various additives, and even from the sizes and conditions of the leaves (Table 6.1). Orange pekoe, for example, now refers to a grade of black tea, although the name came from the Chinese who had added orange blossoms to the leaves. Since most commercial teas are blends of different grades and district products, some names refer to mixtures put together for famous people, for example, Earl Grey or Queen Mary.

The leaves, twigs, and bark of many plants can be infused in hot water and are called teas by the same process of product generalization that plagues companies producing trademarked products. Thus we

Table 6.1
TEAS

Tea Grade	Type	Description	Country
Orange Pekoe	Black	Thin leaves, lightly colored, pale brews	Ceylon, Japan, China, Java
Pekoe	Black	Leaves broader, brew with more color	Ceylon, India, China, Java
Pekoe Souchong	Black	Round leaves, pale brew	Ceylon, India, China
Oolong	Oolong	Medium leaves, delicate brew, pale color	Taiwan, China
Fannings	Black Oolong	Small leaves, quick brewing with good color	India, Taiwan, Java
Dust	Black	Very small leaves, strong, aromatic, dark brew	India, Taiwan, Ceylon, Java
Imperial	Green	Small leaves, delicate, pale brew, faintly aromatic	Japan
Gunpowder	Green		Taiwan
Hyson	Green		China

Figure 6.39
Flowering branch of mate,
Ilex paraguariensis.

have mint, camomile, and sassafras teas whose pharmacological effects are due to sugar and the soothing effects of hot water. Among the steeped plant products that do contain caffeine is *yerba maté* of South America (Figure 6.39) brewed from leaves of *Ilex paraguariensis,* one of the hollies. Its caffeine content is about half that of black tea, and its essential oils confer a delightful aroma to the brew.

Each year, North Americans drink 35 billion servings of tea of which 18 billion are iced; 150 million pounds are imported from India, Ceylon, East Africa, and Taiwan. Aside from green teas served in Chinese restaurants, most are black teas. The average "cuppa" contains only about half as much caffeine as a cup of coffee. Sam Johnson notwithstanding, caffeine is neither addictive nor narcotic. Fortunes have been made from the tea trade. Our clipper ships participated in the tea runs of the nineteenth century, some bankrolled by John Jacob Astor who worked out a deal with the Chinese government to trade furs for tea and obtained a deferment of import duties from a complacent U.S. administration. The Great American Tea Company (A & P) (Figure 6.40), the National Tea Company, and the Union Pacific Tea Company have all become supermarket chains. Tea financed Thomas Lipton's unsuccessful efforts to restore the crown of yatching to Britain. It is still the sales base for manufacturers of fine china—the very word reflecting the history of this most social of drinks.

Figure 6.40
One of the first A&P's. Courtesy Great Atlantic and Pacific Tea Co.

ADDITIONAL READINGS

Eden, T. *Tea,* 2nd ed. London: Longman, 1965.
Harler, C. R. *The Culture and Marketing of Tea.* New York: Oxford University Press, 1964.
Okakura, K. *The Book of Tea.* Tokyo: Kenkyusha, 1919.
Sadler, A. L. *Cha-no-yu. The Japanese Tea Ceremony.* Rutland, Vt.: Charles Tuttle, 1933.
Shalleck, J. *Tea.* New York: Viking Press, 1972.
Ukers, W. H. *All About Tea.* New York: Tea & Coffee Trade Journal Co., 1935.

A Mug of Coffee

Recipe for good coffee:
Black as the devil, hot as hell,
pure as an angel, sweet as love.
C. M. TALLEYRAND (1815)

Gemaleddin, grand mufti of Aden, a city-state on the Persian Gulf during the ninth century of the Hegira (the fifteenth century of the Christians), had occasion to travel to Persia. While in the city of Kahvah, he saw people drinking a dark-brown, aromatic beverage which did not stupify, but seemed to lessen the cares of the world. Returning to Aden, he found himself indisposed and decided to try this drink. His scribe, Schehabeddin Ben, wrote of the recovery of the grand mufti. Observing that the drink lessened the desire for sleep, Gemaleddin recommended it to the dervishes, an especially religious sect, to allow them to pass the night in prayer with greater zeal and attention. Coffee had entered the consciousness of the Arabian world.

COFFEE OF ARABIA

There are many legends of how we first learned of this precious brew. It was said that goats near a monastery in the remote city of Kaffa returned one evening much friskier than usual and that a shepherd boy found that the flock had eaten of a shrub with dark red berries (Figure 6.41). The monks mixed the berries with hot water, tasted the brew, and praised Allah for the wonderous flavor and aroma. Another story is that Mohammed was visited in a dream and was told to go to a remote spot where he knelt down next to a tree with red fruit. He ate of the fruit and was suddenly filled with energy. Since we know that Abyssinian warriors of the second century A.D. made pellets of roasted, pulverized coffee beans with grease to be carried into battle, the history of coffee is certainly older than the Moslem faith.

From Aden to Mecca and then to the rest of the Arab world, the coffee-drinking habit spread as religious men of letters and laws took it up. The Imans and others in the religious hierarchy tried to restrict its

Figure 6.41
Fruiting branch of coffee, *Coffea arabica.*

use, but the pleasant taste and mild stimulation so captured the fancy of people that it moved down the social ladder to night watchmen and finally to the whole population. In about 1475 the inhabitants of Mecca began drinking coffee publically in small *kahvah khaneh*—coffeehouses. Businessmen, students, and workers gathered there in the afternoons to escape the heat and the press of trade and study. Dancing and gambling became part of the entertainment. In 1554, the opulent Café of the Gates of Salvation opened in Damascus, a model of luxury taken up by other establishments. The strict religious sects, outraged that this holy gift was secularized and thus profaned, petitioned the great Soliman to outlaw this impiety on the grounds that roasted coffee was a sort of burned wood denied the faithful. Soliman ruled that coffee was, indeed, contrary to law and the coffeehouses were shut down. This prohibition against coffee drinking in the sixteenth century was about as successful as was the prohibition against alcohol in the twentieth century; bribery, payoffs, and the equivalent of speakeasies prevailed. The question of charred wood was reconsidered, and the houses were reopened to general rejoicing, but hours were carefully regulated and the establishments were heavily taxed. Only once more were Arabian coffee houses closed. During the reign of Mohamet IV, his grand vizier received reports that local politicians were leaking information to fellow coffee drinkers and what was worse, were denigrating his administration. In spite of large tax losses, he closed the houses, but allowed street vendors to sell it. If found bootlegging coffee, the proprietors were bound head to tail on a donkey and flogged through the streets. When the sultan heard of this, he cancelled the edict, much to the gratification of the people.

The coffee was served thick and very black, without sugar, and sometimes with added cloves, anise, or cardomon. Turkish coffee, with the grounds left in the small cups, is identical to that prepared in the sixteenth century. Not allowed in the coffeehouses, women drank it at home, and it was also used during childbirth. If a man refused to keep his wife supplied with coffee, she had grounds for divorce.

COFFEE IN EUROPE

Coffee entered Europe in about 1615 when a Venetian brought home a supply to start a coffeehouse. It moved to France by 1640 and to England by 1650, although the English were not especially enamoured with it at first. By 1660, however, there was enough sipped in London to tax it at the rate of four pence the gallon. The revenue-generating potential of the coffeehouse was recognized throughout Europe, and various governments placed excises on the product. By 1700, there were 3000 coffeehouses in London serving as trade centers where businessmen, lawyers, and civil servants could gather for discussions. Daily newspapers and desks with writing materials were available, and the customers could receive their mail there. Lloyds of London, an international underwriting and insurance firm, started in a coffeehouse in 1690. Men spent so much time in the houses that in 1750 a group of women met with the city fathers to present a petition "representing to public consideration the grand inconveniences accruing to their sex from the excessive use of this drying and enfeebling liquor." The words "drying and enfeebling" suggest that there were rumors flying about London that intemperate use of coffee might interfere with the sex lives of the men. The very existence of the houses was a political matter as well, for tavernkeepers demanded protection for their product through higher taxation on coffee, and the medical profession was raising questions about indiscriminate use of coffee without strict medical supervision. The rabble preferred gin, and tradesmen had their own coffeehouses, called "penny universities" because entrance cost a penny and one could read the *Tatler* and the *Spectator* and engage in learned discussions on politics. Coffee was popular in France, rivaling wine among the aristocracy. Roasted beans were sold in apothecary shops, ostensibly under medical supervision, but by 1715 there were close to 50 *cafés* in Paris, and over 600 in 1750, and medical aspects were forgotten.

The Arabian city of Mocha, whose name is memorialized in coffee-flavored ice cream and cake frosting, was the center of the coffee trade, extracting higher and higher prices from Europeans who were drinking more and more coffee. It was obvious that this exceedingly valuable product could not remain a monopoly of the infidel, and attempts were made to cultivate the plant under European, and hence Christian, auspices. The Bourgermeister of Amsterdam, who was also governor of the East India Company, told the governor of Java to obtain trees—by any means possible—and establish a plantation in the hills above Batavia. This was accomplished by 1690. Louis XIV of France was given a coffee tree for his private gardens in 1714, and, charmed by the gift, loaned money to the East India Company to expand their plantations. In 1718, the Dutch planted trees in the New World in Surinam, the French followed suit in Martinique, and the English planted trees in Jamaica in 1727. Within ten years, the first plantings were made in Brazil, and by 1780 coffee trees were growing throughout much of Central America and Southeast Asia, with the bulk of Europe's supply under Dutch control. Ceylon became British in 1797 partly because of England's coveting of the Ceylon coffee plantations, but, ravaged by disease and neglect, these declined seriously, and tea again replaced coffee as the mainstay of the British home. Nevertheless, coffee was, throughout Europe, the preferred beverage of intellectuals and the aristocracy, and paens of praise were written in its honor; J. S. Bach even composed a coffee cantata.

Today, the world thinks of coffee as a North American drink, and we consume about 30 gallons per person per year as compared with less than eight gallons of tea. Although the first coffee house opened in Boston in 1670, the habit really got started in New Amsterdam because of the Dutch influence and continued when the city became New York. The nascent Americans obtained their beans from British Jamaica and, aping London, established coffeehouses in the business district. The King's Arms on Broadway near Trinity Church and several others played roles in the intellectual ferment leading to the Revolution.

BOTANY OF COFFEE

The coffee tree or shrub is a member of the Rubiaceae, a botanical family which also includes the fever-bark tree (*Cinchona*), the gardenia, and the common bedstraw (*Galium*). There are many species in the genus, but only two, *Coffea arabica* L. and *C. canephora* are of much economic significance. The plant grows well in tropical areas where rainfall exceeds 100 cm/year, and, because it does best at stable temperatures close to 20° C., mountain environments in South and Central America, East Africa, and Java produce superior beans. The fruit is technically a two-seeded berry, each seed surrounded by a sweet, yellowish edible flesh and a red skin. A tree will flower once or twice a year, producing about six pounds of beans. A hectare of a well-managed plantation will yield close to a metric ton per year. Berries are picked by hand (Figure 6.42), subjected to a floatation process to separate defective fruit, leaves, and stems, and then depulped by machine. Residual pulp is removed by allowing the beans to ferment in tanks. The beans are thoroughly washed and dried. The flavor of the final product is the result of many interacting factors, including the variety of the plant, growing condi-

Figure 6.42
Coffee berries are hand-picked so that only fully ripened fruit will be processed. Courtesy Pan-American Coffee Bureau.

tions, location of the plantation, and, above all, the roasting process. Clean, dry beans are heated to 250° C. for five minutes to develop the essential oils and other compounds that provide aroma and flavor and to dry the cell walls to facilitate grinding. Roasted beans from different areas and from different harvests are blended to obtain a uniform, distinctive product and to produce a distinctive product for each of the companies which sell coffee. Prices are established through an International Coffee Agreement signed by 60 producing and consuming countries to stabilize prices and to establish quotas. The United States and Canada, the largest consumers, do not grow coffee and are understandably interested in keeping prices low to protect their consumers and their international balance of payments; producing countries (Brazil, Angola, Kenya) want high prices for the same reasons. Agreements are hammered out after negotiations that last for months.

Probably the greatest threats to stable supplies of coffee are bad weather and disease. In 1974, close to 8 percent of Brazil's coffee trees were killed by frost, and a single frost in July, 1975, damaged 70 percent of the trees. It takes three to five years to bring damaged plantations back into production, and such frosts invariably increase the cost of coffee throughout the world. Several plant diseases plague the plants including bacterial, viral, and fungal pathogens. The most serious and dramatic pathogen is the coffee rust, *Hemileia vastatrix*, a fungus. British-owned plantations in Ceylon, which at one time accounted for a significant portion of world trade, were afflicted with the disease. It appeared in Ceylon in 1850–1860 and by 1875 had spread throughout the plantations. In 1892, coffee production in Ceylon was at an end. The Central and South American plantings were free of rust until 1970, but the problem is now receiving anxious attention from growers and consumers.

ADDITIONAL READINGS

Ellis, A. *The Penny Universities: A History of the Coffee Houses.* London: Secker & Warburg, 1956.
Jacobs, H. E. *The Saga of Coffee.* London: G. Allen, 1953.
Mayer, H. *Old English Coffee Houses.* Emmaus, Pa.: Rodale Press, 1954.
Ukers, W. H. *All about Coffee,* 2nd ed. New York: New York Tea & Coffee Trade Journal Company, 1935.
Uribe, A. *Brown Gold: The Amazing Story of Coffee.* New York: Random House, 1954.
Wellman, F. L. *Coffee.* New York: Wiley, 1961.

Hot Chocolate,
Devil's Food Cake, and Cola

> Howard Johnson's sells chocolate ice cream
> and 27 alternates for the real thing.
> RICHARD M. KLEIN, 1978

As part of the rituals of coronation and important religious ceremonies, priests of the Aztec empire would reverently sip liquid from a
golden goblet reserved solely for this libation. Legend said that the
plant from which the drink was made had been a gift to mankind from
the god of the air, Quetzalcoatl, who was punished by other gods for
giving a thing so precious to lowly beings. Quetzalcoatl believed, however, that only through partaking of this beverage could humans communicate with Him and tell Him of their needs. *Chocolatl* was invested
with sacred powers due in part to the relatively high theobromine alkaloid content. Because trees were scattered in the tropical rain forests in
the Mexican highlands, they were rare and, hence, sufficiently expensive to warrant reserving their fruit for the emperor and the priests.
Western man first saw the *cacahuatl*, the pod-shaped fruits, when Columbus was given several on his first voyage to the New World; he included them in the gifts laid at the feet of Ferdinand and Isabella when
he returned to Madrid in triumph. The significance of this gift was lost
on the Spanish court, for no one had any idea of what they could be
used for, and, eventually, as the enthusiasm for what Columbus had
discovered died, the pods were simply discarded.

CHOCOLATE IN MEXICO

Only when Cortez reached Mexico in the early sixteenth century
did the whites learn the delicious secret contained in the pods. The
Aztecs believed that a pale-skinned god would come to their land in a
vessel with white wings, so when the Spanish fleet hove into sight in
1519, the ships' crews and bearded chief god clothed in a shiny skin
were greeted with great joy. Coming as they did, just at the time of the
coronation of Montezuma II, it was a most auspicious concordance.
Cortez was led to the royal palace at Temistitian, and with all the pomp
the priests could muster, the white god was presented with a golden
goblet containing the ritual drink from the hands of the young emperor
himself. Cortez tasted a very bitter, very hot liquid compounded of
dried, coarsely ground beans of the revered tree mixed with unknown
spices and whipped to a foam in hot water. Not being used to it, Cortez
became somewhat dizzy, but not too dizzy to note that the goblet was
solid gold. According to Bernal Diaz, scribe of the expedition, Cortez
thought that Mexico must be laden with gold and, then and there, the
decision was made to subjugate the land and its people and take what

gold could be found; the fate of Montezuma had been sealed with his own hands.

CHOCOLATE IN EUROPE

By 1550, Spanish galleons were carrying precious metal and many plants back to Spain, among which were pods and seeds of the chocolatl plant. The court of Spain developed its own version of the drink of the gods:

> Take 100 cocoa beans, two pods of chili pepper [also a New Spain import], a handful of aniseed, six roses of Alexandria, a pod of logwood, two drachmas of cinnamon, a dozen almonds, a dozen hazelnuts and one half pound of loaf sugar. Pound these things into a paste, add one pint of boiling water and stir vigorously over a fire. Drink while still very hot.

Word of this new drink soon reached the courts of other countries of Europe. Demand from France, Portugal, Italy, and as far away as the court of the tsar for the cocoa beans reached a level high enough to warrant putting the conquered Indians, now virtually slaves, to work clearing land and starting plantations of chocolatl. Methods used to prepare the beans for making a drink was a Spanish secret for about 50 years, until an unknown Spanish monk told the method to others, thereby allowing factories to be started in France and Italy. The raw material was, however, still a monopoly of Spain. Spain advertised the virtues of chocolate throughout Europe, asserting that it was the religious drink of the Aztecs, whereupon the Eastern branches of the Roman Catholic Church tried to ban its use as pagan. The wealthy commissioned chocolate sets from Meissen to match the coffee services that were "all the rage." Royal approbations and the tendency to want to keep up with the Joneses assured an interest in chocolate among the upper classes and the upper middle classes; the lower classes couldn't afford it. By the 1650s there were chocolate houses just as there were coffee and tea houses in London and Amsterdam; however, although the cocoa habit spread throughout Europe, it never achieved the popularity of the other drinks. And no wonder. Public acceptance was limited not only by the price, but by the very high amount of fat in the beans. This cocoa butter, a semisolid fat which constitutes up to 40 percent of the weight of the beans, was a severe problem; it made grinding difficult, became rancid, and would develop an unpleasant gray color as it oxidized. Cups of hot chocolate had the annoying habit of coagulating as the liquid cooled. In spite of these drawbacks demand rose, and Spain developed plantations in Trinidad, the Antilles, and Venezuela, maintaining their monopoly on production. They left the processing business which was taken over by the Dutch.

PROCESSING

With the availability of relatively inexpensive sugar, the bitter taste of chocolate could be modulated, but the hand-grinding of the beans still limited production and kept prices high. By 1800, the power of the steam engine was harnessed to stones used to grind cocoa beans and a much smoother paste could be made in a much shorter time. The steam engine's power also allowed the use of presses to squeeze out part of the cocoa butter, resulting in a more stable product with fewer annoying characteristics. In 1876 a Swiss, Daniel Peter, added milk to dark chocolate paste to produce milk chocolate candy, now a significant fraction of the international cocoa business. The Peter process was further modified by the Hershey Company to give a much smoother product in which nuts, raisins, and fruit could be incorporated. At about the same time, chocolate with a controlled cocoa butter concentration was developed, and cooking processes were modified to produce dark, sweet, or milk chocolates which were smooth and velvety. These fondant chocolates completely replaced the less satisfactory product, lowered the price, reduced rancidity, and world-wide interest in chocolate reached new heights. From fondant, the production of a stable powdered product, cocoa, involved only a few simple manufacturing steps, steps developed by the Dutch, whose plantations of chocolate trees in Java had broken the Spanish monopoly.

The cocoa tree is a member of the Sterculiaceae, a small family that contains the cola-nut tree, whose fruit is also a source of caffeine alkaloids. Like coffee, cocoa grows best in regions of high humidity with stable temperature—conditions much like those found in a warm greenhouse—areas which include only tropical zones within 20 degrees north or south of the equator. Cocoa beans develop in a pod (technically a berry) from small, odorless, pink-to-white flowers that develop directly from the trunk and main branches (Figure 6.43). Each pod is 10–12 cm long and contains about 40 bean-shaped seeds, each sur-

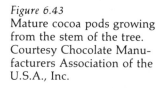

Figure 6.43
Mature cocoa pods growing from the stem of the tree. Courtesy Chocolate Manufacturers Association of the U.S.A., Inc.

Figure 6.44

Figure 6.44
Cocoa pod opened to show seeds. Courtesy U.S. Department of Agriculture.

rounded by a white, sweet, and juicy pulp (Figure 6.44). When removed from the pod, the pulp ferments and the seeds turn brown. These seeds are placed on racks in the sun for a week, during which time they lose water and begin to develop the color, odor, and flavor we associate with chocolate. Large plantations now substitute mechanical drying to save time and reduce the chances of decay. After cleaning and grading, beans are bagged for shipment to chocolate factories where they are cleaned and prepared for the critical roasting step. Heating to 140°C. for several hours converts the beans into a product recognizable as chocolate.

The production of industrial chocolate involves grinding beans in steel rolling mills with sufficient heat to melt cocoa butter, to form a dark-brown, oily mass called cocoa liquor. The liquor can go through cocoa manufacturing steps or into the production stream leading to chocolate. In cocoa production, liquor is pressed to remove cocoa butter (reserved for other uses including cosmetic manufacture) leaving cocoa press cake to be crushed into a powder. The product contains up to 20 percent fat and is very bitter, still unsuitable for drinking. The bitter taste and some of the alkaloid is removed by treating the powder with alkali, the Dutch process discovered in the late nineteenth century by two Dutch firms, Droste and van Houten (Figure 6.45).

Baking and eating chocolate are manufactured from cocoa liquor by continued grinding and blending to produce an extremely smooth product with controlled amounts of cocoa butter. Emulsifiers are added to prevent separation of the chocolate and fat, and other substances may be added to impart special characteristics to milk chocolate, semi-sweet chocolate, etc.

Figure 6.45
Droste cocoa box advertising
the "Dutch Process."

A new era in world chocolate production was ushered in at the end of World War II when the Dutch plantations were destroyed. A few seeds, casually picked up by an African visitor to a West Indian plantation, were brought to Ghana to start the cocoa industry in Africa. With the development of plantations in Brazil, these two countries came to dominate the world market. Well over 85 percent of the world's supply comes from these countries, with Ecuador, Venezuela, and the West Indies supplying the rest of the crop. Many plantations in Java were destroyed by a virus disease which struck in the 1930s, and the remaining plantations were decimated by monkeys whose numbers soared after 1945 as leopard coats became fashionable. Current production is close to 900,000 metric tons, with North America the major consumer. World prices have increased several fold since 1970 due to demand, poor growing conditions, disease, and the insistence of African producers that they obtain a fair price for their major export product.

Nutritionally, chocolate is a fine food with about 200 calories, 3 gm of protein, and 16 gm of fat per ounce. A cup of hot chocolate contains about the same concentration of caffeine-type alkaloids as does a cup of tea and is an effective, mild stimulant.

OTHER CAFFEINE SOURCES

Although tea, coffee, and cocoa are the major caffeine-containing beverages, several other plants contain enough alkaloid to be eco-

Figure 6.46
Fruiting branch of cola, *Cola acuminata*.

nomically and socially important. Most familiar to North Americans is the cola-nut tree, *Cola nitida*, a member of the cocoa family. Several other species contain caffeine including *C. acuminata* and *C. verticillata*, but these are used locally in western Africa as a masticatory—something to chew on—and have not gained popularity in the Western world. The plant from which we get cola is a large shrub with glossy leaves and a fruit containing reddish seeds (Figure 6.46). The seeds contain about 2–3 percent caffeine plus essential oils, giving a pleasant flavor, and the extract serves, as do all other caffeine-containing products, to relieve fatigue and hunger. Dried seeds are produced on small plantations in tropical Africa in a belt extending from Ghana to Sierra Leone. The caffeine and flavorings can be extracted with hot water to make a syrup whose main economic use is in the fabrication of cola drinks.

The most famous of these products is Coca-Cola which, with blue jeans and the electric guitar, may be among America's more important contributions to modern civilization. Invented by an Atlanta, Georgia, pharmacist, John S. Pemberton, it was first marketed in May, 1886, through the new-fangled soda fountains popularized at the Centennial Exposition of Philadelphia in 1876. The original syrup, mixed only with plain water, contained sugar and caramel colorings with extracts of cola for caffeine and flavoring and an extract from the coca leaf which provided additional flavoring and a very small amount of cocaine.

The matter of supplying the public with a product which might contain even traces of a dangerous drug, cocaine, was of concern to the federal government. The first federal food and drug act became operational in 1906 and the first federal narcotic act in 1912. Shortly after 1900, the Coca-Cola Company asked the U.S. Treasury Department, who administered the narcotic acts, to supervise the extraction of the flavoring from coca leaves to ensure that no cocaine entered the mixture used to make the basic syrup. This control is still being carried out. The basic formula, known as 7X, is still company property, but all ingredients have been approved by the Food and Drug Administration.

The cola beverage of Brazil and other countries in South America is made from the fruit of the Guarana shrub (*Paullinia cupana sorbilis*). The small, black seeds are pounded into a paste with tapioca flour and mixed with hot water. The caffeine concentration is close to 5 percent, among the highest of caffeine drinks, and there is a large quantity of tannin, giving the drink a strong taste. People in eastern Equador and parts of Peru drink a water extract of Guayusa leaves, a plant in the holly genus (*Ilex guayusa*) closely related to the Agentine maté. The leaves were originally used by the head-hunting Jivaro Indians to keep them alert for night battles, and when the Society of Jesus found out about this, they started selling the leaves as a cure-all. Although the Jesuits were expelled from Ecuador in the middle of the eighteenth century, the plantations established at the Catholic missions are still being harvested and a few new ones planted.

ADDITIONAL READINGS

Chatt, E. M. *Cocoa Cultivation, Processing, Analysis.* New York: Wiley, 1953.
Mandulay, L. S. *The Chocolate Tree.* Washington, D.C.: Pan American Union, 1966.
Urquhart, D. H. *Cocoa,* 2nd. ed. New York: Longman, 1961.

Spices and Savory Herbs

Are you going to Scarborough fair?
Parsley, sage, rosemary and thyme.
ENGLISH FOLKSONG

Authorities on spices and herbs suggest that the desire for these substances in the Middle Ages was due to the wish to control the development of strong odors and tastes of meats that had been sitting around too long. It is reasonable to believe that they were also employed just as we use them, to make food more appetizing and less boring. Consider a stew made without bayleaf, pepper, an herbal bouquet, a dash of mace, and another dash of ground cloves—dull and uninteresting. Pickles made with only salt are merely salted cucumbers, and steamed spinach without a sprinkle of nutmeg has turned millions of children against this succulent green vegetable. North American supermarkets display well over 50 spices and herbs, but we use few of them because of lack of imagination. We also have the North American ignorance or distain for the wonderful cooking of non-European cultures.

There is considerable nomenclatural confusion about spices and herbs. The word herb is itself misleading. Botanically, it refers to a plant that is either an annual or that dies down to the ground each fall. Medically, it refers to plant material from which drugs can be extracted. Gustatorially, it means the leaf of a plant used in cooking. Spice is an even more obscure word; Christopher Morley defined spice as the plural of spouse, and under happy circumstances a spouse can be the spice of one's life. For our purposes, we will operationally define a spice as a plant part such as a seed, stem, or root. The two terms are not mutually exclusive; the confusion can be left to lexicographers and ignored by botanists and chefs.

SEASONINGS IN EUROPE

Since our food habits are those of Europe, a look at the seasonings available in the eighth century can be instructive. The northern countries had few seasonings. Meats were dipped into pulverized mustard or into horseradish root pounded with vinegar and salt (Figure 6.47).

Figure 6.47
Pounding horseradish root. Taken from a German woodcut published in 1493.

Figure 6.48
Dill, *Anethum graveolens.*
Leaves and stems are dill-
weed; the seeds are used as
spice in soups, meats, and
salads. Courtesy U.S. Department of Agriculture.

Dill seed (Figure 6.48) and young shoots, celery seed and leaves, wintergreen and peppermint leaves, and garlic, onions, and leeks were
available. Winter savory and thyme were not generally used, although
the plants grew in Germany and Scandinavia. The flavoring spectrum
in southern Europe was considerably broader. In addition to peppermint, the Mediterranean basin supported the growth of many members
of the mint (Labiatae) family, including basil (*Ocimum basilicum*), marjoram (*Marjorana hortensis*), oregano (*Origanum spp.*), rosemary (*Ros-
marinus officinalis*), sage (*Salvia officinalis*), savory (*Satureia hortensis*) and
thyme (*Thymus vulgaris*). It is interesting to note that all these go exceedingly well with tomatoes introduced into Italy in the sixteenth century. The parsley (Umbelliferae) family was also well represented in
southern Europe. Caraway (*Carum carvi*) was carried by Roman legions
from its home in Asia Minor throughout the known world for stews, as
seasoning for northern Europe's breads and sauerkrauts, and for addition to the Scandinavian aquavit and the German *kümmel* schnapps,
either of which is sipped with beer while eating herring and boiled potatoes. Chervil (*Anthriscus cerefolium*) was used as an alternate for parsley, and in England "its tender tops . . . are never wanting in our
sallats." Coriander (*Coriandrum vulgare*) was used in sausages and our
hot dogs still contain it, while cumin (*Cuminum cyminum*), initially from
North Africa where it is used in *cous cous*, was so valuable that it was
used to pay taxes in Roman Palestine: ". . . you tithe mint and dill and
cummin and have neglected weightier matters . . ." (Matt. 23:23).
Anise (*Pimpinella anisum*) was used in baking, and Romans hung a plant

in their bedchambers to prevent nightmares. Charlemagne required that the plant be grown on imperial lands as his personal condiment; its licorice flavor was used in cordials, to cover up the bad taste of medicines, and in the preparation of cookies. Its addition to cottage cheese and to the water used to boil shrimp is recommended. Fennel (*Foeniculum vulgare*) and celery (*Apium graveolens*) were grown as vegetables, and their seeds were and still are used in sausages and stuffings for fowl, and both provide a most agreeable addition to boiled vegetables.

Yet, with the exception of mustard and horseradish, these are relatively mild seasonings, and the palate-stimulating and saliva-generating flavors and odors of pepper, bay, clove, cinnamon, nutmeg and mace, ginger, and others were almost lacking from the foods of Europe. One has to say almost, not absolutely, because small and very precious supplies of the spices of the Orient were known as far back as the time of ancient Egypt.

In 1874 George Elbers, a German Egyptologist, discovered a long papyrus roll now known as the Elbers papyrus. It was a medical treatise in which was listed herbs and spices used to treat disease and in the medico-religious art of embalming. The body cavity of the corpse was treated with ground cinnamon, cumin, anise, and marjoram plus cloves and nutmeg. Cinnamon and cloves were soaked in sweet oil for anointing, and stick cinnamon was burned as incense to purify sick rooms. Since these plants did not grow in the Nile Valley, they were transported from Madagascar, Indonesia, and India to the shores of East Africa, through the Gulf of Aden and the Red Sea, and finally overland to Thebes. The same route was also used to carry cotton cloth from India to the land of the pharaohs. From Egypt, cargoes of spices could be carried across the Mediterranean to Greece. Herodotus reported in the fifth century B.C. on cinnamon sold by Arabic merchants. Pepper, cinnamon, ginger, and other spices were added to wine and bread in addition to their use in medicine. Hippocrates included saffron among the effective herbal medicines and recommended black pepper in honey and vinegar for "feminine disorders." We know that in addition to the sea route, there was at least one overland route. Consumed with envy, Joseph's brothers got rid of Jacob's favorite by selling him to a caravan of Ishmaelites bearing spices to Egypt (Gen. 37:25). When the Queen of Sheba visited Solomon, she brought with her "a very great retinue and camels bearing spices" (2 Chron. 9:1). Some were not known to the Israelites for 2 Chron. 9:9 states that ". . . there were no spices such as those which the Queen of Sheba gave to King Solomon."

SPICE CARAVANS
BY LAND AND SEA

By the third century B.C., caravans of camels were carrying the vaunted spices of the Orient to the civilized world of the Mediterranean. Some of these spice and silk routes were very ancient. One led from the port of Muziris through the Persian Gulf to Charax, then up to Aden and through the Red Sea. A land route went up the Indus Valley and along the Kabul River, climbed the mountain passes of the Hindu Kush, and entered the Middle East at Bactra. Caravans based in Bactra transported the goods to modern Iran, thence overland to Antioch or by sea to Egypt where merchants repacked them and shipped them across the Mediterranean to Europe. A caravan would usually spend two years on a trip, and the dangers of dust storms, bandits, thirst, and starvation served to keep prices so high that a bit of cinnamon bark or a peppercorn could be worth its weight in silver. For the most part, control of the spice trade was an Arabian monopoly. Rome almost broke the Arabian monopoly when a merchant discovered that the wind systems in the Indian Ocean reversed their direction twice a year. The April-October monsoons favored the journey from Aden to India, and the October-April winds permitted the return voyage. It still took a full year for the round trip, but this was half of what the journey formerly required, it was much safer, and it allowed bigger payloads than the caravans. Prices began to drop with a corresponding increase in spice utilization. As the ultimate in conspicuous consumption, Romans piled their funeral pyres high with cinnamon, cloves, and nutmeg, anointed their bodies with fragrant oils, and added "heat" to their foods and wines with numbers and quantities of spices that modern people would find unpalatable.

The extension of Roman civilization to northern Europe and to Britain was accompanied by the introduction of spices to these lands and also contributed to the downfall of Rome. The semicivilized northern Goths tired of paying tribute and envied the foods, wines, and silks of the Romans. When Alaric appeared at the walls of Rome in 408 A.D., he demanded gold, silver, silk, and 3000 pounds of peppercorns in exchange for his promise not to sack the city. Although the tribute was paid, Alaric returned two years later and took Rome, thus ending the hegemony of the empire. Constantinople had been built in 330 A.D. to serve as a trade center for goods coming from the East. Cloves and nutmeg were brought by Indonesian sailors from the Molluca Islands to Aden where they reached Constantinople via the Red Sea. In the early sixth century a monk, Cosmas Indicopleustes, traveled to India and Ceylon and described the spice industries, including the cultivation of pepper and the smell of the clove trees that wafted on ocean breezes a hundred miles from the islands. Europe was interested in this report, but the descent into the dark ages had begun, and it became impossible for merchants to mount the expeditions necessary to exploit these resources.

Mohammed (570–632), in addition to being the prophet of Islam, author of the Koran, and founder of a legal system, was himself a spice merchant. He worked as a boy for spice merchants in Syria and became a camel driver and a caravan leader before marrying his employer, the widow Khadija. After his death, the crescent swept from Spain to the borders of China and south into Ceylon and Java, again establishing the monopoly on spices held in pre-Roman times. Basra, situated at the head of the Persian Gulf where the Tigris and Euphrates met, became the primary trade center between East and West. Arabic apothecaries invented distillation to permit the extraction of the essences of the spices, resulting in the development of perfumes and essential oils like oil of cloves and oil of peppermint. The wealth accumulated from these activities paid for missionary work, facilitated the development of arts, sciences, and medicine, and permitted the flowering of a sophisticated culture system. It also increased the hate and envy of Christian Europe which found itself completely dependent upon the infidel for life's necessities such as silk, pearls, jewels, and spices. The desire to wrest Jerusalem from the Muslims was not due solely to heightened religious sensibilities, but included a large component of economic self-interest on the part of the nobility and the Church. With the slow demise of feudalism in the ninth century and the development of centers of commerce, craftsmen and artisans organized into guilds. The merchant class, the bourgeoisie, recognized that it could make common cause with the nobility and Church if it had direct access to the goods of Cathay—a term that included both China and India.

The first Crusade was mounted in 1096 and Jerusalem was freed in 1099, with concomitant exposure of Europeans to luxury that they had not dreamed existed. Venice and Genoa were the main ports from which supplies of men and material flowed to Palestine and these two city-states became wealthy. During the fourth Crusade of 1204, Venetian merchants financed the destruction of the power of Constantinople, thus eliminating a hated economic rival from competition in the spice and silk trade. Venetians converted Constantinople into a supply depot with financial control centered in Venice. New spice routes led to Baghdad, to Trebizond on the Black Sea, and then to Constantinople where the goods were transshipped to Venice. In exchange, Venice bought European grain, glass, wine, and woolen cloth, and shipped them to Constantinople for transfer to India and China. Venice developed economic power never before experienced in Europe, power and wealth that matched that of the principal cities of China and India. This wealth, and the availability of leisure that accompanies wealth, culminated in a patronage of the arts resulting in Titian, Tintoretto, Veronese, and Giorgione; in the construction of Venice as we see it today; and to the magnificence of many of the art objects that constitute much of our Western heritage and contribute to our standards of beauty. Merchants in Nuremberg, Antwerp, Bruges, Paris, and London established ties with Venice through the good offices of the Vatican and secondary centers of art and culture developed. By 1180, during the reign of Henry II, a pepperer's guild of merchants who "sold in gross" was organized in London, later to become the powerful Grocer's Company which financed the British East India Company in the sixteenth century. Apothecaries used the Arabian distilling apparatus, the alembic, to make medicinal essential oils including oil of cloves for toothaches, rubs for sore muscles, and unguents.

In the middle of the thirteenth century, Nicolo and Maffeo Polo, businessmen of Venice, traveled east to establish connections with the great Khan, Mongolian emperor of the Yüan dynasty of China. In 1269, Nicolo's son Marco accompanied his father and uncle on a second trip, and 26 years later, three ragged men returned to Venice carrying pearls, diamonds, sapphires, and emeralds in the seams of their clothing. While Marco, then a middle-aged man, was a prisoner of war in Genoa, he wrote his *Travels* which whetted Europe's interest in things Chinese. Marco described the mouth-watering flavor of ginger-spiced pork, Peking duck rubbed with nutmeg, star anise, and honey, and other foods and drinks unknown to Europeans. He provided detailed descriptions of pepper, nutmeg, clove, and cardamom plants and told almost unbelievable stories of marble palaces, paper money (Chapter 8), and a thick, black liquid that burned with a hot blue flame. Venice was opportuned to bring more and more of these wonders to the wealthy of Europe, and merchants happily obliged.

The safety of the caravans was, however, a matter of concern. Constantinople had fallen to the Turks in 1453, and Muslims were reasserting control over the caravan routes through Persia and the fabled cities of Damascus and Samarkand. Each caravan was being harassed and tribute exacted. This increased costs, and although the costs were being passed on to the customers, demand was falling off. Nevertheless, Europe kept Venice rolling in wealth—and this was translated into economic, social, political, and religious power. The doges of Venice could and did dispute the pope, and their arrogance and greed was unfavorably noted throughout Europe.

AROUND THE CAPE OF GOOD HOPE

Under the leadership of Prince Henry the Navigator, Portugal established a naval college at Sagres to train ship's officers. Henry believed that the economic future of his small country depended on the exploration and development of new sea routes for trade and commerce. Navigators, astronomers, and geographers staffed the school; new and better charts were prepared, and more efficient and accurate instrumentation was developed. Working their way slowly down the western coast of Africa, Portuguese ships found the Madeira and Cape Verde Islands. In Africa Guinea, they found the aromatic spicy seeds of *Amonum melegueta*, the Melegueta pepper—a plant related to ginger but one that could partly replace true pepper and could be sold cheaper than the black pepper controlled by Venice. In 1486, Diaz circumnavigated the Cape of Good Hope, demonstrating that the Indian Ocean could be reached by sea without need to force the Red Sea controlled by the Arabs. Stimulated by the reports that the Genoese admiral Columbus had found a western route to India, King Manuel I ordered Vasco da Gama to find a southern route to the Orient. Da Gama charted the eastern coast of Africa, and in 1498 reached Mozambique before sailing northeast across the Arabian Sea to Calicut on the west coast of India. Waiting almost six months for the winds to shift, da Gama retraced his outward journey and in spite of an outbreak of scurvy, reached Lisbon in August of 1499 with precious jewels, spices, and an agreement with the zamorin of Calicut to exchange spices and sapphires for gold, silver, and coral. Halfway around the world, Columbus's crews discovered allspice in the West Indies, red (*Capsicum*) peppers, and tobacco.

Anxious to secure all trade rights, the Portuguese mounted another voyage to Calicut and in 1505 sent out 16 ships under the command of Don Francisco de Almeida. Capturing and staffing naval supply bases at Quiloa and Mombasa, the fleet subdued Ceylon, Malacca, and other spice ports, established factories for the processing and packing of spices, and—through agreements with local rulers—assured the control of all the seaport-based spice trade of Asia with the exception of China. The Portuguese East India Company monopolized the pepper trade and drove up prices to the point that many spices were reserved solely for the very wealthy. Ferdinand Magellan, trained in Prince Henry's naval college, hired himself out to Charles V of Spain to find another route to the Spice Islands, since Spain realized that Columbus had merely discovered a New World devoid of cinnamon and black pepper. Although Magellan was killed in the Philippines, one ship of his fleet completed the circumnavigation of the world in 1522. Cloves, nutmegs, mace, and cinnamon carried by the survivor more than paid the Spanish crown for the cost of the trip. Captain Sebastian del Cano, master of the surviving ship Victoria, was rewarded with a handsome pension and a coat of arms that included two crossed cinnamon sticks, three nutmegs, and twelve cloves.

EAST INDIES COMPANIES

With the success of the Portuguese and the Spanish as a goad, Britain, Holland, and France formed East India companies to find new sources for spices and new routes. Sir Martin Frobisher discovered Labrador while searching for a northwest passage to India. Hudson and Cabot for England, Champlain for France, and the founders of New Amsterdam were actively probing the northeast coast of America for inland waterways, while Sir Francis Drake visited the Spice Islands on his voyage around the world. Spain was exploiting Mexico, Peru, and Central America. Portugal established bases in Brazil, and red pepper, allspice, and vanilla were being packed into the galleons sailing back to the Iberian peninsula. Venice, being too conservative to mount expeditions beyond the Mediterranean, was reduced to the status of a tourist's museum. By the beginning of the seventeenth century, Britain and

Holland were—by guile and force of arms—undercutting Portugal in the Orient. By 1630 the Dutch had seized Malacca and secured control of most spice production in the east. England bullied her way into control of India and established tenuous trade relationships with the Ch'ing dynasty through Canton and Singapore, driving the Portuguese from Macao and other off-shore islands. Britain and Holland vied economically and militarily for hegemony over the lands of the Indian Ocean. After more than a century of sometimes bitter fighting, the two countries signed a treaty in Vienna in 1824 under which Holland received the Malay archipelago except for North Borneo, and England controlled India, Ceylon, Singapore, Hong Kong, and much of the coastal areas as far west as Siam. These islands were, as it turned out, not only valuable for spices, but also for rubber, tea, and other plantation crops.

During the fourteenth to sixteenth centuries, spices were collected from wild trees scattered throughout forested areas. Pepper vines were cultivated, ginger planting was a cottage industry, and poppy seed was imported from modern Pakistan. The Europeans recognized that yields could be improved and production efficiency increased if spice trees were grown in plantations. Cinnamon was introduced into the Seychelles by France, an example of what was considered theft and piracy by the British and Dutch, who imposed the death penalty for anyone caught smuggling spice-planting stock. Even today, authorities of Zanzibar are empowered to kill anyone caught smuggling cloves into Kenya, but such restrictive measures have little meaning; indeed North America is supplied with many oriental spices now being grown on plantations in Central and South America.

PEPPERS

Although it is not feasible to discuss all or even the majority of spices and herbs available to us, certain of them require some discussion. Among these, black pepper is vital (Fig. 6.49); in terms of volume consumed, it is our most important spice. Apparently native to the Malabar coast (also the home of cardamom, bananas, and sugar cane), pepper cultivation moved eastward to the Malay peninsula, Indonesia, India, and China, and has been cultivated throughout Asia and Micronesia for untold centuries. Our name for the spice is derived from the Sanskrit *pippali*. Botanically, the plant is *Piper nigrum* in the Piperaceae, a small family with few other economically important plants except the peperomias grown as house plants. Although now grown in South America, the bulk of the world's supply comes from India, Indonesia, and the Malagasy Republic (Madagascar). Grown from cuttings to ensure genetic uniformity, the fruits are harvested just before they ripen, and they darken into the typical black peppercorn as they dry. When the entire fruit is ground, black pepper results, but if the outer hull is removed, white pepper is produced. White pepper is less pun-

Figure 6.49
Pepper, *Piper nigrum*. Taken from *Historia Generalis Plantarum* by Jacques d'Alechampes published in 1587.

gent than black. The sharp taste of pepper is due to the presence of a resin, an essential oil and an alkaloid.

The fruits of two species of the genus *Capsicum* in the tomato family are almost as important a seasoning as is true pepper. The genus contains the familiar bell peppers which are not hot at all, but the fruit of *Capsicum frutescens*, the red or cayenne pepper, can blister the insides of your mouth. The genus is native to Peru, where seeds have been unearthed from prehistoric burial sites in the Andes. Peter Martyr, who accompanied Columbus on his second voyage in 1493, brought back to Spain "peppers more pungent than those from the Caucasus," and, he added, "there are innumerable kinds . . . the variety whereof is known by their leaves and flowers." From Spain, the plants spread throughout Europe and the Near East for, since they could be easily grown as far north as Denmark, they were an inexpensive substitute for black pepper. Our name chili pepper is Spanish, derived from the initial collection of seed in Chile. The chili con carne we associate with Mexican cooking is actually a nineteenth-century invention of the Texas *gringos;* commercial chili powder contains red pepper pods, plus the European herbs cumin, oregano, and garlic with the addition of salt, cloves, and allspice. In addition to its use in Tex-Mex dishes, scrambled eggs, tomato juice, and even soups can be spiced up with judicious additions of chili peppers either as cayenne powder or chili powder. The Chinese started to grow chilis in the sixteenth century when a few pods were brought to Canton by the Portugese. They were gratefully accepted in Szechwan and Hunam provinces where the people liked "hot" foods. Szechwan paste is a fiery mixture of ground chilis and garlic in oil and is a major ingredient in Kung Pao chicken, a dish named for a Pekinese official who settled in Szechwan.

KUNG PAO CHICKEN
(FOR TWO PEOPLE)

1 chicken breast boned, skinned and cut into ½-inch strips
½ egg white beaten lightly
1 tsp cornstarch
2 tbsp Chinese brown bean sauce
¼–½ tbsp Szechwan paste
1 tbsp brown sugar
½ tbsp rice wine or dry sherry

1 tbsp white vinegar
3 peeled and crushed garlic cloves
2 tbsp oil
½ cup fresh, unsalted peanuts
2 whole chili peppers (more if you can stand it)
¼ tsp salt
⅛ tsp MSG (Accent)

Combine chicken strips with egg white, salt, and MSG. Refrigerate for 1 hour. Combine bean paste, Szechwan paste, vinegar and garlic; reserve. Heat oil in wok or skillet and when very hot, add peanuts and cook until they are medium brown (about 30 seconds). Remove peanuts with slotted spoon to absorbent paper. Add chicken and egg white to hot oil and stir rapidly for no more than 1 minute. Remove chicken, place chili peppers in oil, and heat for no more than 10–15 seconds until brown. Return chicken to wok, add bean paste—Szechwan paste mixture and heat briefly. Sprinkle with peanuts. Serve with steamed rice and cold beer.

Two other important seasonings are from the fruits of plants in the same genus. Pimentos decorate potato salads and are used to stuff olives. Roasted peppers, a necessity in antipasto, are thick-fleshed peppers (*Capsicum annuum*), a species that also includes the red or green bell peppers. Cultivars may also be dried, deseeded, and ground to a powder—the paprika of commerce. Two kinds of paprika are available, the milder, bright-red Spanish types which are used primarily to garnish foods, and the lighter and spicier Hungarian types. The latter were introduced into Hungary when the Turks invaded that country in the late sixteenth century; the word paprika means "Turkish pepper." With paprika as a base, Hungarians developed the distinctive goulash, essentially beef or veal stews thickened with sour cream with lots of paprika.

PRAGUE VEAL GOULASH
(FOR FOUR)

1½ lb shoulder veal cut into 1½ inch cubes	1 tsp caraway seed
4 medium onions, sliced thinly	1 tsp dry mustard
1 clove of garlic, minced	1 cup sour cream
¼ cup shortening (one-half butter)	1 cup beer, chicken broth, or water
2 tsp salt	2–3 tbsp Hungarian paprika
4 oz can of tomato paste	pepper to taste

Sprinkle meat with salt and pepper and half the paprika. Heat oil in a skillet and brown the meat on all sides. Remove meat to heavy casserole. Add onions and garlic to the oil and cook until onions are translucent, then add them to the meat. Add liquid, tomato paste, caraway, mustard, and the remainder of the paprika. Bring to a boil, cover, and simmer until the meat is fork-tender. Just before serving, add sour cream, but do not allow to boil. Serve with noodles or dumplings, rye bread and sweet butter, and a properly chilled bottle of white wine.

MUSTARD

Mustard is a "hot" spice, and its use as a dressing for meat dates back as far as records on cooking are available. Seeds of several species in the genus *Sinapis* (*Brassica*) are used. *S. alba* (yellow mustard seed) and *S. nigra* (black mustard seed) originated in Europe and *S. juncea* (brown mustard seed) is native to the Himalayan Mountains. Pythagoras, about 530 B.C., recommended it as an antidote to scorpion bites; Hippocrates prescribed it as an emetic; and Diocletian, the Roman emperor, placed a heavy tax on it knowing that his subjects would pay without too much complaint. Greeks, Romans, and the northern barbarians ate the leaves as a boiled green vegetable and ground the seeds into a paste. Matt. 13:31–32 refers to the parable of Jesus in which the kingdom of heaven is likened to a mustard seed, in that, in spite of its small size, it grows into a large plant; this demonstrates—if nothing else—that the plant was known in the Near East.

Its place on Europe's tables led Anatole France to remark that "a tale without love is like beef without mustard; an insipid dish." Powdered mustard is used in cooking and mustard seed in pickling spice, but most is consumed as prepared or salad mustard in which the ground seed is mixed with vinegar, salt, and other spices, including turmeric, to enhance the yellow color. French and German mustards utilize brown or black seeded species and may include wine or beer in the sauce. A magnificent dressing for boiled beef or smoked tongue consists of prepared mustard, horseradish, and mayonnaise—proportions adjusted to taste.

SAFFRON

Saffron, the most expensive seasoning, consists of the stamens of the saffron crocus (*Crocus sativus*), a species closely related to the garden crocus. Over 220,000 hand-picked stamens are required to make a pound of saffron. The word is derived from the Arabic *za'faran*, meaning yellow, for the stamens were used as a source of yellow dye for woolen and linen cloth. Ancient Minoan goddesses are figured with sacred snakes and the equally sacred saffron crocus flowers. Its expense and the delicate flavor imparted to foods is noted in the Song of Solomon. (4:13–14). The Chinese particularly esteemed saffron, not as a flavoring, but as a cosmetic used by women. Even a few stamens impart a yellow color and a delicate flavor to the rice that complements the classic Spanish paella:

PAELLA
(FOR FOUR)

3 slices bacon, chopped	generous pinch of saffron
2 medium onions, chopped	½ lb peeled shrimp, parboiled 1 min
3 cloves of garlic, chopped	½ lb scallops, parboiled 1 min
3 sweet Italian sausages, sliced	12 small clams, carefully washed
2 hot Italian sausages, sliced	paprika
1 chicken cut into serving pieces	olive oil
1½ cups dry rice	1 cup peas
1 tbsp salt	sliced pimiento
¼ tsp pepper	

Fry onions, garlic, bacon, and sausages until onions are golden and translucent. Remove to heavy casserole and add ½ cup water. Dust chicken pieces with salt, pepper, and paprika and fry in olive oil until brown on all sides. Add to casserole. Add rice and 2 cups of water. Cover and bake at 450 °F. for 30 min until the rice is almost done.

Steam clams separately until they begin to open before adding them to the rice. Add shrimp, scallops, peas, and pimiento. Return to heat, uncovered, until seafood is done. Serve with a green salad, crusty bread, and a dry red wine.

CINNAMON, NUTMEG, AND MACE

Several spices are so firmly associated with baking that most people forget that they were formerly used in other ways. Cinnamon is a case in point. There are two species: *Cinnamonum zeylandica* is true cinnamon, and *C. cassia* is the spice cassia, more familiar in Europe than in North America. Both are imported from Ceylon and the Seychelle Islands where the small, bushy trees are grown in plantations. In ancient Egypt, Greece, and Rome, the bark was steeped in oil to make a fragrant unguent, in addition to being used in cooking. It was used on funeral pyres to disguise the smell of burning flesh; when Nero's wife Poppaea died in 65 A.D. a full year's supply of precious cinnamon was burned. The Portuguese controlled cinnamon until the Dutch captured Ceylon in 1656. They continued the brutal practices of enslavement of the collectors and established a price monopoly that was maintained by destroying cinnamon when prices began to fall. Their control was broken when the British took Ceylon in 1796, and by the early part of the nineteenth century, plantations were in several countries and prices dropped to roughly their present levels. Both cinnamon and cassia are members of the Lauraceae, the same family to which the bay leaf tree (*Laurus nobilis*) and the avocado belong. The cinnamon sticks we use as swizzles in hot mulled wine are curls of the peeled bark of young twigs, the cinnamon quills of commerce. In addition to its use as cinnamon powder in baking and as cinnamon sugar on doughnuts, the addition of a small piece of stick cinnamon to a pot of spaghetti sauce should be experienced, and a dash of cinnamon powder on fish is a pleasure most North Americans have not had.

Apple desserts would not be the same without cinnamon, nutmeg, and mace; and a touch of nutmeg or mace in baked squash, sweet potatoes, or even the lowly boiled carrot transforms these into gourmet foods. Both nutmeg and mace are from the same plant and, indeed, from the same organ. The nutmeg tree, *Myristica fragrans*, is a large, evergreen member of the Myristicaceae. The tree is dioecious, that is, male and female flowers are borne on different plants (the term means "two houses"), so that plantation planters were careful to include a few male trees scattered throughout the plantings. Recently, horticulturists have found a way to graft male branches onto the female trees. The fruit looks like a small apricot with an orange-yellow pulp (Figure 6.50). Surrounding the seed is a lacy, scarlet layer called an aril. This, separated from the nutmeg seed and dried, is mace. It goes well with tomatoes, including juice and catsup, and adds an unusual flavor to fish dishes, including tuna and chowders. Nutmeg is a necessity in baking;

Figure 6.50
Nutmeg and mace, *Myrstica fragrans.* One part of the fruit coat is lifted to show the aril on top of the seed (mace) and the seed itself (nutmeg).

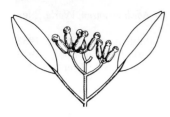

Figure 6.51
Inflorescence of the clove tree, *Eugenia aromatica*, and an enlarged view of an individual flower, the "clove."

it is the secret ingredient in Swedish meatballs and lamb stews, and no self-respecting eggnog neglects a sprinkling of nutmeg.

CLOVES

The French call them *le clou*, the nails, an apt description of the dried, unopened flower buds of the clove tree (*Eugenia aromatica*) in the Myrtle family (Figure 6.51). In Greece, Easter is marked by the baking of light, delicious cookies, each containing a clove to represent the nails used to fix Christ to the Cross. A native of the Spice Islands, it has been transplanted to other tropical countries where it is grown on plantations. Records from the Han dynasty in China (ca. 200 B.C.) show that it was much appreciated, being known as the "chicken tongue" spice. It, along with star anise, anise pepper, fennel seed, and cinnamon comprise the "five spice powders" used in red-cooked roast pork and other meat dishes. Along with peppercorns, cinnamon, and nutmeg, cloves were imported into dynastic Egypt and were known in Europe by the fifth century. The Dutch maintained a rigid monopoly on cloves for more than a century, but smugglers obtained planting material and plantations were started in Malaya and on the West Indian island of Grenada. Today, much of the world's supply comes from Zanzibar. In both Africa and in the Orient, cloves are used to flavor tobacco and are chewed with betel nut. In the West, its use has generally been restricted to studding baked hams and for cakes and cookies. It adds flavor to chocolate and, during the Middle Ages, cloves were stuck in oranges to construct the pomandors ladies carried to avoid smelling the bodies of the great unwashed masses and, they hoped, to ward off the plague.

CHINESE PORK IN FRUIT SAUCE
(FOR TWO)

1 orange, peeled and sectioned	½ tbsp brown sugar
1 tangerine, peeled and sectioned	1 tbsp rice wine or dry sherry
1 pear, cored and sliced	½ tbsp steak sauce
½ lb lean pork cut into 1-inch cubes	⅛ tsp anise
¼ tsp lemon juice	⅛ tsp ground cloves
¼ tsp salt	⅛ tsp ground cinnamon
2 cloves of garlic, minced	

Heat small amount of oil in a wok or small pan and thoroughly brown the meat and garlic. Add sugar, wine, ginger, steak sauce, spices, and salt. Add ½ cup of water and simmer gently for 30 minutes. Add fruit and lemon juice and simmer 3–4 min.

ADDITIONAL READINGS

Clair, C. *Of Herbs and Spices.* New York: Abelard-Schuman, 1961.
Clarkson, R. E. *Herbs and Savory Seeds.* New York: Dover, 1972.
Hayes, E. S. *Spices and Herbs around the World.* Garden City, N.Y.: Doubleday, 1961.

Lopez, R., and I. Raymond, *Medieval Trade in the Mediterranean World.* New York: Columbia University Press, 1955.

Parry, J. W. *The Story of Spices.* New York: Chemical Publishing Company, 1953.

Rosengarten, F., Jr. *The Book of Spices,* rev. ed. New York: Pyramid Books, 1973.

Schery, R. W. *Plants for Man,* 2nd ed. Englewood Cliffs, N.J.: Prentice-Hall, 1972.

Stein, A. *On Ancient Central-Asian Tracks.* University of Chicago Press, 1964.

CHAPTER 7
Booze

Wine

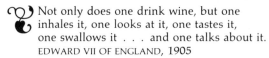

Not only does one drink wine, but one
inhales it, one looks at it, one tastes it,
one swallows it . . . and one talks about it.
EDWARD VII OF ENGLAND, 1905

In a book on wine published in 1923, the author described wines
he had drunk. He spoke of one as "a wine of senatorial dignity with the
mysterious haunting appeal found only in very old wines that may be
compared with the sound of the harmonic on the violin." And of another: "This wine breathed forth a perfume worthy of the Gods. It was
compounded of a multitude of subtle fragrances, the freshness of sun-
ripened grape, etherealized by the patient work of Nature into a quin-
tessence of harmonic scents. The full organ swell of a triumphal march
might express its appeal." These overwrought statements do not tell
one just what wine is. As a start, wine is the fermented juice of grapes.
But wine is considerably more than alcoholized grape juice, and its
preparation is worthy of study as part of the liberal education of the
intellectual (Figure 7.1). If one were to state that the world's finest wines
are French, the uproar from other countries would shake the walls of
Jericho, but techniques developed in France over 2000 years of wine-
making are the standards by which all wine is judged. To discuss the
science of oenology, the French model will be used.

Figure 7.1
Punishment for drunkenness:
the stocks. Taken from *Spiegel
Menchlicher Behaltnis* by
Berger published in 1489 in
Augberg.

LE VIGNE

The wine grape, *Vitus vinifera*, originated in the Caucasus Mountains at the southern end of the Caspian Sea as a self-fertile (mone-
cious) mutant from the normally two-sexed (dioecious) wild plant. It
was domesticated before 5000 B.C., was grown in Greece by 3000 B.C.,
and "The blood of the grape" was known in Egypt by 2400 B.C. (Figure
7.2). Homer sang of wine in the *Iliad.* According to *The Bacchae* of Eurip-
ides, Dionysus gave wine to man: "Filled with that good gift, suffering
mankind forgot its grief." As libation and beverage, wine was known to
the Israelites; there are over 160 references to the vine and its product in
the Testaments. Rome, too, was well acquainted with wine. Pliny observed that it refreshed the stomach, sharpened the appetite, blunted
care and sadness, and was conducive to slumber. Vines were planted in
France about 600 B.C. and Rome introduced wine grapes to the Rhine
Valley in 100 B.C.

The wine grape is one of 50 species of *Vitus* in subtropical and
temperate zones around the world. It is a climbing vine, capable of at-
taching to upright supports by tendrils. Flowers are small and inconspicuous. The fruit is botanically a berry, produced in clusters, and
contains two or more hard, bitter seeds. Through selection of natural
mutants (what horticulturists call "sports") and by cross-breeding,

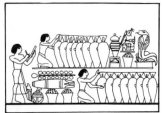

Figure 7.2
Steps in winemaking. Taken
from an Egyptian tomb
painting.

there are now several thousand recognized cultivars or varieties, although fewer than a hundred provide the bulk of the world's fine wines.

In France, two dozen cultivars are used. In Burgundy, the Pinot types and the Gamay, brought to France as tangible results of the crusades, are the primary wine grapes. Within each of the cultivars there are subtypes, the French *cépages,* so that a black pinot grape of one grower is not exactly the same as the pinot noir of another vineyard. The tokay of middle Europe is a pinot, as are many others under different names. The white-skinned semillion and sauvignon grapes are used for most white wines in Bordeaux including both dry wines and sweet sauternes, while the dark-skinned merlot and cabernet sauvignon grapes are used for rich, red clarets. The reisling and sylvaner grapes produce rhine and moselle wines while two red-skinned grapes, the sangiovese and the malvasia, predominate in the chianti of Tuscany.

Of the six or eight North American species of *Vitus,* the most important is *V. labrusca,* the northern fox grape. It differs from *vinifera* in that the skin is easily separated from the pulp because of a mucilaginous layer just below the skin. Wines made from *labrusca* grapes have a characteristic taste due to the presence of methyl anthranilate. The fox grape so impressed Leif Erikson in the tenth century that he called the shores of Newfoundland "Vinland." Labrusca was first bred by the Rev. W. W. Bostwick of upper New York state to yield niagara, and the famous concord was a mutant of niagara isolated in Massachusetts. The catawba cultivar was isolated from the wilds of North Carolina. Labrusca cultivars are disease resistant and cold-tolerant, but the grapes are low in sugar, and produce soft, fruity wines frequently fortified with sugar and additional alcohol; kosher wines are examples of this processing.

LE CLIMAT

Climate is almost as important as is the cultivar. Vinifera grapes can be divided into two major groups. Those that do well in the hotter, drier climates of southern Europe are called Mediterranean types and include grenache, syrah, muscat, and cabernet grapes. Temperate climate grapes include gamay, semillion, sauvignon, all pinots, the merlots of Bordeaux, and the white grapes used for rhine and moselle wines. The success of California's wine industry is due to the wide variations in climates found in the state. Dry, very hot Mediterranean climates are found in semidesert valleys with cooler, moister areas of mountain slopes planted to temperate-zone cultivars.

Climatic preferences are, however, more subtle than seen on a temperature chart. Vines grown on north- or south-facing slopes of the same valley will produce grapes varying in sugar, acid, and pigmentation. The wine of Chateau Latour and that of Chateau Leonville–Las Cases are produced from the same grapes grown in Bordeaux vineyards

Figure 7.3
Breaking the soil to plant grape vines. Taken from *Opus Ruralium Commodorum* by Piero Crescentio published in 1493 in Venice.

separated by a narrow gully. Latour's slopes face south, receiving perhaps an hour more sunlight a day than do the vines on the north-facing Leonville–Las Cases, but this adds up to several hundred more sunlight hours for the photosynthesis of Latour's vines. Latour is a great wine, while Leonville–Las Cases is merely a very fine wine.

Variations in weather may strongly affect the character of wine. Low night temperatures favor the conversion of sugar into acids resulting in a more long-lasting wine. Rain, or lack of rain, a too hot or too cold period of only a few days, hail, or a multitude of slight changes may so alter the grapes that the best skills of the vintner are required to utilize the grapes properly. Differences in weather are reflected in the vintage dates that appear on wine labels, and "perfect" years are few and far between, resulting in prices which make a wine-lover weep.

LA TERRE

Soil composition and structure are important (Figure 7.3). Soils used for wine grapes are generally rapidly draining, poor in organic matter, and relatively poor in nutritive minerals. Even slight differences in soil can greatly alter the character of the wine. In the St. Emilion district, the highly rated wines of Chateau Cheval Blanc are from grapes grown on a stony soil overlying solid limestone. The neighboring Chateau La Gaffelier–Naudes has a slightly better soil with more humus, but its wines are not as highly rated. Fertilization of the vines is done cautiously. Addition of too much nitrogen fertilizer results in more growth of the stems and leaves, and the photosynthate that would otherwise go into the grape is used for vegetative growth.

LA CULTIVER

In general, five years are needed to bring a new planting into full production and up to ten years may be required to determine whether a planting will have the potential to produce a superior wine. All grapes are propagated by cuttings or graftings and remain true to type. Vines are trained along wire trellises with row distances calculated to allow maximum exposure to the sun. Only rarely are there more than 1500 plants per hectare. Pruning is done after the harvest (Figure 7.4) to control the number of flower-bearing branches laid down the previous summer and early fall. The grower must balance maximum grape production against grape quality.

Although it was their own fault, the French have never forgiven Americans for the introduction of *Phylloxera* into French vineyards. Europeans took grape vines from North America to France in 1855–1860 as breeding stock, unaware that the roots carried this tiny insect. Wingless females of *Phylloxera vastatrix* insert their mouth parts into roots, sucking up the sweet sap, and causing disruption of root structure. Females produce more wingless females and also lay eggs which

Figure 7.4
Pruning vines. Taken from *Opus Ruralium Commodorum* by Piero Crescentio published in 1493 in Venice.

312   Booze

Figure 7.5
Harvesting the grapes. Taken from *Opus Ruralium Commodorum* by Piero Crescentio published in 1493 in Venice.

develop into winged females. In less than 15 years, phylloxera had almost destroyed the vines of Europe. Charles V. Riley, a U.S. entomologist, noted that the phylloxera then ravishing Europe's grape vines was relatively benign on American species of *Vitus* and suggested that the vinifera grapes be grafted to *V. labrusca* rootstocks. This saved the wine industry of Europe.

LA VENDANGE

There is usually 80–100 days between fertilization and fruit maturity. Sugars accumulate only slowly in young fruit, as we know from Aesop's fable of the fox and the young, sour grapes. In the few weeks before full maturity, sugars accumulate rapidly. When ripe, 70 percent of the weight of the fruit will be the juice (cell sap) and between 20–24 percent will be the two hexose sugars, glucose and fructose. Organic acids such as malic (also found in apples) and citric (in oranges) plus tartaric acid, vitamins, proteins, and other cell constituents are present. Acid levels are important; if the acid constitutes less than 0.5 percent, the resulting wine will be insipid; levels above 0.7 percent, will allow the production of a tart, long-lasting wine. Aroma and flavors are determined by many different compounds including aldehydes, esters, and alcohols, and decisions on when to pick the grapes include evaluations of all these factors.

Depending on the grape and the wine to be made, picking may be done all at one time or may be spread over several weeks (Figure 7.5). The sauvignon grapes used to make sauternes are picked at full maturity to yield a relatively dry, fruity wine. At Chateau Yquem, the grapes are allowed to become overmature to the point that they shrivel and begin to rot. The consequent reduction in water content and increase in sugar plus the flavors imparted by the rot fungus interact to produce a sweet dessert wine. In the Rhine wines, grape overmaturation determines the name of the wine. *Spatelese* wines are made from grapes picked late; *auslese* wines are made from individually selected bunches of overmature grapes. *Beerenauslese* wines are made from the juice from individually selected grapes from bunches showing both drying and some rotting while the most expensive Rhine wines are the *trockenbeerenauslese* made from individually selected, rotted grapes which have dried out almost to the raisin stage.

LE PRESSOIR

Following mechanical removal of stems, grapes are pressed to yield their juices. Traditionally done with human feet, the screw press of the fifteenth century and its modern descendants are now used (Figure 7.6). Pressing must be thorough enough to crush the grapes, but not so harsh as to break the seeds and release compounds which can ruin the wine. For white wines, the grape skins are removed immediately

Figure 7.6
A windlass wine press. Taken from a sixteenth-century woodcut.

after pressing to avoid getting any skin pigments into the fermenting vats. Both white-skinned and red-skinned grapes can be used to make white wines. For rosé wine, the skins are allowed to remain with the juice during the initial stages of fermentation.

LA FERMENTATION

Grape juice plus a pomace of pulp, seeds, skins, and some stems is transferred to large fermentation vats traditionally made of oak (redwood in California), but now of concrete, ceramic, or stainless steel (Figure 7.7). Temperatures close to 30° C. are best for the initial fer-

Figure 7.7
Primary fermentation of wine. Vintner checking temperature of red wine. Courtesy Wine Institute of America.

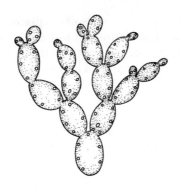

Figure 7.8
Yeast, *Saccharyomyces ellip-soides.* Asexual reproduction by budding with more than one daughter cell per mother cell.

mentation of red wines and 18° C. is optimum for white wines. Yeasts are killed at temperatures above 38° C., and, since active fermentations can reach this point, cooling the vats is important. Although a few vineyards still utilize natural yeasts which form a whitish bloom on the surface of grapes, the bloom also contains bacteria, other yeasts, and other fungi which can spoil the fermentation. It is now standard practice to kill or inactivate all microorganisms in the juice (now called the *must*) in the vat and then to inoculate the must with a pure culture of the desired strain of *Saccharyomyces ellipsoides* (Figure 7.8), the wine yeast. This concept of using a pure culture of yeast was initially developed by Louis Pasteur, a loyal wine-drinker whose scientific life revolved around the wine industry. Pasteur also found that when fermentations went "bad," resulting in bitter, acidic or evil-smelling wines, they could be restored by heating the very young wine briefly to kill all microorganisms and starting again with active "good" yeasts. This Pasteurization process was subsequently adapted to kill pathogenic and spoilage bacteria in milk, an adaptation no Frenchman would have anticipated.

Fermentation of the must starts almost immediately and continues at a high rate for several days. The amount of alcohol formed depends on the strain of yeasts used, the temperature, the level of acid and the initial concentration of sugars.

AN ESSAY ON FERMENTATION AND RESPIRATION

A physical chemist, looking at living cells, would suggest that life is thermodynamically impossible. Living systems are so complex, so dynamic, and so delicately poised that they should lose their integration and "decay" to reach the lowest possible energy level. Entropy, a measure of this "decay," should be high. In order for this not to occur, there must be a continual input of energy into the system. As an analogy, consider a house. If neglected, it will weather, decay, and collapse. The repair and maintenance of a house requires the input of materials and of energy to stave off the inroads of entropic chaos. So it is with a cell. Formation of new cell walls, replacement of worn out protein, and revitalization of structure and function requires not only the synthesis of the building materials, but also the availability of energy needed for the requisite construction. The biochemical processes of fermentation and respiration supplies the cell with both energy and building supplies.

Basically, respiration is the oxidation of a carbohydrate formed by photosynthesis to water, carbon dioxide, and energy. Again using an analogy, when one lights a fire, the overall oxidation equation can be expressed as:

$$\text{Wood (cellulose)} \xrightarrow{\text{O}_2} \text{H}_2\text{O} + \text{CO}_2 + \text{energy (heat and light)}$$

Except for minor modifications, the equation also summarizes respiration. The substance burned—the substrate—is not cellulose, but another carbohydrate, the simple hexose sugar glucose. Instead of all the energy being given off as heat and light, some of it is converted into the energy-containing compound adenosine triphosphate (ATP). The reactions occur through intervention of enzymes which sequentially facilitate the conversion of the initial hexose substrate into carbon dioxide and water plus ATP. It is not surprising that a good deal of the initial research on the energy metabolism of cells was done in France, occasioned by interest in wine fermentations. Joseph Gay-Lussac reported in 1810 that the formation of alcohol in the wine was due to conversion of sugar to carbon dioxide and alcohol, but it was Louis Pasteur who asked and then answered the important question, "But how to account for the working of the vintage in the vat?" Pasteur's studies led not only to the isolation and identification of yeast as the vital factor in fermentation, but paved the way for his studies on spontaneous generation.

The biochemical processes of fermentation and respiration are initiated with a molecule of glucose or from the glucose derived from starch. Through the intervention of 11 enzymes, each molecule of sugar is converted into two 3-carbon molecules called pyruvic acid. This initial process, called glycolysis (sugar dissolving), occurs in all living cells. At this point, there are two major routes for the further metabolism of pyruvic acid, the fermentative and the oxidative. Which pathway is followed is determined by the concentration of oxygen available to the cells. When oxygen is low, the fermentative route is used; yeasts are among the few organisms that can actively ferment pyruvic acid in the presence of oxygen, and they ferment more actively when oxygen levels are low, as they are in large vats of grape juice. Under fermentative conditions, pyruvic acid is converted to the 1-carbon carbon dioxide and the 2-carbon ethyl alcohol. Fermentation is energetically less efficient than respiration, and in order to survive and grow, organisms using this route must ferment actively.

When, however, ample oxygen is available, cells convert pyruvic acid into citric acid to initiate a cycle conversion process, the tricarboxylic acid cycle (citric acid is a tricarboxylic or three-acid-group molecule). Glycolytic reactions occur freely in the cytoplasm, but the TCA reactions occur on and in the mitochondria (Chapter 1); this packaging of a set of enzymes is notably efficient since enzyme and its substrate are always in intimate contact. Citric acid is metabolized through a series of steps in which molecules of carbon dioxide are excised and in which hydrogen ions and electrons are shifted from substrate to an oxidase system. The electrons activate the formation of ATP, and the hydrogen ions combine with atmospheric oxygen to form water. The fermentative and respiratory systems can be summarized as:

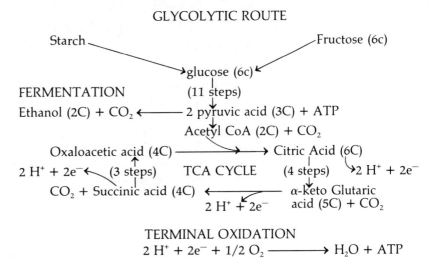

GLYCOLYTIC ROUTE

TERMINAL OXIDATION
$$2 H^+ + 2e^- + 1/2\ O_2 \longrightarrow H_2O + ATP$$

The efficiency of the conversion of the energy in glucose into the energy stored in ATP is high, about 20 percent.

Respiration provides not only the ATP energy required by cells, but also the compounds needed for cellular repair, maintenance and growth. For example, α-keto glutaric acid or pyruvic acid can be converted into amino acids needed for protein synthesis, and from these amino acids others can be made. Pyruvic acid can be converted into glycerol, the backbone of fats and oils, and the fatty acids can be made from acetyl CoA. There are hundreds of these reaction pathways, all utilizing one or more of the intermediates of the glycolytic, TCA, and fermentative routes. The accumulation of compounds in plants also involves the respiratory route. If the enzymes which convert citric acid into α-keto glutaric acid work slower than those converting acetyl CoA and oxaloacetic acid into citric acid, citric acid will accumulate in cells, as it does in citrus fruits.

Within a few days, the rate of wine fermentation in the vats slows down and the *pomace* of pulp, skins, and seeds floats to the top. This thick layer reduces the penetration of oxygen into the young wine and reduces the danger of conversion of alcohol to vinegar by aerobic bac-

teria. Fermentation stops when the alcohol reaches 15–18 percent by volume, a concentration that inactivates yeasts. This level is usually reached in about two to three weeks. In the preparation of sweet wine, fermentation is stopped by adding sulfur dioxide before all the sugar is converted to alcohol, by pasteurizing the wine, or by the addition of alcohol. The wine is allowed to remain in the fermenting vats for several more weeks to allow solids to settle out. At this point, the young wine has a distinct "grapy" or fruity flavor and aroma.

RACKING

Racking consists of drawing the wine from the fermenting vat into clean barrels (Figure 7.9). These barrels are held in cool cellars (*les caves*) over the winter. Fermentation may continue at a very slow rate until the last trace of sugars is converted into alcohol. Sugar-free wine is called "dry." By spring, the wine has a precipitate of dead yeast cells, proteins, and other components of the grape, and the wine is again racked into clean barrels. For red wines, racking is repeated the following fall to allow removal of precipitated tannins and tartaric acids (*lees*). The final racking may last one to three years depending on the wine. This period, *la marié*, is the period of the marriage of the flavors and aroma in the wine (Figure 7.10). Many subtle changes occur, including complex chemical unions of acids and alcohols to form aromatic esters that provide bouquet to the wine. The time of racking is critical: bottling too early will delay wine maturation and bottling too late will result in a dull-tasting wine.

LE COLLAGE

Prior to bottling, wines are clarified (*fined*) to remove suspended solids. For premier wines, collage is done by adding egg whites, gelatin, or clay to racked casks. These additives slowly sink to the bottom of the cask carrying impurities with them. Fining takes two or three months.

Figure 7.10
Testing racked wine. Courtesy Taylor Wine Co.

Clear wine is drawn from the top of the barrel into a cask where it is chilled to precipitate tannins in solution and a final "polishing" or filtration is done to make the wine very clear. Inexpensive wines may be pastuerized or are sterilized by passage through bacterial filters, but these wines will never show any further development in the bottle.

MIS EN BOUTEILLES

Finally, the wine is bottled (Figure 7.11). Different types of wine are placed in bottles of traditional shapes, each major wine-making area having evolved its own bottle shape; the raffia-covered *fiaschi* for chianti and the tall slender bottles for Rhine wine are immediately recognizable. The bottles are capped with cork, the bark of the cork oak tree. Corked bottles should be stored on their sides to prevent the cork from drying out and allowing oxygen to enter the bottle. It is also a good idea to open a bottle of wine an hour or so before drinking to allow the bouquet to develop; part of the pleasure of wine is its aroma.

Where the label indicates mis en bouteilles followed by the name of a chateau, estate, or domaine (the German equivalent is *original abfüllung*), it means that the wine was bottled where it was made. In many instances, wines from several vineyards are blended by the vintner or by a shipping company, and the label can bear only the designa-

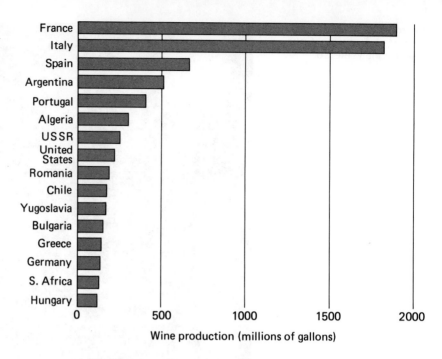

Figure 7.11
Major wine producing countries of the world.

Wine production (millions of gallons)

tion of the district from which the wine came. This is the *appelation controlée* in France, and wines simply called medoc or burgundy are usually blends. This is not necessarily bad. In years when none of the wines are outstanding, judicious blending can provide an excellent wine. In 1855, the French government classified all the vineyards in Bordeaux using a complex rating scale that included assessment of climate, soil, cultivars, and so on. Although this classification, later extended to other districts in France, is debated endlessly, classifications are still generally accurate. Outstanding vineyards are called *grand cru* and, in descending order of excellence, wines from specific vineyards are called first, second, or third growths. Until recently, most California wines were not regulated, and the consumer was told only that the wine was a burgundy or a rhine. Since California is not in French Burgundy nor in Germany's Rhine district, such names were misleading and, under international law, were illegal. Reputable vintners in California are now providing the name of the area plus the name of the grape cultivar used. Such "varietal" wines must contain 51 percent of the wine from the listed grapes and proposed federal legislation is designed to raise this to 75–85 percent.

Most red, rosé, and white wines are "still" or table wines and contain 12–14 percent alcohol by volume. If not sterilized, they will continue to age and mature in the bottle. Red wines, usually high in acids and tannins, may age for 30 or more years, but white wines rarely age well for more than 2–4 years. The vintage year refers to the year the wine was made, not necessarily when it was bottled.

Figure 7.12
Drinking champagne. Taken from *Opus Ruralium Commodorum* by Piero Crescentio published in Venice in 1493.

VIN AROMATIQUE

Thus far, our discussion has centered about table wines. There are, however, many other grape vine products, and three categories are so common that brief discussion is warranted.

The manhattan or martini cocktail is an important part of the social life of North America. The manhattan consists of whiskey plus sweet Italian vermouth, and the martini of gin or vodka plus minute quantities of dry French vermouth. Both vermouths are made from white wines in which herbs and spices including coriander, wormwood (*Artemesia*), cinnamon, cloves, nutmeg, and many other flavorings are steeped. Sweet vermouth has sugar plus caramel coloring added and, for both vermouths, the alcohol level is raised to 20 percent by volume. May wine—best served chilled with a garnish of fresh strawberries—is a Rhine wine flavored with woodruff (*Asperula odorata*).

VIN MOUSSEUX

Champagne (Figure 7.12) is a sparkling (carbonated) wine produced only in the Champagne district of France from both white and black pinot grapes, but other wines can be made by the same process (sparkling burgundy, crackling rosé, cold duck, etc.). Traditionally, the invention of champagne is ascribed to Dom Perignon, a Benedictine cellarmaster in the monastery of Hautvilliers. The vineyards surround the towns of Épernay, Ay, Gramant, and Avise. Growers sell their grapes to large companies which press, ferment, and bottle the white wine. Each of the two or three grape cultivars are pressed, skins immediately removed, and each is fermented separately. The young wines are racked over the winter and, the following spring, the cellarmaster blends the different wines to form a *cuvee* the success of which depends on his skill and vision. The blends are bottled and undergo a second fermentation over the hot summer. The wine rests for two or three years during which time each bottle is slowly turned and eventually upended to allow the sediments to collect in the neck of the bottle, the *remuage* process. The neck of each bottle is then frozen and the cork removed. The plug of sediment gushes out of the bottle, and the wine is ready to be made into champagne. A small amount of sugar solution is added to each bottle, it is corked, and the cork wired in place. Since yeasts are still present, a third fermentation occurs with the carbon dioxide remaining in solution under pressure. The amount of sugar added determines the final dryness of the champagne. *Brut* champagne has less than 0.2 percent residual sugar; *extra dry* has less than 0.5 percent sugar; *sec* may have up to 2 percent sugar; and *doux* is a cloyingly sweet 3–4 percent. Vintage champagne is made from wine produced in a single year; *sans année* champagne may have wine from several years' pressings. A Frenchman, M. Charmet, who should have had more native pride, invented the "bulk champagne" process in which the third fermentation is done under pressure in large tanks, and the sparkling wine is bottled under pressure. The Russians, who enjoy a very sweet sparkling wine, invented a continuous process in which, from grape to bottle, the wine is made in only three weeks.

LE VIN CUIT

Wines to which alcohol is added are generally very long lasting, do not develop in the bottle, and are called aperitif or dessert wines. The most famous of these are the sherries. The vineyards are dispersed around the city of Jerez de la Frontera near Cadiz in southwestern Spain where the primary grape is the palomino. The first fermentation lasts the unusually long time of three months. During the latter part of this fermentation, a film of yeasts forms on the surface of the vats. This *flor* or flower of natural, wild yeasts confers a special flavor to the young wine. For heavy, sweet oloroso sherries used as dessert wines, the fermentation is stopped by the addition of grape brandy when there is still considerable sugar left in the wine. For the drier, lighter fino sherries, fermentation is stopped later, and the final product is used as an aperitif. This fortification raises the alcohol concentration to 19–21 percent by volume. All sherry is aged by the solero process in which the young wine is racked into the top row of stacks of barrels. As older, aged wine is removed from the bottom row of barrels, younger wine is racked down and fresh wine is placed in the top row. The hot weather of southern Spain causes some residual sugar to carmelize, providing the brown colors and the "nutty" flavors characteristic of sherry. Finos (amontillado and manzanilla sherries) are less caramelized than the oloroso and cream types. Our name "sherry" is an English corruption of Jerez, and when Falstaff, in Henry IV, spoke of ". . . a good Sherris sack," he corrupted not only the name of the city, but also the Spanish word *saco*, which means for export.

In the 1930s the agricultural experiment station in Geneva, New York, invented a quick method of obtaining a fortified, brown wine sold as sherry. Wine from *labrusca* grapes is fortified with sugar and alcohol and aged at temperatures of 40–60° C. for several months to carmelize residual sugars. Madeira is a fortified wine aged at an elevated temperature for three to six months. Port, muscatel, tokay, and others are fortified wines whose fermentation was stopped before all the sugar was converted to alcohol.

ADDITIONAL READINGS

Amerine, M. A., and M. A. Joslyn. *Table Wines.* University of California Press, 1951.
Grossman, H. J. *Grossman's Guide to Wines, Spirits and Beers,* 5th rev. ed. New York: Scribner, 1974.
Johnson, H. *The World Atlas of Wine.* New York: Simon & Schuster, 1971.
Lichine, A. *Encyclopedia of Wines and Spirits.* New York: Knopf, 1966.

Schoonmaker, F. *Encyclopedia of Wine*, 6th rev. ed. New York: Hastings House, 1976.

Simon, A. L. *Wines of the World*. New York: McGraw-Hill, 1972.

Younger, W. A. *Gods, Men and Wine*. New York: Harcourt Brace Jovanovich, 1966.

Whiskey

See what the boys in the backroom will have.
AMERICAN BALLAD

On May 4, 1964, cognizant of its awesome responsibilities in guiding the nation, the Congress of the United States of America passed Senate Concurrent Resolution 19 introduced by Senator Thruston Morton and Representative John C. Watts, honorable gentlemen of the great state of Kentucky.

> That it is the sense of Congress that the recognition of Bourbon whiskey as a distinctive product of the United States be brought to the attention of the appropriate agencies of the United States Government towards the end that such agencies will take appropriate action to prohibit the importation into the United States of whiskey designated as Bourbon. . . .

It was said that Congress adjourned immediately after this exhausting act of statesmanship to demonstrate their whole-hearted support for the product under consideration. This was far from the first act referring to distilled beverages. One of the earliest was a law enacted by the first general assembly of Virginia in 1619.

> Against drunkenness be it also decreed that if any private person be found culpable thereof, for the first time he is to be reprooved privately by ye Minister, for the second time publically, the thirds time to lye in boltes 12 howers in the house of ye Provost Marshall and to paye his fee.

These two quotations demonstrate the sinfulness and profitability of distilled spirits; twin themes that run throughout the entire liquor experience.

A BRIEF HISTORY

Although production and consumption of low-proof alcohol predate recorded history, Aristotle mentioned that one could obtain the spirit of wine. Greek and Egyptian alchemists in the second century obtained rectified liquids and apparently stumbled upon ardent spirits. Distilling entered the thinking of Europe only when crusaders learned the technique from North Africans. A treatise on distilling appeared in

Adapted from "A Nation of Moonshiners," *Natural History* 85 (1976): 23–31. Copyright American Museum of Natural History.

Figure 7.13
A brick still with a metal alembic. Taken from a woodcut of the sixteenth century.

France in 1310 and, by 1500, there were books and practical manuals on preparation of brandy from wine, *aqua vitae* from beer, and the fabrication of medicinal cordials from herbs infused in grain alcohol made from wheat, barley, and millet. Our English word "whiskey" is derived from the Gaelic *uisa—beatha* or water of life (aqua vitae). Originally, it was medicine containing anise, cloves, nutmeg, ginger, caraway, licorice, sugar and saffron—an all-purpose cure.

Preparation of alcohol is very simple. Basically, a carbohydrate source is fermented with yeast and the 8–13 percent alcoholic solution is rectified by distilling the alcohol away from the water. Both in Europe and in its colonies, the medieval alembic (Figure 7.13), a closed vessel to which heat was applied, was filled with the ferment and the vapors, high in alcohol, were passed through a pipe in the top of the alembic to a twisted metal "worm" immersed in cold water to condense the vaporized alcohol. In addition to alcohol, distillates contain a wide variety of volatile compounds which conferred specific odors and flavors to the brew. When grain is the carbohydrate substrate, fermentation is initiated either by adding sugar or supplying an enzyme (amylase) which converts starch to the sugar fermented by the yeast. Barley is an excellent source of amylase. The grain is wetted, allowed to sprout, dried with gentle heat, and the seedlings ground into malt. Scotch whisky's distinctive taste is due to the fact that the malt is dried over peat fires. Grain is ground into meal, boiled into a gruel to solubilize its starch, malt added, inoculated with yeast, and allowed to ferment. Most yeasts stop working when the alcohol content reaches 15 percent and the alcohol can be distilled away from the beer. The resulting mash can be refermented several times and spent mash used as animal food. It was a source of much innocent amusement among the colonists to see how drunk their pigs would get on used mash.

SOLACE OF THE AMERICAN REVOLUTION

For reasons lost in the idiocies of history, the early settlers of North America failed to bring the alembic and worm with them and thus were restricted to beer, ale, and wine. Thomas Hariot of the Roanoke colony reported that barley malt was unavailable and that he had some success with malt made from Indian maize. George Thorpe, a kind-hearted soul, gave the Indians back their corn in the form of strong beer and was killed by them for his trouble. Jamestown was making corn beer by 1608, derived from maize planted by Captain John Smith in 1607. With Yankee ingenuity, just about all the colonist's produce could be used. Pumpkins, potatoes, plums, Indian corn, carrots, and even turnips were made into beer. New Englanders used honey from wild bee trees and sap of the maple. Along with other necessities, seedling fruit trees were carried from Europe and orchards were planted as soon as the land was cleared. "Hard syder," about 4–5 percent alcohol, was made from "crabb trees." The alcohol content of these ciders was increased to about 25 percent by putting the jugs out in the winter cold and carefully removing the ice. In the Virginia colonies, peaches were favored because they fermented into a somewhat stronger product with a delightful odor and taste. Elderberries, currants, and wild *V. labrusca* grapes were used for wine. Ardent spirits were not used routinely. Brandy was standard in medicinal ship's stores, and rum made from molasses on the sugar plantations of the West Indies had to be transshipped to the colonies via England.

Of course, distilling apparatus was soon in use. At first, a blanket was put over a kettle and the condensed vapors simply wrung out into a pail. Each village had a mechanically inclined inhabitant who constructed an alembic and worm from bits and scraps of wood and metal, and applejack and grape, peach and pear brandies were in use by 1630. The fruit brandy of New Jersey was considered the most potent, producing an "apple palsy" after just two drinks. Wilhelm Keift, director-general of the Dutch colony on Staten Island was distilling grain in 1640 with imported barley malt mixed with imported rye and native maize. Brandy was made from imported grapes by 1650 and aged for two to four years in barrels made from native white oak instead of the limoge or gascon oaks of France. Gin, too, was made in America by 1660, but never attained the popularity it did in England. A mixture of gin and applejack called "strip and go naked," because this behavior was observed after several mugs, was used by the poor in northern cities. It was tamed for children and women by adding beer and blackstrap molasses.

The puritans of the Massachusetts colonies favored rum, usually served on social occasions as a rum flip. A mixture of rum, beer, and sugar was stirred with a logger—a heated iron poker—and after a half dozen, the guests were frequently at loggerheads. Rum is a distillate of yeast-fermented sugar cane syrup or the molasses left after the crystallization of the sugar. Blackstrap molasses, a dark brown, caramel-colored endproduct, is the most common starting material. The rum first consumed in the colonies was made in the West Indies, but by 1660 the Salem, Newport, and Medford branches of the Massachusetts colonies were importing molasses to make their own dark, high-proof product. Benjamin Franklin, first president of the American Philosophical Society, devised a whalloping drink for membership dinners, and for his lady friends, consisting of high-proof medford or demarara rum, loaf sugar, and orange juice. The sugar refineries were mainly in French and Spanish hands. Britain, preferring to keep this lucrative trade entirely English, attempted to limit the profits of the foreigners by requiring that medford rum be shipped to England or, if kept for local consumption, be taxed at home country rates. The Molasses Act had two effects. It increased smuggling and it was as distasteful to the colonists as the stamp, navigation, and tea taxes; the seeds of revolution were planted early. As a substitute for rum, New England fermented and distilled honey. "Old metheglin" was a dark-brown, sweet drink with about 60 percent alcohol. In Vermont it was asserted that one glass was enough to allow—even in winter—the buzzing of bees to be heard.

RYE

Wheat and barley grew poorly in the colonies; wheat was needed for bread and barley was needed for making beer malt. Rye, not a native plant, formed the base for the first whiskeys of the colonies. It flourished in western Maryland and in eastern Pennsylvania, land settled by Scots who had previously lived in Ulster and by Germans from Moravia. These settlers had the skills and the equipment to make whiskey, and they set to it with a will and a fervor true to their heritage. Rye whiskey was originally called tiger spot, but soon became known as either Maryland or Monongahela. There were good economic reasons for the industry of the distillers. A pack horse could carry no more than four bushels of grain, but the same horse could carry two kegs of whiskey representing 24 bushels of rye. Spoilage was minimal unless the horse or driver stumbled, and the two kegs brought the equivalent of 40 bushels of grain. The fermentation and distillation of rye and barley malt produced a whiskey with a heavy, intensive flavor, in great contrast to the blended whiskey "rye" which the modern unenlightened mixes with ginger ale. We have been sold a bill of goods; such "rye" is little more than a small amount of whiskey mixed with pure alcohol, water, and caramel coloring.

BOURBON

In the South and on the western frontier, rye was not a successful crop. The weather was too warm with too much rain in midsummer. The settlers turned to a cereal grain that did grow well, Indian maize. As Virginians moved west, they established villages that were sufficiently remote to require their own legal structures. By 1770, Harrodsburg, Harwood's Landing, and Boosfort were settled agricultural communities and were growing corn for food, fodder, and fermentation. In 1780, Fayette, Jefferson, and Lincoln counties were established in the western Virginia lands, a judicial district of Kentucky was formed in 1783, and Bourbon County was cut out from Fayette County in 1786. The population of Kentucky at that time was over "thirty thousand souls," of which a goodly proportion were growing corn. Just who first started making "corn likker" in Kentucky is unknown, and there are several claimants for this exalted title. Marsham Brashears purchased land in 1782 for 165 gallons of whiskey, Bartlett Searcy willed his 96 gallon still to his son John in 1784, and the Kentucky District court had nine cases of illegal whiskey retailing in 1783, including one presentment to the grand jury against a woman. For convenience, the Rev. Elijah Craig is credited with making the first bourbon whiskey. Elijah's brother Lewis, also a Baptist minister, was a whiskey dealer who supplied the cargo for the flatboats which followed the rivers down to New Orleans. There was serious debate on the propriety of ministers engaging in the liquor trade, but since most of the congregation was similarly employed, no pot could call any kettle black. The Baptist debate raged until about the time of the Civil War when a firm antiliquor stand was taken. Drunkenness was, however, another matter and occasionally led to outright expulsion from the congregation; a man and his wife were supposed to be able to handle their liquor.

As Maryland and Pennsylvania rye and the corn whiskey from Bourbon became widely distributed, liquor's role in social, business, and political affairs became all-pervasive. Payment for debts, mortgages, and purchases of necessities was often in liquor. Jugs of whiskey and distilling apparatus were formally parts of estates; births, weddings, and funerals were paid for and celebrated with whiskey; and no business arrangement could be considered consummated without pledges sloshed in a cup or glass. A sovereign remedy for summer complaints of children and adults was an infusion of rhubarb, caraway seed, and orange peel in whiskey. Copperhead or rattler bites were treated both internally and externally with bourbon.

Most distillers were the farmers themselves, taking advantage of the fine, pure limestone water and the abundant corn. None of the products were trademarked; they were all bourbon or rye. A James Beam and a J. W. Dant had moved to Kentucky from North Carolina by 1760 and a John Ritchie and a Jacob Beam were licensed distillers by 1780, but it wasn't until after the Revolutionary War that specific names were attached to the product of a distillery or a district. Prior to and even after the revolution, whiskey was collected at Louisville or Cincinnati, then barged down the Ohio and Mississippi rivers to reach the Southwest and Southeast via New Orleans. John James Audubon made a stake for his then-ignored painting when as a merchant in Henderson, Kentucky, he barged a load of kegs to Missouri where he sold the whiskey at two dollars a gallon.

The whiskey of the colonists was sold and consumed in its God-given natural, water-clear state. Occasionally a small amount of caramel coloring was added to give the amber color associated with European brandy. The colonists knew that aging in charred oak barrels imparted a golden-brown color and smoothed out the product by the marriage of various organic compounds, but this took valuable time, and no one cared a whit what the color was so long as it was strong. The complex of harsh, skull-popping volatiles (fusel oils, aldehydes, and wood alcohol) could be absorbed by barrel aging or by filtering the raw whiskey through a layer of maple charcoal. Peach brandy was similarly treated with charred peach pits, which also contained a minute amount of prussic acid. One serious problem was the tendency for the alembic to become too hot and scorch the mash, and some genius figured out a way to introduce live steam into the pot still to volatilize the alcohol. Only much later did the continuous Coffey still replace the pot method.

By the beginning of the eighteenth century, copper stills were in general use, imported from either England or Holland. Most native stills were iron which imparted an off-flavor and sometimes colored the water-clear liquid. Because of British restrictions on local manufacturing, rolled copper for alembic, worm, and boiler was impossible to obtain. Paul Revere, silversmith and sometimes horseman, was best known for being the only manufacturer of rolled copper in the colonies, a monopoly that he held until 1802.

Proof was determined by the gunpowder method. Equal volumes of gunpowder and whiskey were mixed and set afire. If it flashed up, the whiskey was too strong, and if it didn't burn, it was too weak. When it burned evenly, the whiskey was 100 percent perfect; it was "proved out" and the proof was 100, that is, 50 percent alcohol within about 2–3 percent.

Rye whiskey was made by inoculating the mash with fresh yeast, but bourbon utilized the sour-mash process. Here the mash was charged with a small amount of the liquid from the previous batch. This spent beer contained a small amount of lactic and other acids that promoted development of yeasts and repressed the growth of bacteria which might otherwise spoil the taste of the whiskey. Colonists were more particular about taste than spelling, and the -y or -ey ending was used interchangeably until about 1850. Frances Trollope used both spellings in her *Domestic Manners of America* of 1839. After the Civil War, the -ey ending was restricted to the rye, Canadian, and bourbon whiskey of the Western Hemisphere.

George Washington learned the value of whiskey very early. He stood for the Virginia House of Burgesses from Frederick County in 1785 and won with 307 votes . . . at a cost of £38 of which £34 was for liquor. At the time, rum was 16 shillings per gallon, rye whiskey was 8s per gallon, and corn liquor was 4s per gallon. It was money well spent. James Madison refused to supply refreshments for the voters and consistently lost elections. Washington's plantation was noted for its peach brandy and as his liquor became more famous, he branched out into making whiskey from rye and corn and imported a Scotsman to oversee the business. During the war, Washington insisted that the army be supplied a liquor ration: ". . . in many instances such as when they are marching in hot and cold weather . . . it is essential that it [liquor] not be dispensed with." He recommended to the Continental Congress that public distilleries supply liquor to the troops. The young navy had its daily tot of rum following the British model. The spirit ration to sailors was ended by act of the 37th Congress in 1862, and the adjutant general of the army supinely went along in 1880. With many men away from their farms during the war, grain was in short supply. Washington noted with approval the restrictions imposed by the states on excessive use of grain for whiskey-making and insisted that the liquor ration be of peach or apple brandy or of rum. This didn't sit too well with Southern and Western troops weaned on corn and rye whiskey, but their complaints were subdued by the high proof rum passed out by supply sergeants. War-time profiteering was evident as the price of rum rose fourfold in a year.

THE WHISKEY REBELLION

All governments have taxed liquor, and all citizens have attempted to evade payment. The young republic, desperately searching for means to pay debts resulting from the war and to cover the growing obligations of statehood, considered a whiskey excise. Secretary of the Treasury Alexander Hamilton and his Federalist party rammed through the Excise Tax Act of 1791 empowering the federal government to impose duties on all spirits "distilled within the United States, from any article of the growth and produce of the United States, in any city, town or village." Progressive duties on each gallon according to its proof and a yearly tax on each still were mandated, together with the onerous responsibility for each producer to maintain complete records and to permit inspections of distilleries, warehouses, taverns, and even private homes. Indignation ran high. Kaintucks, Virginians, Tennesseers, and Pennsylvania "Dutch" rye distillers were hauled into court for back excises; Reverend Elijah Craig had a liability of $140 assessed against him and nearly went bankrupt, as did many others. The excise was, they all agreed, an invasion of rights won with their blood; they fought the British for freedom from excisemen, and they'd be damned if they were going to go through that again. Although the excise was softened in 1792 and 1794, distillers joined with shippers and retailers to oppose the tax, and meetings and resolutions gave way to direct opposition including tarring and feathering the collectors. It was to be a test of the power of the federal government to tax the people directly without the interposition of the states, and Hamilton convinced Washington that the whiskey rebels of the Monongahela must be utterly and permanently crushed.

With a penchant for military overkill that seems to mark the United States' defense establishment, a massive army was gathered. In the autumn of 1794, 13,000 troops with artillery, mortars, supplies (including whiskey), and other appurtenances of a punitive force had been mustered. General Henry (Light Horse Harry) Lee, then governor of Virginia, volunteered to lead troops through Cumberland Gap, General Howell and the New Jersey contingent moved through Carlisle, and Washington himself made an appearance on a white horse to show that the government really knew what it was doing. Happily, there was no battle; the rebels dispersed. It cost about $1.5 million, but excises were collected. Jefferson, who never supported any of Hamilton's ideas, repealed the tax when he assumed the presidency, and it wasn't imposed again until the Civil War. Since that time, whiskey has become a favorite target for state and federal legislatures.

BRANDY

Although the oversized glass is often considered an affectation comparable to a lorgnette or handlebar moustaches, there are good reasons for using one when drinking fine brandy; half the pleasure of brandy is in its aroma. True brandy is defined as a grape wine, distilled when young, and then aged in oak casks for 5 to 50 years. The wine is distilled to yield a clear, slightly aromatic spirit at 25 percent alcohol (about 50 proof), and then is distilled again to produce raw "heart of spirit" at 70 percent alcohol (140 proof). During the aging process, the alcohol content drops to 85 proof, esters and other flavorings develop, and the brandy becomes a rich, reddish brown. Any white wine can be used; German brandies use Rhine wine and *V. labrusca* cultivars are used in North America. By law, only brandies made in France's Cognac district can be called cognac. Different brandies are aged separately and are then blended to produce a uniform product. Brandies from the Champagne areas of Cognac produce the finest brandies, a source of some confusion since these areas are named for the local chalky soil and have nothing to do with sparkling wine. Names like "Napoleon" or a system of stars means very little except to advertisers. VO (very old) cognacs are usually less than 10 years in the cask, while VSOP (very superior old pale) brandies have been aged for more than 10 years.

France also produces other fine grape brandies, among the most notable being armagnac, produced in the area near Spain where Dumas placed the birthplace of d'Artagnan and Louis XIV completed the cathedral of Auch. In most wine-growing districts of France, wine grapes are pressed a second time to yield juice called *marc*. Too strong and too full of tannins to produce a table wine, marc is fermented separately, distilled, aged, and sold as marc brandy. This varies in taste depending on the grape cultivar used and the subsequent handling.

Distilled spirits are made from many fermentable products in addition to grapes. These water-clear alcohols contain flavorings of the fruit used. Kirsch is derived from cherry pits, fraise from strawberries, and mirabelle from plums. Wood-aged brandies are made from many fruits: calvados is from Normandy apple cider and slivovitz from plums grown in eastern Europe. Peaches, apricots, pears, and other fruits are fermented into a beer, distilled, and then aged. In addition, many liqueurs include brandy as the flavoring and alcohol base: Chartreuse is a brandy with herbal essences, creme de menthe has mint flavorings, Cointreau, Curaçao, and Grand Marnier are flavored with oranges. Many liqueurs have sugar added for taste and to provide the "heavy" syrupy flavor.

UNAGED SPIRITS

Of the unaged distilled spirits, the two most used are gin and vodka. For both, and for the Scandinavian akavit, the base is usually grain, fermented into a beer, distilled several times, and purified of all other compounds by fractional distillation and filtration through charcoal. To produce gin, a variety of herbs are steeped in the 95 percent alcohol (grain neutral spirits) for several weeks followed by a series of distillations to yield a high proof alcohol containing essential oils and other flavorings. For gin, the major flavoring ingredient is the fruit of Juniper (Figure 7.14). Other plant extractives include coriander seeds, angelica roots, and citrus peels. Akavit uses caraway seeds and some Slavic vodkas include buffalo grass, sorrel, or other herbs. Most should be taken neat and very cold; a bottle chilled in a block of ice is recommended.

Figure 7.14
Juniper, *Juniperus spp.* The "berries" are a major flavoring in gin.

ADDITIONAL READINGS

Baldwin, L. D. *Whiskey Rebels.* University of Pittsburgh Press, 1939.

Carr, J. *The Second Oldest Profession: An Information History of Moonshining in America.* Englewood Cliffs, N.J.: Prentice-Hall, 1972.

Crowget, H. G. *Kentucky Bourbon: The Early Years of Whiskeymaking.* University Press of Kentucky, 1971.

Gray, J. H. *Booze.* New York: Macmillan, 1946.

Packowski, G. W. *Alcoholic Beverages, Distilled.* In W. Kirk-Othmer (ed.). *Encyclopedia of Chemical Technology,* 2nd ed., vol. I. New York: Wiley, 1963, pp. 501–531.

Wilke, H. F. *Beverage Spirits in America: A Brief History.* Downington, Pa.: Newcomen Society in North America, 1949.

Beer

Beer! Beer! Beer! cried the privates, Merry,
merry men are we. There's none so fair as can
compare to the fighting infantree.
SOLDIER'S SONG

From cans, bottles, and kegs, by the glass, stein, or pitcher, beer is sipped, drunk, chugged, and guzzled in quantities greater than that of any other alcoholic beverage. In contrast to distilled spirits, whose development can be dated from the Arabic invention of the still, beer has been a primary product of grain from the time people became farmers (Figure 7.15). Royal breweries existed before the eleventh dynasty in Egypt (ca. 2000 B.C.), and the Code of Hammurabi of 1750 B.C. regulated the quality and price of dark, light, pale, and red beers. The Chinese

Figure 7.15
Beer sippers. Taken from a
Babylonian seal of 1900 B.C.

used both rice and millet beers, called *kui*, by 2000 B.C. In all cases, the
people believed that beer was a gift from their gods. Egyptians credited
it to Osiris, Babylonians to Siris, and the Chinese simply stated that it
was a gift from heaven. Greek tradition has it that Demeter fled Mesopotamia in disgust because the people drank beer instead of wine and
chose to settle in Greece where the wine-drinkers were more civilized.
Rome dedicated beer to Ceres, goddess of the corn, and their name for
beer, *cerevisia*, is the specific epithet for yeasts (*Saccharyomyces cerevisiae*)
used for both baking and brewing. The role of beer in ritual results
from its attribution to a female deity. Women, closer to the corn goddess than men, were priestesses of these goddesses and the brewers.
Rameses III modestly noted that he sacrificed 30,000 gallons of beer
each year to Egyptian gods, and credited his country's prosperity to
these libations. Norse Vikings asserted that beer was the drink of
Valhalla.

A BRIEF HISTORY

From the beginning of the Christian era, brewing was a household
art, with every girl instructed in baking and brewing, both processes
requiring the same ingredients, both invested with the same mysteries,
and both exalting woman as giver and sustainer of life. Beer was "liquid
bread," and a meal could consist of bread, beer, and cheese—a combination not unfamiliar in North America today. Since water was suspect,
beer was the thirst quencher, and people demanded high quality.
Roman legions carried beer with them and mixed it with vinegar to
avoid catching the diseases of the barbarians. When they conquered an
area, they introduced Roman yeasts to ensure getting a decent drink.
The *Senchus Mor*, a compilation of ancient Celtic or pre-Celtic law, contained regulations on beer quality. William of Orange invented British
bureaucracy when, in 1067, he appointed an ale coiner to regulate quality and price. Official state breweries were established in Bohemia in
1256, in the town of Budweis, and, in 1384, Pilsen's breweries were
under the personal control of Charles IV. Thomas à Becket, then chan-

cellor and later archibishop, brought casks of good English ale to France when he crossed the channel to conclude a peace treaty, and Elizabeth I sent courtiers to see if the local beer was drinkable before she visited towns. In Germany, if the local beer went bad, city fathers would import it from another town and sell it at cost in the basement of city hall—the *rathskeller*. Medieval marriages were toasted with brides-ale, now "bridal," and the name ale is derived from the medieval "hael," meaning "good health."

Government regulated beer not only in the public weal, but also to ensure the collection of taxes. Since beer made at home was impossible to regulate, taxes were imposed on the ingredients, the containers, and on alehouses frequented by the public. Monasteries and nunneries were exempt on the grounds that they consumed what they brewed, but many bishops enriched the Church coffer by requiring that the community buy beer from the Church. The resentment in England was a factor in the overthrow of "popery."

BEERS OF NORTH AMERICA

The invasion of the New World by Europeans found beer already in situ. Columbus drank a corn beer on his fourth voyage in 1502. Beer was brewed from English barley in Jamestown. The Pilgrims carried kegs in the Mayflower and worried near the end of their voyage that "our victuals being much spent, especially our beere" whether they would survive the winter. The Dutch established the first commercial brewery in New Amsterdam in 1623. A Captain Sedgwick was licensed to brew by the Massachusetts Bay Colony in 1637. Roger Williams authorized a communal brewery for the Rhode Island Colony. William Penn brought Quakerism and a brewhouse to Pennsylvania, and every family maintained its own equipment for making "small beer," consumed immediately after fermentation. Heroes of the American Revolution were brewers and beer merchants: Samuel Adams managed the family brewery, General Israel Putnam was part-owner of a brewery, and Thomas Jefferson imported a Bohemian brewmaster. Virginia planters all followed the advice of Dr. Benjamin Rush who wrote of the "friendly influence upon life and health" of malted beverages.

Colonial tavernkeepers-brewers were respected men, who inherited the esteem of members of European brewer's guilds to whom medieval people bowed. Usually retired military figures, they accumulated political and social power, and their taverns were the center of a town's activities. Because it was the town's gathering place, the British rightly viewed the tavern with suspicion; all too often seditious words were uttered. Patrick Henry frequented his father-in-law's tavern, and Captain Stephan Fay was the proprietor of the Catamount Tavern where Ethan Allen and his Green Mountain boys met. When Great Britain required that Redcoats be quartered in private homes, indignation over this breach of privacy led to refusal of householders to supply beer to the hated "Lobsterbacks." Washington joined Patrick Henry and Richard Lee in the Declaration of the Association of Williamsburg which stated "that we will not hereafter import . . . from Great Britain any one of the goods hereafter enumerated including . . . beer, ale or porter." Beer joined sugar and tea as symbols of the colonists' resolve to be free. Washington's farewell address to his officers in 1783 was, as would be expected, given in Fraunce's Tavern in New York's Battery.

THE BREWER'S ART

Basically, beer is merely fermented carbohydrate. Depending on temperature and other conditions, yeasts will produce carbon dioxide, ethyl alcohol, and their own energy supply from almost any carbohydrate. Wheat, rice, corn, pumpkins, beets, potatoes, and other starchy plant tissues can all be converted into beer. But starting from the Mesopotamians, people noticed that the cereal grains produced better beer and also noticed that barley was the best cereal. When barley grains were wetted, allowed to sprout, and then dried and ground to a powder, the resulting malt fermented more quickly than ungerminated grain. Barley malt seemed to cause the fermentation of other grain powders to occur more quickly. Only in the nineteenth century was it discovered that during germination of cereal grains, an enzyme is activated which breaks starch down into its constituent sugars—the primary substrate for yeast's fermentation processes. Not only did this enzyme (diastase or amylase) hydrolyze barley starch, it could break down wheat, corn, potato, or any other starch, thus speeding up fermentation. Barley soon became the preferred malt because the resulting beer has a lighter, more pleasant taste. Today, especially bred strains of barley are steeped in water for several days, germinated on a temperature-controlled malting floor or in large drums for 5–6 days, dried, and ground into the light brown malt powder also used in malted milkshakes. Most modern beers mix the malt with "adjunct" starches from wheat, rice, or corn, but malt liquors use only barley malt as the substrate for the yeasts.

The malt and adjunct starch are mixed with water to make a mash. Pure water is essential; the best beers are from iron-free water, preferably from springs that flow through limestone. The mash is allowed to saccharify, that is, to allow the enzymatic conversion of starch to sugar, resulting in the formation of the wort, an aqueous mixture of starch, sugar, and many soluble substances from the cells of the grains. The clear wort is transferred to a mash kettle where it is boiled for several hours with hops.

Were beer to contain only alcohol, wort, and the flavor of malt and grain, it would be insipid, slightly sweet, and would spoil quickly. By the eighth century, brewers in central Europe found that the addition of the unopened conelike inflorescence of the hop plant (*Humulus lupulus*) preserved the beer and provided a slightly bitter taste which

Figure 7.16
Hops, *Humulus lupulus,* showing the flowers used to flavor beer.

increased the palatability of the product (Figure 7.16). Hops was not the only bitters used; *Tanacetum balsamita,* a chrysanthemum, is still used in England to make alecost, tannins from oak and ash trees were used in Scandinavia, cinnamon was used in southern Europe, and American colonists dosed their beer with sweet fennel (*Foeniculum vulgare*), licorice, or sassafras. Hops was first recorded in the *Kalevala,* the Finnish saga of 3000 B.C. but was unknown in the rest of Europe until the father of Charlemagne donated his hop garden to the Monastery of St. Denis. Addition of hops was accepted in central Europe, but since Henry VIII liked ale made without bitters, it was not until the end of the fifteenth century that Flemish brewers were brought to Kent to make hopped beer. Parliament denounced "this wicked weed which will spoil the taste of our drink and endanger the people." Bohemian hops from Czechoslovakia, Wisconsin, and Oregon account for the supply used in North America.

Boiling in the mash kettles transfers the bitter flavor of hops to the wort, sterilizes it, and causes the precipitation of proteins. Filtered again, cooled wort is placed in fermenting tanks where an appropriate strain of brewer's yeast is added to initiate alcoholic fermentation. Each brewmaster guards the yeast carefully; subtle differences in fermentation result in beers with different flavors. Beer differs from ale in several respects; the fermentation temperature is lower for beer (45° C.) than for ale (60° C.) and the yeasts used for beer tend to remain at the bottom of the tank instead of floating to the top as in ale brewing. The result is a higher alcohol content in ale, tarter taste, and a paler color.

Young beer is separated from the yeasts and aged for several weeks before bottling. Some brewers add chips of beechwood to the aging tanks to serve as a nucleus for yeast growth. A second, slow fermentation occurs at low temperatures in which remaining sugar is fermented and carbon dioxide accumulates under pressure to yield a carbonated beer. This *lagering* process ("to store" in German) was discovered in the eighth century when it was noticed that the flavor of young beer was improved by aging it over winter in cold cellars in tightly stoppered wooden barrels. The second fermentation is sometimes eliminated and the beer carbonated with the carbon dioxide saved from the first fermentation. To increase shelf life, most beers are pasteurized following the recommendations of Louis Pasteur in his classic *Studies on Beer* in 1876. Draft beer is unpasteurized or sterilized by filtration to avoid the flavor changes which occur upon heating.

Virtually every step in the brewing process can be modified to give a somewhat different product. For bock beers, the temperature at which malt is dried is higher and the caramelized sugars give a darker and sweeter beer. Porter and stout use different malts. Variations in time, temperature, water quantity and quality, and yeast strain all play roles in providing a range of beers and ale. Many Mexican beers use corn both as malt and adjunct starch; south Germans make a *weissbier* entirely from wheat, and Japanese *sake* and Chinese *samshu* are made from rice. Birch beer, ginger beers, a spruce beer from young needles and twigs of *Picea*, true root beer from sarsaparilla (*Smilax*), and spikenard (*Aralia*) are but a few of the world's beers.

Like other alcoholic drinks, beer has come in for its share of both taxation and vilification. Over $600 million are collected in the United States, not including local taxes, and the tax burden is proportional in other countries. Never popular, beer taxes have sometimes been collected at gunpoint and have occasioned minor peasant revolts which have been put down violently as attacks on government itself. Public and private drunkenness have long been ascribed to beer drinking. Assuming an average alcohol content of 5 percent in beer, there will be 50 ml of alcohol in a liter (roughly a quart). A 30 ml "shot" (roughly 1 oz) of bonded bourbon will contain 15 ml of alcohol. It is possible to get very drunk on beer.

TEMPERANCE—THE NOBLE EXPERIMENT

People of Europe and North America have been modestly heavy drinkers for centuries and the interaction between drinking and the moral standards of organized religions have been with us for as long. The phrase of poor Bill Sykes in *The Beggar's Opera* that "gin was mother's milk to me" was literally true in seventeenth-century London. The Holborn district boasted of 7000 homes in 1617 and 1300 of them sold gin. Gin mills advertised "drunk for a penny" and made good on their promise, even supplying straw piles on which to sleep it off. Between 1785 and 1787, the foreign exchange of the young United States was reduced $12 million for importation of heavy, dark rum with alcohol levels about 140 proof. In 1826, the American Society for the Promotion of Temperance was founded in Boston, the major port for rum importation. Proper Bostonians raised the rallying cry against "the demon rum" in spite of America's permanent love affair with whiskey and beer. The society first approved of cider, beer, and wine, declaring that all they wanted to do was to "suppress the too-free use of ardent spirits." In view of the high level of American alcoholism, this was a social good. It was, in fact, 30 years before the Society decided that "the only logical basis for the temperance movement is entire abstinence from all intoxicants." The Society's leaders were religious people, mostly Baptists, Methodists, and Congregationalists. Catholics and Jews did not support the movement, but since members of these religious groups did not include political leaders, their voices were unheard in congress.

Temperance was very popular in less urban parts of the United States in the mid-nineteenth century when most of the population was rural, religious, and righteous. Three years after its founding, the Society claimed 100,000 members and this number rose to over a million by 1833. The convention of the society in 1836, when the doctrine of total abstinence was enunciated, also gave rise to the word "teetotaler," because, as the story goes, one who fully abstained (even from beer), was marked down in the register of the Society with a "T" for total. If this story isn't true, it should be. Prior to the 1860–1865 Civil War, the antis mounted a massive propaganda campaign. Edward Delevan, their spokesman, asserted that water used in brewing contained dead rats and also claimed that the delightful, nutty taste of Madeira wine was obtained by suspending a bag of cockroaches in the casks. Delevan also said that General U. S. Grant was a member of the Society, most unlikely in view of Grant's ability to consume over a pint of bourbon a day and still fight a major war. Maine passed a temperance law in 1846 which became the model for laws in Rhode Island, Massachusetts, New Hampshire, and Vermont by 1852, later to be introduced into the middle western states and the then territories of Minnesota and Nebraska. Southern legislatures assiduously avoided even discussing such bills in committee. Canada, with its large French and Irish populations, also avoided the question in spite of active temperance groups in the Maritimes and Ontario.

The Civil War, subsequent expansion of settlements in the West, growing affluence in the eastern states and grinding poverty in the South held the country's attention and prohibition laws were either repealed or ignored. Yet, the postwar period of 1870–1890 was the golden age of American religiosity, when preachers on horseback moved their revival meetings and evangelical fervors from village to hamlet. "Sin! No Heaven for the tippler! What will happen to the drunkard's child?" Rum was, in literal terms, the drink of the demon, and Satan was a palpable reality. With tears streaming down their cheeks, men took "the pledge," and it frequently took several drinks to get over it. The country ignored its growing alcoholism rate and the organization of the Anti-Saloon League in the 1890s went almost unrecorded. A small cloud in an otherwise clear sky, the League soon became a thunderhead. The antis were clever. They cut formal church ties, although their financial and moral support was still Baptist and Methodist, and, very wisely, they organized the women of America into a formidable coalition, the Women's Christian Temperance Union. With a flair for publicity which could be profitably emulated by modern television personalities, superstars like Carry Nation wielded her shiny hatchet.

Rather than a direct frontal assault on the last bastion of American malehood, the Anti-Saloon League and the Temperance Union decided to work quietly, to "nibble away at the Devil." Sunday blue laws were enacted in many states; even the ice cream sundae was banned. Candidates for office were sure to be asked their "stand on morality," and when the antitemperance governor of Ohio was defeated in Ohio in 1904 by an avowed prohibitionist—and this in spite of the fact that women couldn't vote—American politicians became very circumspect when it came to this burning issue. President Taft caught holy hell when he authorized the attendance of his secretary of agriculture at the International Brewer's Congress in Chicago in 1911. Taft was defeated in 1912 and, while the loss of the presidency was not due to his stand on liquor, the Anti-Saloon League made political hay over its "victory." While a lame-duck president, Taft vetoed a bill prohibiting liquor shipments into or through the dry states of Kansas and Oklahoma, but Congress overrode the veto.

Woodrow Wilson, concerned about the possibility of war, came down hard on both sides of the issue, but American military intervention in 1917 provided the temperance groups with powerful arguements. With food in short supply, why use precious stocks of wheat, corn, and barley for liquor? If, as the president proclaimed, the role of the United States was that of a moral crusader, no moral issue was as immediate as temperance. After all, most brewers and distillers had German names, and it was not unlikely that they were poisoning the youth of America to prevent "our boys from defeating the kaiser." With society dislocated and preoccupied, with men in uniform effectively disenfranchised and with weary and overworked legislatures wrestling with major economic and political decisions, the Eighteenth Amendment to the U.S. Constitution was passed by Congress in December 1917, President Wilson's veto was overridden, and the amendment passed by the states. The Volstead Act of October 28, 1919, defined an intoxicating beverage as one containing more than 0.5 percent alcohol by volume and authorized the formation of appropriate sections of the department of justice to enforce the law. Alcohol produced for medicinal or industrial use was *denatured*, a curious word which meant that it was made unfit for drinking by adding poisonous compounds.

In spite of dire predictions that the agricultural economy of the United States would collapse, the drop in demand for wheat, corn, and barley for beer and whiskey was more than matched by the desperate needs of people in Europe. Farm income rose as barley fields became wheat fields and corn was fed to hogs and cattle. Brewers, distillers, and their employers were badly affected, and many survived economically by entering other fields or brewing "nearbeer" and industrial alcohol which was in great demand. The effect on the mores of the people was, however, less happy. Bootlegging—the smuggling of liquor from Canada, Scotland, the West Indies, and South America—was soon dominated by criminal leaders like Al Capone of Cicero, Illinois. Injury and death rates caused by drinking denatured alcohol rose. Disregard for government and law resulted in the wild twenties and thirties when speakeasies, smuggling, home brew, and public drunkenness were fairly common behavioral modes.

Europe's economy recovered in the 1920s with corresponding decreases in farm incomes in the United States. The American people became heartily sick of weak beer, expensive imported wines, adulterated spirits, a quantum jump in violence, and the massive, calculated federal disregard for civil liberties, President Hoover appointed a commission in 1929 to study the impact of prohibition. The commission reported that the Eighteenth Amendment should be retained and Hoover agreed. With the stock market crash in late 1929, depression, the subsequent election of Franklin Roosevelt in 1932, and the feeling that the times had changed, Congress passed the Twenty-first Amendment to the Constitution thereby repealing the Eighteenth Amendment. The states quickly confirmed the amendment. Liquor could again be sold, as long as it was in bars or taverns—the Anti-Saloon League was able to prevent the reintroduction of this hated name, saloon. No battle over a plant product, not even that raging about marihuana, has so polarized the people of any country in several thousands of years. The sanctimonious zeal of organized religion and its political power within the United States was broken by the repeal, leaving a legacy of distrust and suspicion toward government that is still in evidence. Prohibition was not a "noble experiment"; it was an unmitigated disaster.

ADDITIONAL READINGS

Astbury, H. *The Great Illusion.* Garden City, N.Y.: Doubleday, 1950.

Baron, S. B. *Brewed in America: A History of Beer and Ale in the U.S.* Boston: Little, Brown, 1962.

Braidwood, W., et al. "Did Man Once Live by Beer Alone?" *American Anthropologist* 55(1953):515–526.

Cook, A. H. *Barley and Malt.* New York: Academic Press, 1962.

Sinclair, A. *Prohibition: The Era of Excess.* Boston: Little, Brown, 1962.

CHAPTER 8
Plants That Changed History

Paper—The Memory of Man

Papyrus sheets preserve the thoughts and deeds of man.
LEONARDO DA VINCI

Each of us acquires the morality and culture of our group through example, verbal signals, exhortation, and from the written record of the ideas and knowledge of the ages. The written word is paramount in education and is no less vital in other activities in which we participate. Writing is information storage and reading is information retrieval. To meet the need for a cheap, permanent storage-retrieval system, methods of recording speech led to the invention of writing and to the invention of paper.

EARLY WRITING SYSTEMS

Man was recording long before paper was invented (Figure 8.1). Cuneiform clay tablets of ancient Babylon, incised sheets of metal, and virtually any other surface from cave walls to public monuments graced with graffiti attest to the insatiable drive that humans possess to preserve their thoughts for a posterity that may not want them. But such surfaces are expensive and not easily transportable. Two of the forerunners of paper are noteworthy. Parchment was produced by rubbing skins of sheep and goats with lime instead of tanning them into leather. Today, this expensive surface is used for prestige diplomas—the sheepskin—but in the time of ancient Persia, it was the preferred material for

Figure 8.1
An Egyptian god of life writing a person's fate on a leaf held by a servant. Taken from an Egyptian tomb painting.

Figure 8.2
Papyrus, *Cyperus papyrus*.
Plants grow up to three m
tall.

edicts and religious tomes. The best parchment came from Pergamum, a Greek kingdom on the coast of Asia Minor. Based on the assumption that the method of producing parchment originated there, "parchment" was derived from "Pergamum."

The other writing surface which preceded paper was papyrus (Figure 8.2), a Latin word from an Egyptian word from which the English word "paper" is derived. As early as the third dynasty (ca. 2500 B.C.), writing sheets were made by cutting the pith of an Egyptian reed (*Cyperus papyrus*) into long, thin strips, pressing them flat, and gluing them together into large sheets. As long as the roll of papyrus could be handled on rollers, there was no limit to its size. The Great Harris papyrus, made during the reign of Rameses III (1198–1167 B.C.) is 400 meters long and is in better condition than last month's newspaper. Athens, and later Rome, used it for consular and imperial edicts such as the law scrolls exhibited in the Agora and the Amphitheatre. It was used in Palestine as evidenced by the Dead Sea Scrolls. Its surface was smooth enough and had enough body to serve as the vehicle for paintings. Rolls of papyrus, mostly of Egyptian manufacture, served the Western world as its primary writing surface until the eighth century, supplemented by parchment and, for lesser writings, by wooden tablets.

INVENTION OF PAPER

Paper, as we know it, consists of sheets of matted or felted plant fibers. Its inventor was the Chinese scholar Ts'ai Lun, who in 105 A.D. pounded the inner bark of the *Ku* tree (the paper mulberry, *Broussonetia papyifera*) into short fibers, treated this with alkali (lye) from wood ashes to dissolve other material, mixed this slurry with water, and sieved off the water through silk. This left a sheet of paper which could be pressed flat and dried in the sun (Figure 8.3). Contemporary records proclaim, "Under the reign of Emperor Hi-Ti, Ts'ai Lun of Lei-yang conceived the idea of making paper from the bark of trees. . . . The paper was then used throughout the entire universe." For the Chinese of the Han dynasty, China was "the entire Universe." They guarded their secret process for over 500 years.

During the Han and the subsequent Wei (220–246 A.D.) and Chin (265–520 A.D.) dynasties and into the troubled Six Dynasties period (420–618 A.D.), strips of bone, tablets of wood, and split stalks of bamboo were the usual writing surfaces. These elongated surfaces lent themselves nicely to the unique Chinese style of vertical writing, a style attributed to Ts'ang Chieh who invented calligraphy in 2700 B.C. Postal service was a government responsibility, and "letters" of wooden tablets in a clay envelope stamped with official seals have been found thousands of miles from the cities where they were written. And they were probably delivered on time. Woven silk sized with rice starch or gelatin was another medium, but because of cost, silk scrolls were used only for important edicts and religious paintings. Silk and paper had

Figure 8.3
Paper-making in China. Bark of the paper mulberry, *Broussonetia papyrifera,* is pounded in a tree-trunk mortar and the pulp sieved from the water on a silk screen. Individual screens are dried in the sun and sheets of paper peeled off.

surfaces that permitted calligraphy with camel's hair brushes and carbon-black inks made from animal bones.

The period of the T'ang dynasty was marked by the development of foreign contacts and trade with the "barbarians," all peoples who were not Chinese. Trade routes were established to the west over the mountains and deserts to modern Afganistan, into modern Iraq and Iran, and possibly into Greece and Rome. Although wooden and bamboo tablets of military records and business transactions have been

Figure 8.4
Household ikon of the Chinese gods of life, health, and wealth. These were printed with the red ink of good luck on ceremonial paper.

found along these routes, no paper has been found; it was either not used for such mundane correspondence or was too fragile to survive a thousand years. Attracted by the flowering of Buddhism in China and the sophisticated and effective methods of governance, finance, and the fine arts, Japan sent many delegations to the T'ang court to learn all they could and to bring this new knowledge back to the Asuka and Heian emperors at Kyoto. The Silla rulers of Korea sent scholar-supplicants to China for the same purposes. Among the arts which went to Korea and Japan was that of making paper.

In China, Korea, and Japan, the art and technology of paper-making developed over centuries with other plant materials used to modify the character of the paper and the addition of coatings to improve its ability to hold various inks and paints. Fibers of precious metals were incorporated to yield papers of outstanding visual beauty, and the parallel development of calligraphy as an art form was exploited by scholars, poets, and even emperors. In Asia, these arts were not to be taken lightly; paper and writing were used to transcribe edicts and sacred thoughts for the benefit of posterity. Fibers of paper mulberry bark, bamboo, hemp (*Cannabis*), and rice straw were combined to prepare special sheets upon which were imprinted woodblock cuts of household deities (Figure 8.4). Such printed pictures have the same religious significance as do the ikons of the Orthodox Russian and Greek churches. The "joss paper" used in funeral services was made from these ceremonial papers, the word joss being pidgin from the Portuguese *deus* or god. Sheets of joss printed as "money" were burned so that the spirit of one's ancestors would have funds to buy their way through the beyond.

MOSLEM PAPERMAKERS

Although the Chinese maintained trade routes to the West as early as the latter part of the Ch'u dynasty (ca. 300 B.C.), it was not until the T'ang dynasty that the major routes were extensively used. Caravans of camels and horses followed several routes laid out by Chinese engineers and policed by troops who built walls, forts, and watchtowers for thousands of miles. One route crossed the Gobi Desert, the desert of Takla Makan, and the formidable Tarin Valley to Samarkand, a city in the Uzbek Soviet Socialist Republic just north of Afghanistan. The Chinese established a papermaking factory in Samarkand to utilize the fibers of flax (*Linum*) and hemp (*Cannabis*) that grew in this area. The paper was shipped back to China under heavy guard. In 751 A.D. an army of the Prophet Mohammed clashed with guard troops of the T'ang Emperor Hsüan-tsung on the banks of the Tharaz River, a few miles east of Samarkand. The Moslems captured the city and with it the paper mill and its skilled artisans. Immediately recognizing that this product would provide a vehicle for spreading the faith, the conquerors continued paper production. Arabian artisans were trained by the Chinese, and new factories were established in Baghdad and Damascus by the beginning of the ninth century under the patronage of Harun-al-Rashid, he of *The Thousand and One Nights.*

The techniques used by the Moslem papermakers differed little from those developed by Ts'ai Lun 700 years previously. Today's processes are also basically the same, although many technical innovations have been developed in the past 1100 years. The inner bark (botanically called the phloem; Greek for bark) of plants consists of several different kinds of cells. Some of these, the sieve tubes and companion cells (Chapter 1), are structurally and functionally specialized to carry sugars made in the leaves down the plant to the roots where they are stored as starch. Another cell type is composed of long, thin cells with thickened walls which serve to strengthen the phloem. These are called bast fibers and in some plants the cells are long enough to be twisted into thread. Linen is a thread made from the bast fibers of the flax plant, and hempen rope is made from the very long and very tough bast fibers of *Cannabis.* Since these cells are tough, they are not destroyed by pounding or alkali treatments which can dissolve away the other phloem cell types. When freed of the other cell types and mixed with water, bast fibers tangle up and interlock as a felt. If this suspension is allowed to settle out on a fine-meshed screen, water is sieved out, and the fibers form thin sheets. Drying and pressing will complete the process. Ts'ai Lun used the bast fibers of the paper mulberry, a happy choice since its fibers are tough and resistant to the harsh alkali treatments he used. As Oriental technical workers continued to experiment, they added fibers from bamboo and other plants to give papers characteristics of toughness, smooth or rough surfaces, ability to take ink, ability to be coated with oil for umbrellas, or for windows. Neither the Chinese nor the Japanese solved the problem of pulping the plants; hand pounding was slow and inefficient. Arabian artisans invented a stamping machine worked by foot power, allowing them quickly to obtain uniform, clean fibers. They also found the old linen cloth gave a superior paper, one very similar to the linen paper used today for thank-you notes.

The Mayan and Aztec cultures independently invented paper using fibers from the century plant (*Agave*), the phloem of species of fig (*Ficus*), or bast from wild mulberry (*Morus niger*). The bark was softened by beating, treated with wood-ash lye to dissolve other cell types, washed with water, and laid on boards to dry (Figure 8.5). These papers, like those of the Chinese, were reserved for official documents and as currency. Paper was a precious commodity and Montezuma II received sheets of papers as tribute. Apparently the Mayans were the first to make books by bending long sheets of paper into accordion folds and binding them between boards. Because paper can transmit information, it was involved in witchcraft, and Mexican peasants still cut out paper dolls of evil spirits, sprinkle the effigies with blood, and burn out the devils. The process and even knowledge of Mayan and Aztec papermaking was destroyed by the Spanish conquerors and

Paper-making in Central America. Tree bark is pulled off and pounded on a log into a thin sheet. Sheets are laid on boards to dry, are peeled off, and pleated into an accordion-folded book.

played no role in the development of paper technology in the rest of the world.

Although Baghdad and Damascus attempted to maintain their monopoly on paper, the secret was bound to leak out. In the tenth century, a factory was built in Cairo, the Moors carried the technique throughout North Africa, and a major factory was built in Fez, Morocco. Moslem paper was made primarily from linen and cotton rags including the wrappings of mummies when available. With the conquest of Spain by the Moors, the first paper factory in Europe was established in 1150 in Zativa in southern Spain. Paper-making had arrived in Europe a thousand years after its discovery in China.

EUROPEAN PAPERMAKERS

The early European papers were, however, not well received. The product was more expensive that the parchment used by monastery monks to hand-letter bibles, it was much more fragile, and it was impossible to scrape away an error in transcription as one could do with parchment. Since paper was made by infidels, it was suspected of being a product of the Devil, a belief fostered by the parchment-makers guilds. Slowly, however, paper began to be accepted. When factories were built in Fabriano, Italy, in 1276, in Troyes, France, in 1348, and in Nuremburg, Germany, in 1390, the curse lifted suddenly and paper's acceptance by the Church for religious writings made its general adoption only a matter of time. The development of moveable type printing by Gutenberg and others and the printing of the Bible broke down any residual reluctance. Since the Latin word for bark is *liber*, books of paper were "on liber" and a collection of books formed a library. The pages of a book, bearing a clear resemblance to the leaves on a tree are called *feuillet* in France, *blatt* in german, and *leaves* in English; equivalent words are used in most languages.

By the end of the fourteenth century, paper was in great demand, not only for imperial and religious use, but also for the transmittance of business information and for a modest, but growing, secular literature. With demand so high, the papermakers in Ulman Strömer's mill in Nuremberg went out on what seems to be the world's first factory strike for higher wages in 1391. In general, the quality of European paper was lower than that made by the Moors, and there was a return to parchment during the fifteenth century. Indeed, it was not until the middle of the sixteenth century that better paper caused a return to its general employment, and it was only then that illustrations of the process began to appear. Under one of these woodblock prints is the following poem:

> Rags are brought to my mill
> There much water turns the wheel
> They are cut and torn and shredded
> To the pulp, water is added.
> Then the sheets 'twixt felts must lie
> While wringing them in my press
> Lastly, hang them up to dry
> Snow-white in glossy loveliness.

Ts'ai Lun's process had been supplemented with a wheel replacing hand-pounding, the use of rags instead of raw bark, and drying felts to speed up the removal of water. Only this in 1500 years!

Two additional inventions in the sixteenth century accelerated the shift from parchment—now called vellum—to paper. Gelatin was added to the slurry of water and fibers before drying to prevent excess absorption of ink, and watermarking allowed identification of the producer. These watermarks were usually crosses since most of the paper was used for the bibles, but later the initials or names of the guilds or the factory were used which tended to raise standards and increase quality control.

Elizabeth I granted a royal patent to a John Spilman to build a paper factory at Dartford to supply the court and to promote the establishment of newspapers. As on the continent, the paper was made from linen and cotton rags, beaten to a pulp (a familiar expression whose derivation is now clear), filled with gelatin, and sized with glue to give a hard, opaque sheet, admirably suited for writing with quill pens and the thin inks then available. Printing presses were improved, and the increase in interest in politics led to the establishment of many newspapers. When William Rittenhouse built the first U.S. mill in Germantown, Pennsylvania, in 1690, rags were his sole source of fiber. His factory's output of two reams per day consumed such an enormous bulk of rags that he wasn't sure that he could find enough to remain in business. An issue of the Boston *New Letter* in 1769 announced that a ". . . bellcart will go through Boston . . . to collect rags for the papermill at Milton . . . when all the people that will encourage the paper manufacture may dispose of them." New, more abundant sources of fiber were needed, and, although bast fibers made less satisfactory paper than rags, papermakers in many countries were experimenting with fibers from local plants. The French used the lime tree (our linden or basswood—a corruption of bastwood). In North America, corn husks, wheat straw, and several trees were tested. Bamboo, rice straw, sorghum stalks, fibers of the common stinging nettle (*Urtica dioica*), and other plants were tested as well, but most were not as good as rags and some were, given the technology of the period, almost useless. The Chinese developed a special paper from the pith or central part of the stem of the rice-paper plant (*Tetrapanax papyriferum*) which became the rice-paper of commerce.

PAPER FROM WOOD

The modern paper industry, based on wood, originated with the acute and perceptive eye of a distinguished French scientist and inventor, René de Reaumer, who noticed that wasp's nests were paperlike and contained filaments of wood. In a report read to the French Academy of Sciences on November 15, 1719, de Reaumer suggested that wood might be a fine source of fibers, although he had not experimented with this possibility. Over 50 years later, Dr. Jacob Schaffer, a Lutheran minister, and Mathias Koop, a Dutch citizen living in England, finally were able to make paper from wood, straw, and rags. It was not until 1840 that Friedrich Keller of Germany made the first all-wood paper. Keller reasoned, quite accurately, that the fibers of wood, cell types botanists call xylem (a Greek word for wood) are sufficiently strong (Chapter 1) and long enough to form a felt which would mat well once the water had been removed. The problem was to find some way to separate the wood fibers. Keller patented a process in which blocks of wood were pressed against a slowly spinning, wet grindstone. The first factory using the ground-wood pulping method was built in Nova Scotia in 1841, followed by another in Maine in 1863. In 1870, the *New York Times* led the United States newspaper industry when it converted to newsprint made exclusively from wood.

Ground-wood paper pulp consists of whole or broken xylem vessels and fibers. At maturity, these are dead cells, their contents having been lost during maturation of the cell. The walls of xylem cells consist of a major component, cellulose, whose chemical composition is much like starch except for the arrangement of the chemical bonds. The cellulose molecule is long and threadlike, and a number of these molecules are twisted about each other to form a microfibril, and several layers of microfibrils form the xylem cell wall. Surrounding and coating the cellulose molecules is another compound, lignin, which serves to strengthen and waterproof the xylem cell wall, making these cells well adapted for their primary function of conducting water from the roots of the tree up to the leaves. Pure cellulose is white—cotton fibers are an example of almost pure cellulose—and the lignin is brownish in color. In paper, lignin is undesirable; in ground wood pulp, strong bleaches are used to decolor the lignin, although this reduces the strength of the paper. The groundwood process was a signal advance in paper-making, but it was time-consuming to grind the wood, and the resulting paper was suitable for cheap newsprint and little else.

Benjamin Tilghman, an American working in France, discovered that if wood chips are heated in the presence of sulfur compounds, the lignin is dissolved away from the cellulose and the xylem fibers will separate easily from one another. In large-scale modern pulping mills, the bark of trees is removed by knocking logs together in a revolving drum. The peeled sticks are fed into a chipper where whirling knives cut the wood into thin, short fragments. These are fed into a digester where steam raises the temperature well above the boiling point of water. The quantity and composition of the chemicals added to macerate (separate) the fibers and remove the lignin will vary with the type of paper pulp wanted. In the soda process, lye (sodium hydroxide) is added, and in the sulfite process, an acid solution of bisulfites and sulfurous acid is used under pressure. A third process used a mixture of lye and sodium sulfide. The macerate or raw pulp, now almost pure cellulose cell walls, is washed with water to remove the chemicals, leaving a water-slurry which can be further bleached or dyed. Water is removed and paper is formed by a modification of Ts'ai Lun's inventions. John Baskerville invented a wire screen in 1757 to replace the cloth screens, and in the same year Nicolas-Louis Robert invented a machine which put the wire screen on an endless belt so that, instead of making sheets of paper, continuous rolls would be formed. Because of the disruptions caused by the French Revolution, the continuous process was actually patented in England by Robert Gamble and Sealy Fourdrinier, and, by 1805, a paper-making mill was built in Hertfordshire which could turn out 100 times as much paper as did the hand process. When rolls of paper made from wood were available, in about 1835, the modern paper industry was ready, willing, and able to meet all possible demands for its product.

Demand increased at a tremendous rate. Kraft (German for strong) papers could be fabricated into bags to replace the burlap and cotton sacks in which grain and sugar were packed, cardboard replaced wooden baskets and crates, wallpaper decorated the homes of millions of families, and the availability of inexpensive paper allowed books, newspapers, and magazines to be within the financial reach of everyone. North America is both the biggest producer and biggest consumer of paper with the bulk of the pulping wood coming from Canada. In the United States, well over half the paper pulp is harvested in the southern states, several million acres being devoted to growing the southern yellow pine which constitutes the bulk of southern pulpwood. In the North and in eastern Canada, spruce, fir, some pine, and a small volume of tamarack are used, and in the West, several species of fir, the Douglas fir, western hemlock, and ponderosa pine are utilized (Figure 8.6). Industrial wastes such as straw, the *bagasse* remaining after sugar has been squeezed from the cane, and a variety of grasses and stems of

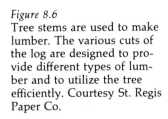

Figure 8.6
Tree stems are used to make lumber. The various cuts of the log are designed to provide different types of lumber and to utilize the tree efficiently. Courtesy St. Regis Paper Co.

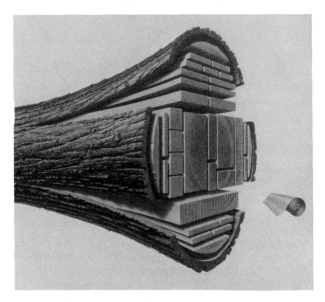

plants like cotton are also used. Genetic engineering of pulpwood trees like poplar has resulted in strains which grow to cutting size in 20 years instead of the 40–50 years it used to take. Recycling of used paper and the introduction of new sources of pulping material may be able to replace the loss of forest lands to industry, housing, and roads that North America has experienced in the recent past. If the rate of paper use continues to increase as it has in the last 50 years, paper supplies may become limited.

ECOLOGICAL CONSEQUENCES

As a consequence of the construction of huge, integrated paper-mills, severe ecological problems have emerged whose solutions are not yet in sight. The assault on one's nose as a mill is approached is well known to most people, but more serious is that some of these sulfur compounds are converted to sulfuric acid in the atmosphere, returning to the earth as acid rain. This alters the alkalinity-acid balance of soils with yet unknown, but potentially serious, effects on microflora in the soil and on plants growing in this soil. The efficient and time-saving method of clear-cutting large stands of forest trees for pulpwood and lumber has resulted in erosion of large areas, with consequent silting-in of streams having severe impacts on the insulted ecosystems (Figure 8.7). Just as severe are the problems of waste disposal in the mills. Until very recently, and in some mills today, mercury compounds have been used in the processing of the wood. These compounds were deliberately or inadvertently (mostly the former) dumped into streams and

Figure 8.7
Erosion and silting of an Alaskan river after clear-cutting the steep slopes of the drainage basin. Courtesy U.S. Department of Agriculture.

lakes along with sulfur compounds, bleaches, and dyes. All these are poisonous to the plants and animals in these waters. In Canada and in Japan, severe human mercury poisonings have been documented in which the mercury used in making pulp is concentrated in food chains, accumulating in the fish people eat. Finally, and not least among this catalog of problems, is the lignin. Virtually undegradable, with no known industrial use, it is routinely discharged into the water courses belonging to *all* of the people. It accumulates in large sludge beds that inexorably reduce the oxygen needed by aquatic animals and plants. The multinational paper companies have a great deal to learn about corporate social responsibility.

ADDITIONAL READINGS

Blum. A. *On the Origin of Paper.* New York: Bowker, 1934.
Butler, J. W. *The Story of Paper Making.* Chicago: Butler Company, 1901.
Clark, T. F. "Plant Fibers in the Paper Industry." *Economic Botany* 19(1965):394–405.
Hunter, D. *Papermaking: The History and Technique of an Ancient Craft,* 2nd ed. New York: Knopf, 1957.
Isenberg, I. H. "Papermaking Fibers." *Economic Botany* 10(1956):176–193.
Perdue, R. E., Jr., and C. J. Kraebel. "The Rice-paper Plant—*Tetrapanax papyriferum* (Hook.) Koch. *Economic Botany* 15(1961):165–179.
Smith, D. C. *History of Papermaking in the U.S. (1691–1969).* New York: Lockwood Publishing Company, 1970.
Sutermeister, E. *The Story of Papermaking.* Boston: S. D. Warren Company, 1954.
Weaver, A. *Paper, Wasps and Packages.* Chicago: Container Corporation of America, 1937.

Cotton—A Despotic King

Look away down yonder in the land of cotton
To Live and Die in Dixie
D. D. EMMETT

In order to utilize a material as fabric thread, several requirements must be met. The fibers must be uniform in diameter, must have the ability to hold together when spun into a thread, must have high tensile strength, durability, and pliability, should individually be long enough to be twisted together, should be light in color, and should have the ability to take dyes. The material should also be relatively inexpensive and readily available. Wool meets most of these requirements except for world-wide availability, and silk is expensive. Thus it came to pass that men and women gratefully gave up wearing fig leaves or animal skins and started using specialized fiberous plant cells. Of these fibers, none has had the social, economic, or political impact of cotton.

THE COTTON PLANT

The genus *Gossypium* was established by Linnaeus within the mallow (Malvaceae) family, along with several garden plants (*Althaea*, *Hibiscus*, and the Rose of Sharon). The genus name is an ancient Greek name for cotton. There are over 30 species found native in Central and South America, Africa, Asia, and Australia. Four species are cultivated and each has numerous subspecies, botanical varieties, and cultivars. *G. arboreum* and *G. herbaceum* form the taxa from which the cottons of India, China, and Africa have been derived. *G. barbadense* (sea-island, pima, and Egyptian cottons) and *G. hirsutum* (upland cotton) are both native to the New World and, because of the superiority of their fibers, are the major species planted throughout the world. The classification of cotton is complicated, not only because man has selected and bred for superior fiber production, but because of the strange geographical distribution of the major species. All Old World cottons have clear affinities to one another, and all New World cottons also show affinities one to another. The difficulty is that there is no obvious geographical connection which would permit a link between the Old and New World species, but botanists are loath to believe that the two groups are not phylogenetically derived from some common ancestral type species of cotton.

ℭ AN ESSAY ON TAXONOMY AND PHYLOGENY

The classification of plants, the science of taxonomy, is an attempt to put like with like—essentially to pigeonhole all known and still-to-be-discovered plants into groups that reflect their structural (usually floral) similarities. Phylogeny, although dependent upon taxonomy, has a different mission. Phylogeny attempts to show how plants in one taxonomic group—anywhere from subspecies up through the entire plant kingdom—are related in an evolutionary sense. Modern evolutionary theory suggests that new species develop over long periods of time by accumulation of mutations that progressively alter the structure and function of an evolutionary line of organisms to the point that the new species can no longer interbreed with its ancestors and it is sufficiently different to be recognizable as being different.

Phylogeneticists attempt to trace relatedness by examination of the structure and function of organisms that might be the ancestors of the species. In many instances, a coherent picture can be obtained. If, for example, several species in a genus are found in one geographical area, it is reasonable to assume that they are all derived from a single species that had, itself, evolved in that area or invaded the area by some means. An isolated area—an island or a valley surrounded by high mountains—may develop a unique flora or fauna because of "adaptive radiation." A few organisms reaching the new area evolve into dissimilar forms adapted to various habitats provided by the geographical area. The most dramatic examples are seen in the variation among the birds (Darwin's finches) on the Galápagos Islands. From a single species, phylogeneticists have been able to trace the adaptive radiation that resulted in finches that look, act, and feed like woodpeckers, seed eaters, or insectivores.

The importance of adaptive radiation as an aid to phylogeny was recognized by the Russian botanist N. I. Vavilov, who was interested in determining where our cultivated plants came from in order to evaluate the botanical resources of the world. Vavilov was able to trace back to the location where the species originated and, by doing so, was able to pinpoint the center of evolution of a species. Since this pioneering research, other botanists have been able to locate the origin of most of our cultivated plants and have used this information to obtain genetic stock for breeding purposes.

There are, however, some puzzling patterns of adaptive radiation. Plants of eastern China are very similar to those of western North America; many plants of Brazil are at least kissing cousins of plants of Nigeria and Angola. Only recently have these discrepancies been resolved as geologists determined that, at one time, our continents were a single land mass that literally floated apart. This new science of plate tectonics has been of great value to phylogeneticists.

There are, however, still unsolved problems. It is not particularly difficult to visualize that even a single species that would be included in the rose (Rosaceae) family could, by mutation, selection, and adaptation, be the direct or indirect progenitor of all members of the rose family. A single type of grass could give rise (over a very long period of time) to genera as diverse as wheat, lawn grasses, bamboo, and sugar cane. Some taxa of plants or animals evolved by simplification of structure and function—internal parasites of animals are examples of this—and other taxa evolved by complications of the basic structure—corn (*Zea*) may be an example of this. There can be instances of parallel evolution in which completely unrelated plants or animals could evolve structures and functions that mimic very closely those of others—the insectivorous plants (Chapter 1) illustrate this quite well. Geographical discontinuities are also troublesome. Plants whose center of origin was near Hudson Bay are also found on exposed ridges of Mount Washington in New Hampshire and in the Great Smokey Mountains. Such plants are called relics, and it is assumed that they were left "high and dry" when glaciers advanced and retreated. Although some scientists assert that phylogeny is a sterile exercise for pedants, the detective work of the phylogeneticist is intellectually exciting and has uncovered useful sources of genetic material for the breeding of economic crops.

Unfortunately, phylogenetic research has not resolved the problem of what appears to be a dual origin of the cotton species. The Old World plants probably originated from a type in southern Africa as a warm-climate perennial and spread north and east into India and southeast Asia giving rise to perennial and annual species. All these species have the same chromosome number, and the chromosomes look the same; indeed many of the Old World species are interbreeding populations and are botanical subspecies and varieties. The New World cottons probably originated in the high mountains of Peru, Ecuador, and Bolivia, the home of the potato, quinine, and tobacco. They have twice the number of chromosomes as do the Old World species. One set seems to be identical with the set in the Old World species while the other set seems to be unique. The presence of the similar chromosome set suggests that both Old and New World species are phylogenetically related, but the geographical discontinuity cannot be explained by a land bridge (such as across the Bering Strait), by plate tectonics, or by human actions. The phylogeneticist sadly shakes his or her head and dolefully walks away from the problem.

COTTON PRODUCTION

In tropical climates, cotton can be a perennial, developing a thick, woody root and stem, and some wild species grow into small trees. The southern United States is too cool for overwintering, and the plants are grown as annuals. The flowers are large, bell-shaped, and look like those of hibiscus. Petal colors range from creamy white through spectacular reds and purples; cotton is a very fine annual garden plant. Cotton is both self-pollinated and bee-pollinated, and after fertilization the ovary develops into a large fruit, the boll, containing many seeds. The outer layer of the seed coat is covered with hairs (Figure 8.8). These

Figure 8.8
Cotton seeds. Courtesy U.S. Department of Agriculture.

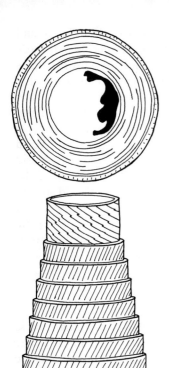

seed hairs, the cotton fibers of commerce, are single cells that may grow to a length of 1 to 2 cm. During growth they elongate and their walls become thickened with additional layers of cellulose fibrils (Figure 8.9). At maturity, the cytoplasm is lost and the cell collapses and twists into a flattened filament, which is almost pure cellulose of considerable tensile strength and durability (Figure 8.10). At maturity, the boll splits open and the cotton fibers expand, covering the plant with white "snowballs."

Cotton is subject to a veritable jungle of insect pests, as well as bacterial, fungal, and viral diseases. Breeding for resistance to soil-borne fungi and bacteria has been important in keeping production costs down, but insecticides and other pesticides are still used in massive control programs. The insect pests have developed increased resistance to pesticides, and there is a continued search for different and more effective compounds. The boll weevil is still a fact of life in cotton fields, causing a five to eight percent loss each year (Figures 8.11–8.12).

Figure 8.9
Diagrams showing the laminations of cellulose fibrils that comprise a cotton fiber. The dark mass in the center of the cross section is residual protoplasm.

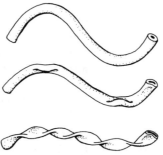

Figure 8.10
As cotton fibers mature, they collapse and twist.

Figure 8.11
Cotton boll weevil on a cotton boll. Courtesy U.S. Department of Agriculture.

Figure 8.12
Infestation of cotton boll. Section of cotton boll with weevil puncturing boll and a larval stage forming its cell. Courtesy U.S. Department of Agriculture.

Figure 8.13
Handpicking cotton in the 1930s. Courtesy U.S. Department of Agriculture.

The growth rate of cotton is rapid, and it is very demanding of minerals and water. In order to supply these needs, cotton must be heavily fertilized and watered and kept free of weeds which compete with the plant. This has traditionally been done by hand-weeding (chopping cotton) but now is accomplished with mechanical cultivators and by application of selective herbicides. Picking cotton, too, was traditionally hand labor (Figure 8.13) since early picking machines damaged the cotton and indiscriminately collected bolls, twigs, leaves, immature bolls, and dirt which reduced the value of the crop. Efficient mechanical pickers were developed in the late 1940s when chemicals were found that could remove the leaves (defoliate the plants) leaving the bolls intact (Figure 8.14). Plant physiologists who studied the responses of plants to chemicals which later became herbicides and who investigated defoliation and leaf abscission as part of their research on plant development were responsible for this modern revolution in cotton cultivation. They were, inadvertently, responsible for the mass migration of the American black to a culturally inhospitable and economically unready North. Since field hands were no longer needed to chop or to hand-pick cotton, there was no longer a living to be made in cotton country, and black families moved North or into the cities of the South where industry was developing. In 1910, 89 percent of the black population was southern and rural; in 1960, three-quarters were urban. Poorly educated (a quarter functionally illiterate), without marketable skills, and subjected to climatic and cultural conditions unfa-

miliar to most of them, they were crowded into ghettos, paid minimal wages, and became the victims of business cycles that resulted in massive unemployment. King Cotton treated his subjects poorly.

COTTON HISTORY

The Western World's first knowledge of cotton was provided by the Greek historian Herodotus who visited India in 484 B.C. and reported that "there are trees on which fleece grows surpassing that of sheep and from which the natives made cloth." In 1350, Sir John Maundaville invented—out of whole cloth—the story of a tree that bore tiny lambs at the ends of its branches, the vegetable lamb (Figure 8.15), and the German name for cotton, *baumwolle*, reflects this legend. Alexander the Great introduced cotton to North Africa in 330 B.C., and when the Moors conquered Spain in the ninth century, they brought cotton with them. A new type of cotton cloth, muslin (from *musselman*), was developed at about this time, and our word for cotton is merely the Arabic word *kutun*. Spanish-made muslin was returned to North Africa and little of it was seen in the rest of the continent. Europe's interest in cotton cloth was aroused by reports of early travelers to China, Japan, and India who saw fabrics "so fine you can hardly feel it on your skin"; Indian poets sang of "webs of woven wind." Venetian merchants obtained a monopoly on Indian cottons and organized camel caravans to transport the cloth across the desert and mountains through Persia to Venice. Some of the cottons from Calcutta were so delicate that "when a man puts it on, his skin shall appear as plainly as if he were quite naked," but most of such cloth was reserved for the harems of the Middle East. These fabrics were a revelation to Europeans who wore heavy woolens or coarse-textured linens and could appreciate lighter fabrics for warm-weather use. Interest was so high that cotton became as valuable as the spices of the Orient which also moved through Venice (Chapter 6).

At the same time, explorers of the New World were astounded to find the natives of Cuba, Mexico, Peru, and Brazil clothed in beautifully woven cotton fabrics rivaling the best India had to offer. Columbus found the people of Hispaniola sleeping in *hamacas* loosely woven from strong cotton cordage. Montezuma wore a *tilmatl*, a girdle and cloak, of delicately dyed and embroidered cotton. Magellan found Brazilian mattresses stuffed with cotton batting, and Piazzaro reported that the peoples of Peru used spinning and looming machinery to weave differently colored cotton threads into complicated and elegant fabrics. Modern archeological research has shown that cotton cloth existed in Mexico and Peru by 3000 B.C. and was used in South Africa for at least 2500 years.

Cotton cloth was, however, a luxury fabric in Europe, more expensive than linen and wool, and it remained so until the beginning of the eighteenth century. In most of Europe, native wool was carded, spun into yarn, and woven into sturdy but rough cloth as a cottage industry, where parents and children worked in their own homes. Spinsters were legally unmarried women who were proficient in spinning woolen or linen yarns. Flemish protestors, feeling persecution by the Catholic majority in the fourteenth century, settled Manchester in Lancashire and introduced new techniques that helped England to achieve supremacy as the wool-cloth manufacturer for Europe. Small amounts of cotton were imported from Egypt and India to be used as bed linens and small clothes (underwear) and for candle wicks. As cotton cloth from India and Egypt became cheaper, parliament passed a law in 1666 requiring that every corpse had to be buried in a wool shroud on pain of a £5 fine levied, it is presumed, on surviving relatives. A few years later, the anticalico laws were amended to provide a fine of £200 on anyone using cotton "for domestic use, either as apparel or furniture." Yet, at the same time, the British West India Company was instituting cotton cultivation in the Caribbean Islands and Indian seed was planted in Virginia. Between 1690 and 1750, the entire cloth manufacturing industry was transformed from a bucolic cottage system into the Industrial Revolution.

THE COTTON GIN

Many factors contributed to this change from a rural to a mechanized city economy in which people became factory workers instead of remaining in villages. One was the development of agricultural practices which allowed smaller numbers of people to supply the food

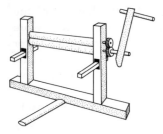

Figure 8.16
Indian churka or roller mills were suitable for separating small quantities of cotton fiber from the seeds.

for city dwellers and the exploitation of the potato as a cheap source of carbohydrates. But among the most important factors were the series of inventions of cloth-manufacturing machinery which were too expensive for a single family and too productive to be managed by a few workers. Prior to the Industrial Revolution, raw fibers of wool, linen, and cotton were carded, spun into threads with the distaff or the spinning wheel, set onto hand looms, and woven with a single, hand-thrown shuttle. Between 1730 and 1760, British inventors devised carding, spinning, and winding machines with tremendous capacities. Arkwright's water-powered loom and the harnessing of James Watt's steam engine of 1769 to the power loom of Edmund Cartwright in 1785 completed the process of mechanization allowing, indeed demanding, that factories be built to make inexpensive cotton cloth. The limiting factor in large-scale manufacture was the separation of the cotton fibers from the seed which had to be done with a simple, inefficient Churka roller gin (Figure 8.16) or by hand. This limitation was overcome in 1793 by Eli Whitney, a Yale-educated tutor to a wealthy Southern family (Figure 8.17).

It is impossible to overestimate the importance of Whitney's cotton gin. The export to England of cleaned raw cotton fibers from the United States increased 800 times between 1790 and 1800 and 2500 times by 1810 (Table 8.1). The parallel increase in the number of slaves is also noteworthy. Equally important was the direct social consequences on the people of England. Thomas Carlyle spoke of the owners of the cotton mills, the coal mines that fed the mill's steam engines, and the railroads and canals that transported the finished cloth as "captains of industry." These he defined as men whose only idea of hell was the

Figure 8.17
Diagram of a cotton gin. The condensed lint is the starting material for thread.

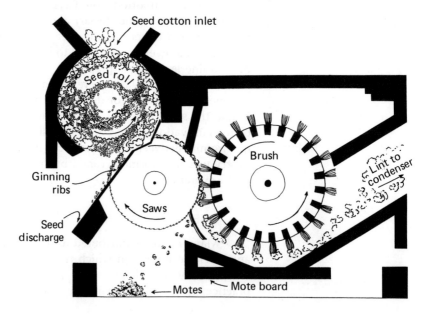

Table 8.1
U.S. EXPORTS OF COTTON TO ENGLAND

Date	Pounds of Cotton	Number of Slaves in Mississippi and Louisiana
1785	5,000	8,000
1790	25,000	8,200
1800	20,000,000	10,000
1810	62,000,000	15,000
1820	125,000,000	32,000
1830	270,000,000	65,000
1840	500,000,000	193,000
1850	930,000,000	300,000
1858	1,100,000,000	450,000

idea of not making money. These men employed hands—not people— who were instruments for the production of capital and who were to be used as needed to make and remake profit. Factories and mills hired women and children who worked 16 hours a day. Most children entered the mills at the age of seven and were frequently chained to the looms, for if they nodded and fell into the machinery, it meant a delay of up to a half hour to extricate their broken bodies.

Wages were pegged below cost of living so that the hands could never escape their just and legal debts. The injured and maimed were summarily dismissed. Rickets, due to lack of sunlight on growing bodies, was the least of a litany of diseases which included tuberculosis, cancer, and bladder or kidney infections. Mill hands married in their early teens and bore many children who could supplement the meager wages of the parents. Death at an early age was common and welcomed. "Charity," declared Lord Brougham, "is an interference with the healing process of nature which acts by increasing the rate of mortality." Adam Smith's *Wealth of Nations* and Karl Marx's *Capital* reached opposite conclusions on the morality of the Industrial Revolution. Thomas Malthus, observing the population explosion that accompanied the start of the Industrial Revolution, reached conclusions embodied in his *An Essay on the Principle of Population,* and this book strongly influenced the later writings of Charles Darwin. Mrs. Elizabeth Gaskell's *Mary Barton* portrayed to a generally unsympathetic middle-class audience the distressed condition of the mill towns. Torys and Liberals fought each other in parliament on the issues of poor relief and child labor. King Cotton treated his subjects poorly.

THE U.S. CIVIL WAR

The colonists of North America wore, as did their English and French cousins, clothing of linen and wool, originally European, but increasingly of local manufacture. By 1660, some cotton was grown in the

Carolinas and by 1700 modest acreages were planted to cotton in Alabama and Mississippi. In 1732, seed of *G. hirsutum* grown in the Apothecaries' Garden in Chelsea (Chapter 5) were sent to Georgia. This upland cotton formed the base for cotton cultivation in the United States. The fibers, although relatively long and smooth, are difficult to separate from the seeds, the Indian Churka roller gin was ineffective, and the fibers had to be plucked by hand by slaves for whom the cotton cloth was intended. Martha Washington wrote that she supervised the weaving of 300 yards of cotton cloth for the slaves of Mount Vernon, and Thomas Jefferson, the Madisons, and others followed the same practices. There were a few water-powered mills in Massachusetts by 1790, but cotton was a minor economic factor in the young United States. The major Southern crop, tobacco, had become a glut on the market. Slave holders were being eaten out of house and home by idled field workers, and they were searching for some way to get rid of their slaves.

All this changed abruptly with the invention of the cotton gin. Land again became valuable, slave labor became "necessary," and cotton moved into all of the South and spilled over into the Southwest. As land was exhausted, planters simply moved to new land until the South was a sea of cotton. In spite of *Gone With the Wind,* there were relatively few large plantations with their white, pillared, antibellum mansions on hills overlooking fields where singing slaves picked cotton. Most planters had less than a hundred acres which they worked with the help of a few slaves. Nevertheless, the increased acreages devoted to cotton caused not only a resurgence in the value of slaves, but a reactivation of the institution of slave-running. The profits of slave-trading were immense, and the laws against these practices—both in Europe and in the United States—were ignored or subverted. Table 8.1 shows the increase in the numbers of slaves which paralleled the increase in cotton sales.

The failure of the American South to industrialize was due to the huge profits that could be made in cotton. This was true in spite of ample supplies of coal and iron, abundant water for power and transportation, and the capital necessary to build factories. Slaves were, it was believed, "unfitted for any other work" and could not be trained to do other than lead a mule through a field. The northeastern states saw and grasped the opportunity to expand manufacturing, and the mill towns of New England date from about 1810. Tensions between an industrial North and agricultural South was fought in congress, with the South favoring free trade and the North demanding a protectionist economic foreign policy. John Calhoun and Daniel Webster spent much of their legislative lives opposing each others views and helping to polarize the country. As their tobacco-planting forebearers had been, Southern farmers soon were in thrall to cotton factors, the bankers who loaned money for seed, slaves, and supplies in exchange for the right to market the crops. Much of the money was from the North, and southerners resented this, believing that they were being cheated—and undoubtedly they were. Economic tensions continued to build as England and New England competed for each year's harvest and the North got richer and more powerful at the expense of the South. Opposition to slavery as a degradation of humanity was fanned by books like Harriet Beecher Stowe's *Uncle Tom's Cabin*, in which Eliza crosses the ice with bloodhounds baying at her little heels. Militant abolitionists like John Brown and the establishment of the underground railway for escaped slaves further worsened relationships, and the slave-or-free issue in Kansas stirred all segments of the country. By the time of the Lincoln-Douglas debates, there seemed little doubt that resolution of the differences was close to impossible and that the South would attempt to secede from the Union.

As Britain continued development of the cotton cloth industry, she became absolutely dependent upon American cotton. Sixty percent of all British exports in 1850 were cotton cloth, and close to 5 million people were directly or indirectly dependent upon the Lancashire mills; a quarter of the cotton cloth in the world was of British manufacture. It was clear by 1855 that trouble was brewing in the United States, and the British Parliament was warning of the "danger of our continued reliance upon the United States for so large a proportion of our cotton." In 1855, David Christy of Ohio published *Cotton is King or Slavery in the Light of Political Economy* in which he proposed that the North accede to all Southern demands or allow secession since, he asserted, the entire economy of the United States was inextricably bound to cotton exports. When Fort Sumter was fired upon and war began in earnest, Jefferson Davis, president of the Confederate States of America, accepted the king cotton idea, believing that the North would soon collapse economically and that Britain would intervene on the side of the South; "Cotton is the King who can shake the jewels in the crown of Queen Victoria." Judah P. Benjamin, Davis's secretary of state, led the move to prevent the export of cotton, a decision fully accepted by Southern cotton growers from the large plantation owner down to the farmer with forty acres and a mule.

The South had, however, made its most serious blunder. Even as war clouds gathered in 1858–1859, millions of bales were shipped to the Lancashire mills, providing a reserve which would last for close to two years and allowing the mill owners time to obtain cotton from India, China, Brazil, and Egypt. The British and French upper classes and the mercantile princes favored the Confederacy, and most newspapers were overtly advocating intervention, but the working classes, familiar with exploitation, were so opposed to slavery that they vehemently favored the Union. Yet, as war continued in 1862 and 1863, raw cotton reserves dwindled, and many of the mills went on reduced schedules or shut down entirely. Linen and wool took up some of the slack as did Indian and Egyptian cotton, but the quality of the Asian product was low and it was expensive. Mill hands were, at best, working only part-time, workers and in subsidiary industries—machine makers, miners, stevedores—were furloughed or fired. Over half the population of Lancashire was destitute. Yet these people supported Lincoln's Emancipation Proclamation to the hilt. Even the captains of industry came around to a noninterventionist position as they found the value of their reserves of cotton cloth rising to unpredictable heights and their profits, undiluted by production and wage costs, continued to increase. The British government, too, saw the wisdom of nonintervention as stores of American grain poured into a country experiencing a period of small harvests. Relief for the poor, late in coming, inadequate, and most grudgingly given, was finally instituted, but the profits reaped by the rich were bitterly noted by the unemployed. The gap between the ruling classes and the workers was widened and has still not been bridged. The present strength of the British Trades Union Council was at least partly based on king cotton.

As the horror of the Civil War (the War Between the States) rose to a crescendo with Sherman's march to the sea and Grant's bulldogging along the line, the South found that the press of the well-equipped Union was more than could be opposed. The end of the war saw a devastated South, stripped of human and material resources and plagued with carpetbaggers who ferretted out every bale of cotton to be confiscated and shipped with large profit margins to the mills in New England or overseas. The South began to rebuild with the only crops it knew—tobacco and cotton—and did so under unbelievably difficult circumstances. Lacking any capital themselves and finding that a vindictive North would not finance industrial expansion, Southern states remained rural, poor, and segregated. A system of tenant-farming or share-cropping developed in which the owner of land would supply seed, tools, and credit to farmers tilling small acreages of cotton land with some agreed-upon distribution of profits. This, too, did little to create an economically viable South. Until the early part of this century, cotton was grown on worn-out soil by worn-out, undernourished, and undereducated people—black and white. A one-crop economy led to economic stagnation, and the monoculture of millions of square miles devoted to a single plant fostered the rapid spread of plant diseases and insect pests. The boll weevil moved from Mexico where it was a minor pest on native wild cotton into Texas by 1900 and into the entire South by 1910, where it reduced yields by over 50 percent in some areas The economic resurgence of the south, its industrialization and crop diversification, did not occur until the period of World War II.

USES OF COTTON

Most cotton grown today is derived from cultivars of *G. hirsutum*, the upland cotton. Close to 12 million metric tons of cotton fibers are produced annually (about 50 million bales of 900 pounds each), with the United States far in the lead among producing countries. Over 24 million metric tons of cotton seed are also produced. Almost all this cotton is grown as an annual, planted from seed each spring, and harvested in the early autumn. Cotton, like tobacco, has high requirements for nitrogen and phosphorus, and the newer, higher-yielding hybrids are even more dependent upon adequate fertilization for their rapid growth. Modifications in the ginning, thread-spinning, and milling processes over the two centuries since machinery was introduced has resulted in a wide variety of cotton fabric types now available. Additional modifications in cotton cloth have been occasioned by the introduction of sea-island cotton (*G. barbadense*). The fibers of upland cotton may rarely exceed a cm while those of sea-island cotton may be double that length permitting the spinning of exceptionally fine, strong threads. Known as Egyptian or Pima cotton (it was first grown in Pima, Arizona), *G. barbadense* is the preferred plant for luxury yarn.

John Mercer (1791–1866) patented the process of soaking cotton threads in an alkaline solution which swells individual fibers making them smooth and lustrous; mercerized cotton has a silky look and feel and slides easily through cloth when used as sewing thread. The first of a host of synthetic fibers was rayon, patented in 1855 by a Swiss chemist who dissolved cellulose from wood chips and forced the viscous solution through small holes into a hardening bath. The filaments are then twisted together in groups of 13 to 270 to form thread or yarn. Commercial production of rayon began in France in 1891 and in the United States in 1911. Viscose thread is made by a similar process. When these and the petrochemical-based plastics were developed, man-made fibers cut deeply into the dominant share of cotton in the world's fiber markets, and cotton growers did little except wring their hands and have their congressmen vote larger subsidies. By 1930, cotton had slipped from 90 percent of the fiber market to 85 percent. By 1940, when the first noncellulose, man-made fibers appeared, cotton constituted less than 80 percent, dropping to 56 percent in 1970. The development of mixed polyester-cotton blends stopped cotton from slipping even further. In 1970, North American consumers were using less than 16 pounds of cotton per year; the cost of producing fine cotton was almost more than its value on the open market. In the past few years, however, the king has found it unnecessary to abdicate. Cost-accounting methods for determining acreage and marketing, some increased affluence in underdeveloped countries, a resurgence of interest in the pleasant feel and flexibility of cotton compared to synthetic garments, and steeply rising petroleum costs have aided cotton production.

FLAX

Cotton is not the only plant fiber used to make cloth. Historically, it is likely that the phloem or bast fibers of the flax plant (*Linum usitatissimum*) was among the first (Figure 8.18). Fragments of linen cloth from

Figure 8.18
Flax, *Linum usitatissimum*.

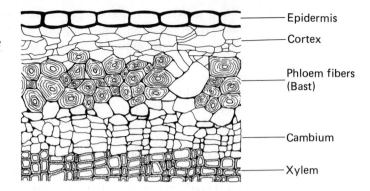

Figure 8.19
Cross section through flax stem showing position of the phloem (bast) fibers from which linen is spun.

Epidermis

Cortex

Phloem fibers (Bast)

Cambium

Xylem

Egyptian diggings have been dated at 4000 B.C. Linen is noted in Gen. 41:42 where "Pharaoh took his signet ring from his hand and put it on Joseph's finger and arrayed him in garments of fine linen. . . ." The name flax is derived from the Anglo-Saxon *fleax,* and our word linen is the Old Celtic *lin* which means thread. The plant was introduced throughout Europe by Romans who found that the cool, damp climate of England and the Low Countries was ideal for the development of long fibers which could be spun into strong thread. Charlemagne (742–815) required all families in Belgium to grow flax and spin linen. By the sixteenth century, linen was so valued that Henry VIII decreed that "every person having in his occupation threescore acres of land apt for tillage shall sow one rood (¼ acre) with flax. . . . upon pain to forfeit three shillings, four pence . . ." Stout linen sails carried the explorers, conquerors, and settlers of Europe around the world. Flax seed accompanied colonists to New France and to the southern and New England colonies where, as in Europe, fibers were spun into linen or mixed with wool to produce the durable linsey-woolsey cloth. Even after cotton became available, linen was used as the warp threads for hand-loomed cloth because it was stronger than the cotton yarns then available.

Like cotton, the bast fibers of flax are almost pure cellulose, but in contrast to cotton, the flax fiber is solid and is stronger (Figure 8.19). Its high cost is due primarily to the difficulty of freeing the fibers from the rest of the stem tissues. It is necessary to *ret* the fibers by allowing microorganisms to digest away the softer tissues. This is done either by placing the stems in ponds or streams for a period of time or, in damp areas like Ireland, by "dew-retting" or placing the stems on damp grass. The retted plants are beaten (skutched) to remove the residue and the bast fibers can then be carded and spun into threads.

HEMP

Hemp (*Cannabis sativa*) was, as noted previously (Chapter 4), originally grown for its fibers rather than for its hallucinogenic resins (Figure 8.20). Like flax, the fibers are bast and are freed in a similar retting

Figure 8.20
Hemp, *Cannabis sativa*, grown for fiber. Courtesy U.S. Department of Agriculture.

and scutching operation. Long used in the Orient, hempen cloth was very long lasting, but it was a rough and scratchy fabric that was not as pliable as linen or wool. This is due to the presence of lignin, the material that strengthens wood, in and around the cellulose fibers. It is, however, possible to obtain single groups of bast fibers which are 3 m long, and hemp has retained its importance as a rope and cord fiber. Both Virginia and Kentucky were deeply committed to the cultivation of hemp; in 1762, Virginia gave cash grants to hemp planters and fined those who would not grow the plant. George Washington and his neighbors grew hemp for cordage and mixed fibers of hemp with those of flax and some cotton to produce cloth. Hemp was a major industry in Kentucky, especially in Fayette and Bourbon counties (the center of the bourbon whiskey country) for export as sail fiber to England and, in the early 1800s, as yarn used to make the burlap that bound bales of cotton. Today, iron bands are used to bale cotton and cordage is derived from jute and other plants.

ADDITIONAL READINGS

Brown, H. B., and J. D. Ware. *Cotton*, 3rd ed. New York: McGraw-Hill, 1958.
Cohn, D. L. *The Life and Times of King Cotton.* New York: Oxford University Press, 1956.
Dempsey, J. M. *Fiber Crops.* Florida University Press, 1975.
Leggett, W. F. *The Story of Linen.* New York: Chemical Publishing Company, 1945.
Scherer, J. A. B. *Cotton as a World Power.* New York: F. A. Stokes Company, 1916 (reprinted in 1969 by Negro Universities Press, New York).
Woodman, H. D. *King Cotton and His Retainers.* University of Kentucky Press, 1968.

Rubber—Sulfur and a Hot Stove

I have seen a substance excellently adapted
to the purpose of wiping from paper the
marks of a black lead pencil.
DR. JOSEPH PRIESTLEY, 1770

In geology, a watershed is defined as a crest dividing two drainage basins. We also speak of historical and cultural watersheds, events which irrevocably alter the course of history. Time is divided into B.C. and A.D. by the supposed date of the birth of Christ. Certainly the heliocentric concept of the then-known universe by Copernicus and Tycho Brache is a scientific watershed, and Leeuwenhoek's peerings through a crude microscope is another. The history and economic botany of rubber is so bound up with the invention of vulcanization in 1830 that one must discuss the topic in terms of pre- and post-Charles Goodyear just as cotton can be evaluated in the light of the invention of the gin by Eli Whitney.

A BRIEF HISTORY

Like many other plant products, rubber was introduced to Europe by Columbus. On his return from his second voyage, he presented Queen Isabella with several small balls that "rebounded as if they were alive." Her Majesty wanted gold, silk, and spices, not children's toys, and was only minimally amused by the gift. A few years later, scribes of Cortez's army reported "there is a tree . . . much appreciated by the Indians. From this tree a kind of very white liquid flows. To obtain this, the trunk is struck with a hatchet and the liquid flows from these incisions." From *caoutchouc*—the weeping tree—latex was collected, smoked over a fire to coagulate it, and molded into balls. Mayans used these to play *tlachler*, a forerunner of basketball with stone rings as baskets. Since the ball weighed 14–15kg and was moved about with head, legs, and shoulders as in soccer, players were selected for strength, not height. In 1735, Charles Marie de la Condamine, a botanist on a collecting expedition for the Paris Academie des Sciences, saw waterproof shoes in the Amazon valley. He brought some of this gum back to France where, dissolved in turpentine, it was used to coat fabrics and as a rubber, or eraser, for pencil marks. Nevertheless, before 1800 there was very little demand for rubber. Smoke-coagulated rubber could be dissolved in naptha, and Charles Macintosh used this mixture to glue two layers of cloth together to make waterproof coats (hence the British "mackintosh" for a raincoat). Thomas Hancock, a London stagecoach builder, made rubberized cloaks for passengers and drivers and in 1819 opened an "elastic works" to make garters, tobacco pouches, and boots. He later expanded the factory to produce rubber bands, cushions, and even air mattresses. Hancock and Macintosh formed a company making hoses of rubber-covered fabric for breweries and attempted to make pulleys and belts for industry. The trouble was that smoked rubber was sticky when warm and brittle when cold; gum boots would literally dissolve in midsummer and crack in winter and they smelled bad in both seasons.

CHARLES GOODYEAR

Recognizing the need somehow to stabilize rubber, many inventors turned their attention to the problem. Most tried mixing rubber with powders to dry out the stickiness, and most used lime, magnesium sulfate (epsom salts), lead oxide (litharge), and elemental sulfur. Charles Goodyear, born in New Haven, Connecticut, in 1800, was a rubber fanatic, convinced that stabilization would be a boon to mankind and the source of a personal fortune. He literally drove himself into physical and mental exhaustion, kept his family in abject poverty, and put himself into debtors' prison several times. In 1839, he discovered that coagulated latex, mixed with sulfur and heated, formed a stable product that possessed all the properties we associate with rubber. It was waterproof, elastic, resistant to abrasion, a nonconductor of electricity, would absorb vibration, and was relatively temperature-insensitive. Goodyear spent the rest of his life attempting to promote the use of rubber. He wore pants, coats, and hats of rubber; his calling cards were rubber; he had his portrait painted on rubber; and he had the story of his life bound in a rubber book. Digging himself deeper into debt, he invented almost all the things for which rubber is now used. Rubber teething rings, bottle nipples, inflatable rubber boats and cushions, elastic shock cords, doormats, and other items were hand-made, and Goodyear tried to sell the processes to manufacturers and even peddled rubber goods from door to door. Although Goodyear patented rubber stabilization with sulfur, Thomas Hancock stole the idea and patented in under his name in England and in France. Hancock chose the name vulcanization (from the Roman god of fire) and successfully beat off Goodyear's feeble lawsuits; Goodyear died a pauper.

Until close to 1850, rubber goods were limited to water-proofings, industrial pulleys, shock absorbers, drug sundries (rubber pants, nipples, hot water bottles), and overshoes. Thomas Hancock presented Queen Victoria with a set of solid rubber tires to replace the iron-banded wheels on the state coach, and with the "patronage of her Majesty" he promoted this more comfortable wheel throughout Europe. In 1845, Robert Thomson coated cotton cloth with rubber and invented the pneumatic tire, later adapted by John Dunlop of Belfast for the bicycle. In 1898, tires for automobiles were made, and the subsequent history of rubber is the history of the automobile industry. The first gasoline-powered horseless carriage in the United States was hand-made by the Duryea brothers in 1893, and in 1905 Ransom E. Olds

Figure 8.21
One hundred years' con-
sumption of natural rubber
in the United States.

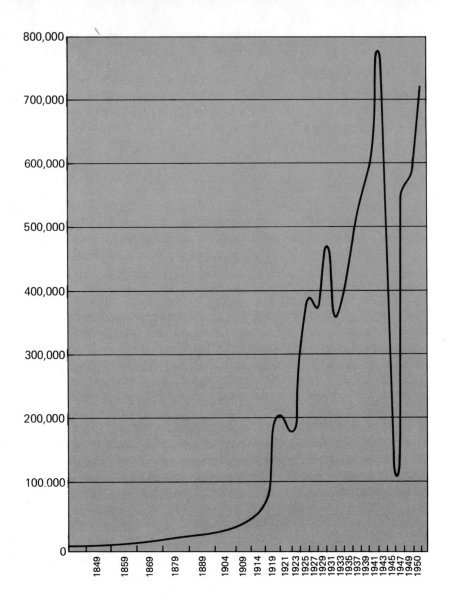

started to produce modest numbers of one cylinder gas buggies. Henry Ford had an assembly line producing the tin lizzy by 1906. There were four automobiles registered in the United States in 1895, 8000 in 1900, 26,000 in 1902, and 2 million in 1908. For better or worse, America—and the rest of the world—was on rubber wheels. Akron, Ohio, a center for bicycle manufacture became "Rubber City" in 1910 when Harvey Firestone invented the nonskid tire. In 1918, the United States had over nine million cars and trucks on the roads (a Ford model T cost $400) and was using 50 percent of the world's rubber production (Figure 8.21).

BRAZILIAN RUBBER

Virtually all of this rubber was collected in the jungles of South America and shipped to the United States through the port of Para (now Belém), Brazil. The center of the rubber-collecting area was the city of Manaos, 1000 miles from the coast. The rubber trees (*Hevea brasiliensis*) were wild, scattered over thousands of hectares of dense tropical rain forest land and were tapped by people who were essentially slaves, so deeply in debt to the companies that they could never work themselves free. Malaria and yellow fever killed thousands, the bush Indians killed other thousands, and the companies also killed thousands who did not meet their daily quotas; it was said that each pound of rubber cost a pint of blood. As demand for rubber increased, unbelievable wealth accumulated in Para and Manaos. During the height of the rubber-tapping period of 1870 to 1890, Manaos built an opera house from imported Italian marble, and individuals built mansions filled with the luxury goods of Europe and America. Jewelry shops and stores filled with French perfume, champagne, and silk clothing sprung up and made fortunes for their proprietors. A few cassandras said that the trees were being overtapped and that prudent management was advisable, but rubber fever so gripped the exploiters of trees and of people that an end to the boom seemed impossible. Yet by 1900, Manaos was a city of decaying houses, an empty opera house, and smashed storefronts. The search for rubber moved elsewhere.

AFRICAN RUBBER

If the exploitation of the people of Brazil was horrible, the rubber experience in the Congo was infinitely worse. In 1850 the American missionary T. L. Wilson had pointed out that when *Landolphia* vines were cut down, a latex flowed from the segment. When rubber production began to fall in Brazil, and as automobile tire manufacture began to climb in 1900, interest in *Landolphia* rubber increased, Leopold II, King of the Belgians, read of the exploits and explorations of Henry M. Stanley (Dr. Livingston, I presume) and decided to "Christianize the negro," spread trade, and exploit the treasures of West Africa. He formed the International Association of the Congo with the assistance of America missionaries who saw souls to save, and the support of industrialists who anticipated a tidy profit. Under the charter of the corporation, Leopold owned tropical West Africa with one-fifth of all profits from ivory and rubber accruing to his personal account. Shares sold initially for $100 were worth $2800 in 1900, and by 1903 dividends were several hundred dollars per year. These profits were obtained by outright enslavement of African natives who were tortured or killed if they did not bring in enough rubber. The Congo was, by 1910, depleted of rubber vines and of people. Supplies of rubber from other latex-synthesizing plants (*Funtumia elastica*) in Angola, Guinea, the Ivory Coast, and the Niger River basin were exhausted by 1914.

By 1918, the British realized that they controlled almost two-thirds of the world's rubber with only trickles coming in from Brazil. There were plantations developed in the Dutch East Indies from seedlings given by the British to plantation owners in Sumatra, Java, and Malaya. Deciding to exploit this financial windfall, Winston Churchill, as secretary of states for the colonies, suggested in 1922 that parliament restrict rubber production to maintain prices; the United States alone was using 50 percent of the natural rubber, and the numbers of automobiles in North America had exceeded 12 million. The price of rubber rose to $1.25 a pound in 1925, and the American tire manufacturers decided that supplies under direct domestic control would defeat the British restrictive practices. Plantations were started in Liberia, a country founded as a home for freed American slaves, in Ghana, and in the Philippines. Henry Ford started a model city-plantation in Brazil which failed when the indigenous leaf blight fungi (*Dathidella ulei*) killed the trees. The African plantations were and still are successful; Liberia produced 50 million pounds of rubber in 1971. As the market for rubber increased, Dutch planters intensified their efforts to develop superior plantings. Following their own example with quinine (Chapter 5), they grafted high-yielding stems to disease-resistant rootstock and increased production to the point that they were in direct competition with the British. The restrictive acts were repealed in 1927, but by that time the Dutch controlled almost two-thirds of the world's rubber, and when rubber became a glut on the market in the early 1950s, a British-Dutch cartel was formed to regulate rubber supplies.

RUBBER CHEMISTRY

The bulk of natural rubber is obtained from *Hevea brasiliensis*, a tall tropical forest tree native to the Amazon basin. Hevea is a genus in the Euphorbiaceae, a family also containing the *Poinsettia*, the crown-of-thorns (*Euphorbia splendens*), and the tapioca plant (*Manihot utilissima*). Several members of the mulberry (Moraceae) family produce latex, including the ornamental rubber tree (*Ficus elastica*) which graces bank lobbies and barber shop windows and a vine (*Castilla elastica*) that sparked a short-lived rubber boom in Panama and Mexico. The mangabeira (*Hancornia speciosa*), a milkweed (*Asclepias*), and the Russian dandelion (*Taraxacum kok-saghyz*) have been used commercially; all told, over 160 species of plants product latex. When removed from the plant, latex is over 50 percent water, 10 percent resins, proteins, and sugars, and 30–40 percent rubber. The "very white liquid" reported by Spanish explorers is an emulsion as are homogenized milk or mayonnaise. Rubber itself is a polymer of a five-carbon compound called isoprene

$$CH_2 = CH - \overset{\overset{\displaystyle CH_3}{|}}{C} = CH_2$$

The polymer may be linear:

$$C = \overset{\overset{\textstyle C}{|}}{C} = C \ldots C = C - \overset{\overset{\textstyle C}{|}}{C} = C \ldots C = C - \overset{\overset{\textstyle C}{|}}{C} = C$$

or cross linked:

$$C = C - \overset{\overset{\textstyle C}{|}}{\underset{\underset{\textstyle C = C - \underset{\underset{\textstyle |}{}}{C} = C}{|}}{C}} = C$$

or it can exist in more complex configurations. Polymer chains may exceed several thousand monomers. Although it has been suggested that latex serves the plant as a sealant of natural wounds, its exact role has never been defined.

In *Hevea*, latex is produced in a series of latex vessels which anastomize through the phloem (Figure 8.22). Trees can be tapped several times a week during the growing season, each set of cuts made successively higher on the trunk (Figure 8.23). Regeneration of phloem is

Figure 8.22
Diagram of part of the stem of the rubber tree, *Hevea brasiliensis*, showing the location of articulated lactifers in the phloem.

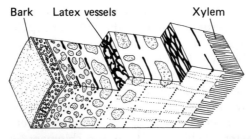

Figure 8.23
Tapping a rubber tree in Brazil.

rapid, and if trees are tapped on alternate years, a single tree can produce rubber for more than 30 years. Each tree can produce three to four kgs of rubber per year. Advances in technology have more than doubled yields with no apparent damage to the trees. A plantation of 1000 trees can provide close to 7000 kg per year. The latex drips into a cup and is brought to collecting stations where the liquid is mixed with weak acetic acid to coagulate the rubber. Water and contaminants are removed by squeezing the rubbery sheets between steel rollers, and the crepe rubber is dried for shipment. At factories, dried rubber is thoroughly mixed, carbon black or other coloring matter added, and then it is vulcanized by the addition of sulfur, litharge (lead oxide), and heat. Used and reclaimed natural rubber and synthetic rubbers are added with additional mixing (Figure 8.24) and heating to produce rubbers for the many different products we use.

AN ESSAY ON BIOSYNTHESIS

As plants evolved from the blobs of organic chemicals we believe formed the first living things, mutation and selection probably resulted in the dominance of one or a few types of organism which were the progenitors of all organisms on earth. Certainly before organisms recognizable as primitive bacteria had evolved, there was the development of an hereditary system based on DNA with protein (and hence enzyme) synthesis much the same as that in existence today. An energy metabolism system involving the single energy-yielding compound adenosine triphosphate must also have been an early development, because it, like the DNA-RNA-protein trinity, is found in all organisms. Virtually identical macromolecules, large starchlike compounds, proteins, lipids, and the complex carbon-rings of porphyrins were synthesized. Insofar as we know, the biosynthetic pathways for these universally found compounds follows the same sequence of chemical steps and require the same starting and intermediary compounds. Biological systems are very conservative; once a successful biochemical system evolved, it was retained and passed along to all organisms derived from that ancestral form.

In evolutionary time, chance mutations continued to occur at slow but calculable rates, and new DNA linkages coded for new protein enzymes which, in turn, direct the synthesis of new compounds. Plants containing these new compounds would succeed or become extinct depending on whether the presence of the potential inherent in the new compound was advantageous or deleterious when the organism was stressed by natural selection. Thus, simple porphyrins have given rise to bile-type pigments, then to more complex cytochromes, chlorophylls, hemoglobins, and phytochromes by mutations which coded for enzymes that added to the chain of biosynthesis. Cytochromes led to advanced respiratory processes and metabolic conversion of cytochrome precursors led to chlorophylls and hence to photosynthesis.

This evolutionary progression is very evident in looking at the isoprenes. A two-carbon intermediate in the respiratory cycle (acetyl coenzyme A) is metabolically converted into a five-carbon compound, mevalonic acid, that is a pivotal or a switch point from which a wide range of biosynthetic pathways start. One route leads to the formation of latexes, sterols like sex hormones and cholesterol, terpenes including turpentine and essential oils, and to the gibberellins—a class of plant growth-regulating substances. A second route from mevalonic acid leads to the synthesis of lycopenes, including those which give tomatoes their rich, red color. A third route results in the formation of the carotenes and xanthophylls, components of the photosynthetic system.

$$
\begin{array}{ccc}
 & \text{Acetyl Co A} & \\
 & \downarrow & \\
\text{Isoprenes} \longleftarrow & \text{Mevalonic acid} \longrightarrow & \text{Carotenes} \\
 & \downarrow & \downarrow \\
\text{Latex / Sterols} & \text{Lycopenes} & \text{Xanthophylls} \\
\downarrow \qquad \downarrow & & \\
\text{Terpenes} \quad \text{GA} & &
\end{array}
$$

It is tempting to think of these branching metabolic chains as developing sequentially in evolutionary time with a respiratory pathway involving acetyl coenzyme A having evolved before the sequence of steps leading to mevalonic acid. Not only do plants have alternate routes for the biosynthesis of various compounds, but there is not enough data to justify such a linear concept of evolution. We are also unable to answer the question of how (much less why) plants in completely unrelated taxonomic groups will develop identical biosynthetic pathways.

SYNTHETIC RUBBER

In 1942, the Japanese invaded southeast Asia and cut off over three quarters of the West's supply of natural rubber. As early as 1939, perceptive governments and rubber corporations realized the dangers inherent in fighting wars that would involve reduction of rubber supplies. Germany, the Soviet Union, Britain, and the United States began stepping up their efforts to produce synthetic rubbers. The first synthetic was based on a discovery of Julius A. Nieuwland at Notre Dame University. Nieuwland sold his patents to DuPont in 1925, and William H. Carothers developed DuPrene in 1931; DuPrene later became Neoprene. Isoprene, the monomer of natural rubber, is difficult to synthesize, and the attention of the German I.G. Farben group and of Standard Oil was directed towards rubber synthesis from polymers made from a related compound, butadiene ($CH_2 = CH — CH = CH_2$) easily obtained from coal tar, coal gas, or, as C. V. Lebedev of the Soviet Union found, from ethyl alcohol obtained from grain fermentation. The Russians called their product SKA and the Germans called theirs BUNA. Neither product was completely satisfactory; abrasion and temperature-resistance characteristics needed for tires running at high speeds were weak. In the early 1940s, nations found that addition of a plastic, styrene, stabilized synthetic rubber. The process of making synthetic rubber is fairly simple in outline:

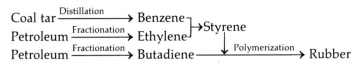

With the end of the war and reactivation of rubber plantations in Africa and southeast Asia, natural rubber was again being collected. Most rubber products are mixtures of natural and synthetic materials.

As the cost of synthetic elastomers (elastic polymers) rises with the cost of petroleum, there has been renewed search for other sources of natural rubber. Attention has focused on guayule (why-oo-lay), a desert shrub (*Parthenium argentatum*) native to northern Mexico, Texas, and southern California. Guayule is a composite (Compositae) related to goldenrod. The latex is a true isoprene polymer. In contrast to *Hevea*, it is not found in vessels, but in the vacuoles of stem and root cells. Since 20 percent by weight of the plant is latex, its exploitation is economically competitive with *Hevea* rubber. The use of guayule is not new; Aztecs used its rubber for balls and for chewing gum. In the early part of this century, processing plants were constructed in Mexico by an American consortium. Between 1900 and 1910, over 100 million pounds of guayule rubber was produced, close to half the rubber used in the United States. Since the plants were destroyed to obtain the latex, and little attention was paid to replacing the bushes, the supply was soon exhausted and the flood of cheap rubber from Asia made the operation superfluous. During World War II, the U.S. government spent

$30 million to develop guayule rubber, and while 3 million pounds were produced in 1941–1943, the development of synthetic rubber made the plantations uneconomical. The federal government, in its wisdom, set fire to 50 square miles of guayule plantings. In 1977, interest was rekindled. A report by the U.S. National Academy of Sciences recommended that otherwise unproductive desert land should be planted with guayule to ensure a domestic supply of rubber and to develop an employment base in the American Southwest.

ADDITIONAL READINGS

Baum, V. *The Weeping Wood.* Garden City, N.Y.: Doubleday, 1943.
Collier, R. *The River that God Forgot: The Story of the Amazon Rubber Boom.* New York: Dutton, 1968.
Polhamus, L. G. *Rubber, Botany, Production and Utilization.* London: Leonard Hill, 1962.
Wilson, C. M. *Trees and Test Tubes: The Story of Rubber.* New York: Holt, Rinehart and Winston, 1943.
Wolf, R., and R. Wolf. *Rubber: A Story of Glory and Greed.* New York: Covice, Priede Publications, 1936.

Tobacco—The Loathsome Weed

A woman is just a woman
But a good cigar is a smoke.
RUDYARD KIPLING

On October 13, 1492, one day after the Santa Maria made its historical landfall at San Salvador, the admiral noted in his journal:

> In the middle of the gulf . . . I found a man in a canoe carrying a little piece of bread . . . a gourd of water . . . and some dry leaves which must be a thing very much appreciated among them. . . .

On November 2, Columbus landed at Cuba and sent his interpreter, Luis De Torres, and able seaman Rodrigo De Jerez with a letter and gifts to the "court of the Grand Khan of Cathay" announcing the arrival of the fleet and expressing his interest in establishing commercial relations with China. The two men met many people on their fruitless journey "always with a firebrand in their hand . . . and a certain leaf after the fashion of a musket [tube]. They light them at one end and at the other they chew or such and take in with their breath that smoke which dulls their flesh and as it were intoxicates, so they say that they do not feel weariness" (Figure 8.25). Gifts of these cigars were given to the admiral who almost casually added them to the growing pile of items to be returned to Spain. One story goes that Queen Isabella became the first Western woman to have smoked a cigar. As other explorers invaded the Americas, a fuller picture of tobacco usage developed.

Figure 8.25
Figure 8.25
The first illustration of tobacco. Taken from *Stirpium Adversaria Nova* by de l'Obel published in 1570.

Figure 8.26
A clay nasal snuffer used in Mexico about 100 B.C.

Figure 8.27
Aztec deity blowing tobacco smoke. Taken from a pre-Columbian stone carving.

Haitians called their rolls of leaves *tabacca.* Amerigo Vespucci visited Margarita Island in 1499 and saw natives with "a green herb which they chewed like cattle to such an extent that they could scarcely talk." Cortex saw Montezuma smoking a hollow reed filled with a mixture of liquid amber and tobacco; Jacques Cartier in 1534 described the ground tobacco of the Algonquins of the St. Lawrence Valley: "after taking some into our mouths, it seemed like pepper, it was so hot." Throughout the Americas nasal snuffers (Figure 8.26) for inserting powdered tobacco directly into the nostrils were in common use. Thus, all of the methods of using tobacco were well established long before the fifteenth century.

In use at least 1000 years before Columbus, tobacco was not only a medicine, but served an important role in ritual. Aztec Gods smoked (Figure 8.27) to tell people of their decisions, and Mayan priests used smoke to intercede for their people (Figure 8.28). Tobacco was used by all North American Indians as money, and the leaf was so important that councils and diplomatic alliances were formalized by passing the calumet, the pipe of peace (Figure 8.29). The tomahawk pipe was a white man's invention, but it fit so well with established cultural pat-

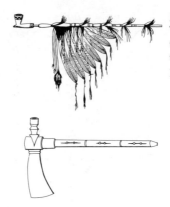

Figure 8.29
The calumet or peace pipe of
the plains Indians and the
tomahawk or war pipe of the
prairie Indians.

Figure 8.28
Mayan priest blowing smoke
as part of a ritual to bring
rain. Taken from a bas-relief
in Chiapas, Mexico.

terns that the Indians immediately accepted it. Quids of folded leaves
were held in the cheeks of men on hunting forays to dull hunger pangs
and lessen the need for water. Women ate powdered tobacco to ease
the pain of childbirth. When in the nineteenth century the settlers of
the West spat out tobacco juice, the Indians were amazed—that was the
best part!

IMPORTATION INTO EUROPE

Explorers noted that there were two different kinds of tobacco
plant, a low-growing type bearing yellow-green flowers that was most
common in the temperate climatic zones of North America, and a taller,
less coarse plant with white-to-pink flowers. The smaller plant, *Nicotiana rustica,* was decidedly less popular since its smoke was biting,
harsh, and stupifying compared with the milder flavor of the taller species, *Nicotiana tabacum.* Columbus's cigars were probably *N. tabacum*
(Figure 8.30), but the tobacco smoked by swaggering Spanish sailors in
the barrooms of Cadiz was probably *N. rustica.* By 1531, the Spanish
imported *N. tabacum* seed from the mainland and started cultivating it
in the West Indies, and by 1540 both species were being shipped to
Spain as dried, pressed leaves. The initial popularity of tobacco was its
supposed medicinal virtues. Spanish ships' crews and the English,
French, and Dutch freebooters who preyed on the Spanish fleet insisted
that it was a potent aphrodisiac; the women in New Spain ports, it was
asserted, were heavy smokers and wonderful lovers. Spanish and Portuguese explorers such as Damiao de Geos reported that "the holy
herb, because of its powerful virtue in many ways, of which I have experienced principally in desperate cases, will cure ulcerated abscesses,
fistulas, sores, inveterate polpys, and many other ailments." That tobacco "improveth the lungs, cleanses the body of excess rheum," cured
headaches, was an "excellent defence against bad air, eased convulsions," antidoted snakebite, and was a "sovereign medicine for female
internal disorders" was avowed with the authority that only ignorant

medical practitioners seem able to muster. As exploration of the Americas continued in the latter half of the sixteenth century, illustrated reports of medical and hallucinogenic uses of tobacco began to appear in Europe. Shamans in both Americas used tobacco smoke (Figure 8.31) to reach hallucinogenic states wherein they could "see the evil" and could prescribe remedies which, in many instances, involved purification of the patient with massive inhalations of tobacco smoke (Figure 8.32).

Although attempts were made to have tobacco dispensed only by medical profession, its "pleasant stupification" spread widely through-

out Europe. Andre Thevet brought seeds of *N. tabacum* to France from Brazil in 1556. Jean Nicot, Ambassador to Portugal in 1560 diplomatically presented *N. rustica*—dubbed by him the queen's herb—to Catherine de Medici and was subsequently honored by Linnaeus when the genus name was applied. Sir John Hawkins, privateer, introduced tobacco to England, but a person with more advertising flair, Sir Walter Raleigh, popularized it in the court of Elizabeth I. Tobacco grew poorly in Europe, and most of the leaf available was *N. rustica*, so harsh that it could comfortably be smoked only in pipes. Gallants in England and on the continent invented the smoke ring, the tobacco belch, and other important social graces (Figure 8.33). Although tobacco sold for its weight in silver, its use spread so rapidly that authorities tried to get rid of it, since money was leaving the country and entering the Spanish treasury. King James I published a *Counterblaste to Tobacco* in 1604 which ended on a high note:

> A custome lothsome to the Eye, hateful to the Nose, harmful to the Lungs, daungerous to the Braine, and in the black stynking thereof, neerest resembling the horrible Stigean smoke of the Pit which is Bottomless.

Figure 8.33
Tobacco drinker and his visions. Taken from *The Melancholy Cavalier* by Cleaveland published in London in 1656.

Unsuccessful with this approach, James decided to impose a duty and excise tax on tobacco. The Portuguese introduced tobacco into China with so much success that the Ch'ing Emperor K'ang Hsi ordered decapitation of anyone selling it; Murad II of the Ottoman Empire regularly executed smokers, and the Romanoff Tsar Michael ordered that snuff-sniffers should have their nostrils slit. Pope Urban VIII issued a papal Bull mandating excommunication of priests who smoked when saying mass, but it was not until the accession of Pope Innocent X in

1645 that the same stricture was placed on communicants who regularly filled St. Peter's with clouds of tobacco smoke. All this was to no avail; people smoked, in the words of one early seventeenth-century writer, "as an excellent defence against bad air . . . it is good for to warm one being cold and will cool one being hot." Jean Baptiste Molière wrote that "he who lives without tobacco is not worthy of living." The hold of tobacco on the population of England was tightened when in 1665 the Great Plague decimated the city and "ye precious stynke" was believed to be the only thing that could ward off the disease.

COLONIAL TOBACCO GROWERS

Although the obligatory purchase of tobacco from Spain or Portugal kept the smoker poor, enriched royal treasuries, and offended the churches, it was obvious that England had to develop its own source of supply. John Rolf, who later married Pocahontas, started a plantation of N. rustica in the Jamestown colony in 1610, and in 1612 planted seeds of milder N. tabacum stolen from the Spanish West Indies. In 1613, leaves were shipped to England and had immediate success. When, in 1620, Elizabeth granted the Virginia Company a monopoly to export leaves to the mother country, virtually every piece of open land was planted to tobacco; even the shoulders of roads were used. Thomas Dale, governor of the company observed that crop land had been planted in tobacco and being concerned that the colonists might starve, ordered that for every acre of land in tobacco, there had to be two acres of Indian corn planted. His order was rarely obeyed since a man could earn £50 sterling per year with tobacco and less than £10 per year with corn. George Calvert, Lord Baltimore, obtained a royal patent to settle Maryland in 1632, and, over the protests of the Virginia Company, planted tobacco as the main crop. The Maryland settlers developed from N. tabacum the cultivar known as Burley which has become a major type of cigarette and pipe tobacco.

By about 1675, Spanish control of the tobacco market was ended, and all of Europe turned to England for its fine tobacco leaf. In order to control prices, the colonists of Virginia and Maryland agreed to restrict production on the understanding that the British crown would reduce their tax burden, an action which the crown used to its own advantage. The British navigation acts provided that all tobacco had to be shipped to England in ships of the British merchant marine. Obviously, the colonists objected to this; they saw that tobacco profits would be reaped by British merchants. They demanded the right to trade directly with other countries, insisting that they had not been consulted about the acts. This was denied, and the growers began selling their tobacco to New England merchants who dodged his majesty's customs officials and sold the tobacco directly to the Dutch out of the port of New Amsterdam (New York). Pointing to this flagrant smuggling as a justification for direct trade, planters of Virginia demanded a new assembly for the colony, one that would stand up for their rights against the collectors of revenue and one that would work towards free trade instead of toadying to the king's governors. Shouting "God dame ye collectors!" and "No duty to ye king!," men like Nicholas Culpepper rebelled, and it took over a year to put down the disturbance. Some of the demands were met, the rebels went unpunished, and the news that the Lion's tail could be tweaked with impunity spread throughout the colonies, to be brought up again a hundred years later when the first Continental Congress met to demand free trade and eventually freedom.

TOBACCO AND THE AMERICAN SOUTH

Tobacco shaped the social, economic, and political life of the central Atlantic colonies from Maryland south through the Carolinas. As land was cleared and slaves became the labor force, tobacco was grown on huge stretches of land. The plantation concept was born, later to be extended to that other staple crop of the American South, cotton. The South was rural, life was gentle, somewhat monotonous, and flowed with the seasons. A small group of landed gentry dominated the area with a very few middle class merchants and a stratum of white people whose job consisted of overseeing black labor. Life on tobacco plantations was, if you were on top, very pleasant with overseers to order and slaves to work. The gentry spent much of their time hunting, gaming, dancing, and being active in politics. Because the navigation acts proscribed trade with other than Britain, virtually everything used in the Southern colonies was imported. Nails, glass, table service, books, brandy and fine wines, and clothing all came from the proceeds of tobacco sales. Tobacco could buy anything; 120 pounds of leaf bought passage of a marriageable woman from England. The local clergy prayed fervently for a good harvest, for they had hogsheads of tobacco rolled to the parish church in payment for their service, and the sons of the planters were educated in England on tobacco money. William Byrd of Westover inherited 43,000 acres and by 1730 had 179,000 acres. William Fitzhugh of Bedford had an income of £6000 from 54,000 acres and 1000 slaves. Robert Carter had 300,000 acres and George Washington's slaves tilled 45,000 acres. Since months or even years flowed by between the time the tobacco was harvested and the accounts of the British and Scotch tobacco merchants were received, planters had no qualms about pledging their crops four, five, or more years in advance and having their bills paid by the merchants. Plantation life was on the installment plan, and as years went by, the planters fell further and further into debt.

Yet, in spite of, or perhaps because of, the easy life, the gentry took their political and public responsibilities very seriously. They served as governors, judges, leaders in the established Anglican Church, formed the local militia, and constituted the legislature. Their political and legal sophistication was high, and, as evidenced by the roster of United States Presidents from Washington to the Harrisons, they dominated the political affairs of the colonies and, eventually, of the United States. The Southern gentry were lords of all they surveyed, and they chaffed under the restrictions imposed by a distant parliament which controlled taxes, duties, tariffs, and enforced the navigation acts. As debts mounted, they recognized that tobacco merchants offered them lower prices than they could obtain on the continent and over-charged them for the Georgian goods they wanted. They objected to being dunned for money by "mere tradesmen."

Yet, the feeling that they were, in fact, aristocrats limited their ability to make common cause with men like Benjamin Franklin, a printer, or John Adams, a businessman, or Alexander Hamilton, a West Indian, a bastard, and a banker. Nevertheless, the desire for free trade was a common bond, a sophistication indicating they could manage their own affairs was another, and a philosophical concept of dignity and freedom (if one were white) was a third. In 1777, the rebel Virginia House of Burgesses passed a law sequestering all British tobacco debts. Jefferson, the Lee family, and the others deposited in the State Loan Bank the sum of £275,000 in worthless paper money to cover these debts; they were gambling that the debts would be uncallable if the revolution succeeded. When, in 1783, the Treaty of Paris provided that the tobacco debts were still outstanding, many planters simply refused to pay them; George Mason rhetorically asked why the Virginians had been fighting if not to get out of the hands of bloodsucking tobacco merchants. The situation was so bad that the congress finally authorized the use of federal funds to cover the debts. Thomas Jefferson negotiated privately and after insisting that tobacco burned by British troops on his Elk Island plantation should be applied to his credit, had a final bill of $10,000, about which he said, "Justice is my object, as decided by reason and not by authority or compulsion." This huge debt, his deep involvement in the affairs of the young country to the exclusion of his personal business, and, it must be noted, his tastes for fine wines and other amenities, placed him in a desperate financial situation. When the British burned Washington in 1814 and the modest Library of Congress was destroyed, Jefferson offered Congress his personal library of 6,500 volumes as the nucleus for a new library. The $23,000 received allowed him to pay his creditors, including the despised tobacconists.

THE BOTANY OF TOBACCO

Botanically, the genus *Nicotiana* is a complex of 65 species, most of them native to the New World. *N. tabacum* is, based on its physiology, morphology, and chromosome complement, likely to be a hybrid of several wild species; there is good reason to believe that N. tabacum is a cultivated species, not found in the wild. Although usually grown as an annual, it is capable of perennial development in tropical climates. For a general description, we can do little better than to present the one given by Dr. Nicholas Monardes of Seville in his *Joyfull Newes Out of the New-Founde Worlde* which was "englished" in London in 1596.

> This hearbe hath the stalke greate, bearded [hairy] and slimie, the leafe large and long, bearded and slimie. The plant is very ful of leaves and groweth in height foure to five foot. In hot countryes, it is nyne or tenne monethes in the yeare laden . . . with leaves, flours and Coddes [fruit], ful of rype graynes. It sproutes foorth neere the roote muche . . . notwithstanding the graine is the least seede in the worlde, and the roote be like small threeds. The leafe of this hearbe being dried in the shadow, and handed up in the house . . . and bee burned, taking the smoke thereof at your mouth through a fonnel or cane . . .

When Linnaeus established the genus to memorialize Jean Nicot, he included only N. rustica and N. tabacum, the only ones known in the eighteenth century. Since then, wild species have been found in Australia, and others have been collected in both North and South America. Many of these have been used as a source of genetic characters for cultivars of N. tabacum.

N. tabacum planting material is all derived from seed discovered by the Spanish in 1530 and transplanted to Haiti and Cuba. The bulk of North American tobacco includes two categories, the Bright tobaccos and the Burley types. Most Bright tobacco is grown in the Piedmont and coastal plain areas of the south Atlantic states characterized by sandy, well-drained soils with low organic contents, acid reactions, and very low reserves of inorganic nutrients. Burley tobaccos thrive on richer, silt-loam soils of Kentucky and Tennessee and in the warmer areas of Ontario. Bright and Burley form the bulk of the cigarette and pipe tobacco. Cigar tobaccos are grown in Maryland, parts of Virginia, and in the Pioneer Valley of Connecticut where the plants are grown under cloth tents which, by reducing solar intensity and increasing humidity, allow the plant to develop large, thin leaves (Figure 8.34). Lovers of fine cigars insist that the best cigars are made from tobacco grown in Cuba under conditions not reproducible on the mainland. Specialized tobacco types, such as the almost black perique cigar tobacco grown only in St. James Parish in Louisiana or the Latakia of Turkey are blended with Burley or Bright.

Cultivation practices vary with plant type, soil, and weather conditions plus the use to which the leaves will be put. High levels of nitrogen fertilizers foster development of large, succulent leaves, while leaves for cigarette manufacturers are smaller and a bit woodier as controlled by restricting the nitrogen supply at predetermined stages in the growth period. Tobacco is, in the words of the trade, "a heavy feeder" requiring extensive additions of fertilizers. In the cigar-leaf areas of Connecticut, close to two tons of fertilizers are supplied per acre per year.

Figure 8.34
Tobacco growing under shade tents in Connecticut. Courtesy Connecticut Agricultural Experiment Station.

ꙮ AN ESSAY ON PLANT MINERAL NUTRITION

In a first course in plant physiology, students learn the memory aid: "See Hopkins Cafe; mightly good with a pinch of salt" which is a listing of the major inorganic minerals—as their elements—needed by plants: C, H, O, P, K, S, Ca, Fe, Mg, Na, Cl. The first three, carbon, hydrogen, and oxygen, enter as carbon dioxide and water. All of the other required elements must be in a water-soluble salt or ionic form in the soil solution, being taken up with the water that all living cells require. These minerals originate in the soil as it is formed from rock, and they are normally replaced in the soil as plants and animals die and decay. Small amounts arrive as rain washes down particulate matter suspended in the air. Thus, there are mineral cycles in which a molecule of, say, potassium phosphate moves from soil to plant to animal to bacterium or fungus and then back to soil. If the cycle is broken when a plant is harvested, a need to replenish the soil is established. Some minerals are present in such large concentrations that artificial renewal is rarely necessary; calcium, sulfur (as sulfate), sodium and chlorine (as table salt) may sometimes be present in excess, as when salts accumulate from irrigation waters. Too much of any mineral element can cause toxicity and poor plant growth.

Of the required minerals, nitrogen is frequently in limited supply. Although 78 percent of air is nitrogen gas (N_2), elemental nitrogen must be transformed into an ionic form (ammonium ion − NH_4^+ or nitrate ion − NO_3^-) before plants can absorb and use it. Natural fixation of the element into ion can occur with the energy of lightning, but the bulk of nitrogen is fixed by the biochemical activities of free-living bacteria and blue-green algae or by bacteria in the root nodules of legumes and a few other plants. In all living cells, nitrogen atoms and ammonium ions are constituents of thousands of different compounds. In order to make a protein macromolecule, several thousand nitrogen-containing amino acids are linked together; 16 percent of protein is nitrogen. Complex rings of oxygen and nitrogen atoms form the purine and pyrimidine constituents of nucleic acids; the genetic code is a nitrogen-containing molecular arrangement. Alkaloids, vitamins, and many other chemical constituents of life are either nitrogen-containing or have their synthesis and degradation controlled by proteinaceous enzymes. Cellular growth and development is thus directly limited by the supply of nitrogen. Nitrogen deficient plants will look stunted, have woody stems and small, pale-green leaves with low chlorophyll contents with reductions of photosynthetic activity (Figure 8.35). If, as in the case of cigar tobacco, large, dark-green leaves are to be harvested, high levels of nitrogen must be supplied to the plants throughout the growing season.

The two other elements in what is called the NPK triad are also required in large amounts. Phosphorus, as the phosphate ion ($-PO_4^{+3}$) is built into nucleic acids and the phosphate-phosphate bond in adenosine triphosphate (ATP) bears the chemical-physical energy needed for every energy-requiring process in cells. Potassium is the third letter in the NPK ratio. A 10:10:10 fertilizer ratio simply means that every

Figure 8.35
Erosion and mineral deficiencies in tobacco. Courtesy U.S. Department of Agriculture.

pound of the fertilizer has 10 percent each of NPK and a 5:10:5 means 5 percent nitrogen, 10 percent phosphorus, and 5 percent potassium. Potassium plays many different roles. It is the primary regulator of the movement of water into and out of plant cells; it is part of many enzymes; and it forms complexes with organic acids (citric, malic, etc.) within the cells.

Few soils are deficient in calcium, magnesium, or sulfur, and few fertilizers contain them, but this does not reduce their importance to the plant. Calcium and magnesium are enzyme activators, serve as carriers for other ions through plant cell membranes, and reduce the toxicity of other ions which may be in excess. Magnesium, as part of the chlorophyll molecule, has a vital role in photosynthesis. Sulfur's main roles are as constituents of certain amino acids, cysteine and methionine, that determine the three-dimensional shape of proteins and as parts of compounds that regulate the oxidation-reduction state of cells. Sodium, needed in higher concentrations in animal cells than in plant cells, acts much like potassium, and the only known role of the chloride ion is as part of the enzyme complex that breaks water in photosynthesis, releasing molecular oxygen into the atmosphere. Iron (Fe) is an ionic constituent of the cytochromes that serve as links in the chain of compounds through which electrons flow to provide cellular energy (ATP) in both respiration and photosynthesis. Iron is also required for the synthesis of chlorophyll so that iron-deficient plants are pale or yellow and grow poorly.

In addition to these minerals that make up Hopkins' Cafe, at least five others are required in very small or trace amounts. Zinc and copper, very poisonous in high concentrations, are absolutely required by all plant cells as parts of certain enzymes; the blackening of a cut potato involves the enzyme tyrosinase which includes copper in its active site. Manganese, like chlorine, is part of the enzyme complex that releases oxygen in photosynthesis. We know virtually nothing about the role of boron except that plant cells will not divide without it, and actively growing tissues of the root and shoot meristem will die without continuous supplies of borate ions. We also know that the difference between adequate boron levels and toxic ones is very small and that even slightly elevated concentrations will result in plant cell death within a few days. All plants require molybdenum in order to carry on normal nitrogen metabolism, and many plants also require silicon, germanium, vanadium, or gallium for reasons which are completely obscure.

That organic additives to soil can supply plants with minerals needed for growth and development is unquestioned; the organic gardener quite properly restores to the soil plant materials or animal wastes. But serious questions can be raised about the assertion that organic materials can supply plants with vitamins and other biologically important molecules. Not only are intact plants fully capable of making these substances, but there is good evidence that the root system cannot absorb and translocate such organic substances dissolved in the soil solution. The main bone of contention between the organic gardener and the plant physiologist is whether a plant can distinguish between a phosphate ion derived from compost and the ion poured from a bag of 5:10:5 fertilizer. Leaving aside the mystical, plant physiologists and plant chemists strongly contend that a phosphate ion is a phosphate ion is a $-PO_4^{+3}$ independent of its source. Be it noted, however, that adequate amounts of organic material in soil provides it with a structure and texture which permits good root aeration, adequate control of moisture content, a balanced soil solution, and other conditions that facilitate the absorption of that phosphate ion by roots. Whether organically grown plants are healthier and freer of disease or whether the fruits and vegetables harvested from these plants taste better than those grown under nonorganic conditions has, surprisingly, received little study under controlled experimental conditions with adequate statistical backup; the antagonists may enjoy shouting at each other.

It is a general biological and thermodynamic rule that the efficiency with which nutrients are used by plants is about 10 percent. The relevance of this fact to recycling via composting must be borne in mind. Of the initial mineral content of composted plants, only about 10 percent becomes available to the plants treated with the resulting compost. This same order of magnitude reduction occurs at every turn in the cycle. Obviously, this does not mean that organic recycling of wastes should not be mandated; our wasting of potentially useful sewage and other organic material is nothing less than sinful. It does mean that even if all of the potential organic material in the world were efficiently recycled, we would still require industrially produced fertilizers. With the need to grow plants to feed the result of enthusiastic human overbreeding, the fertilizer industry is vital in world economy. Any other evaluation of the concept of recycling and the cultish aspects of organic gardening is to wish for a perpetual motion machine.

TOBACCO TECHNOLOGY

Tobacco cultivation is both labor and capital intensive. Plants are started from tiny seeds in seed beds and are transplanted to the field by hand. Subject to many viral, baterial, fungal, and insect diseases, a vigilant spray and dust program is necessary, involving millions of pounds of pesticides. During the growing season, the plants are topped to remove the terminal bud and apical meristem to prevent flowering and fruit production which tend to drain nitrogen and phosphorus from the leaves (Figure 8.36). Once topped, however, buds in the leaf-stem axil are released from dormancy since the growth-regulating chemicals formed in the apex are no longer translocated down the plant to repress bud growth. The resulting axillary shoots, called suckers, must be removed so that the plants do not become bushy and so that the leaves on the single stalk will grow to the desired size. Plants are harvested either

Figure 8.36
Hand removal of sprouts
(suckers) from tobacco.
Courtesy U.S. Department of
Agriculture.

by picking individual leaves or, more usually, by cutting down the entire plant. Leaves are collected and either threaded on stringers or baled for curing.

Curing of tobacco involves a linked series of chemical events. The water content is reduced from 80 percent to 20 percent. At the same time, a complex series of biochemical changes occur (called aerobic fermentations) wherein the leaf color changes from it bright green to various shades of yellow-brown as chlorophyll disappears and yellow leaf pigments become visible. Starch is converted to sugar, some leaf proteins break down to their constituent amino acids, and the alkaloids are modified. Originally, tobacco curing was accomplished by air-drying (Figure 8.37) or by smoking the leaves over a wood fire, and, for some specialized tobaccos, this is still done. Most tobacco is now cured by flue-curing. In modern flue-processing, leaves are heated to 70° C. for a few days resulting in a bright yellow leaf. Further aging may require several months. Leaves are classified by color, by projected use, by the method of curing and aging, and by the cultivar type and where it was grown. In cigarette manufacture, leaves are stripped to remove veins and petioles, shredded, blended with other tobaccos, flavored with sugars, menthol, moisturizers (humectants), and wrapped in special cigarette papers to which filters may be added. Although less popular than previously, plug tobacco is cut, blended with honey, licorice, and other flavorings before being compressed. Cigar-making, originally a highly skilled hand operation, is now usually done by machine. In Eu-

Figure 8.37
Air drying tobacco. Courtesy
U.S. Department of
Agriculture.

rope, small cigars are baked, and these are favored as an after-dinner smoke by women—a solace and pleasure still viewed with surprise in North America.

Although modern cultivars, curing practices, and methods of manufacture have been combined to give a cigarette with a relatively low nicotine content compared with the 35–40 mg of alkaloid per cigarette of the 1920–1950 period, there is still enough nicotine to cause physiological addiction in addition to habituation. Nicotine is synthesized in the roots and translocated up to the leaves where it accumulates, reaching its highest concentration when the leaf is mature (Figure 8.38). Genetic manipulation has resulted in clones of tobacco which are completely free of nicotine, and these plants show no metabolic or reproductive disturbances; it is not known what, if any, role is played by nicotine in the economy of the plant. When nicotine is combined with sulfuric acid, the resulting nicotine sulfate is a potent insecticide. During the burning of tobacco, nicotine is volatilized and may constitute 2–3 percent of the smoke. It is readily absorbed by mucous membranes of the nose, mouth, and lungs, passing directly into the blood stream; unborn human babies can become addicted. Physiologically, nicotine alkaloids act directly on nerve cells, stimulating them at low concentration and paralyzing them at higher concentrations, interfering with the transmission of nerve signals. Whether this dual action can account for the hallucinogenic effects of tobacco is likely, but unproven. And, as with other addictive alkaloids, the pharmacological,

Nicotine

Nicotinic acid

Figure 8.38
Nicotine, the major tobacco leaf alkaloid, and its oxidation product—the vitamin nicotinic acid.

biochemical, and psychological basis for habituation and addiction is unknown. The health hazards of tobacco include not only cardiovascular alterations, but emphysema and cancers of the lung, lips, and mouth caused by ill-characterized carcinogens, the "tars" formed during the combustion of tobacco. Since cigarette smokers do more inhaling of smoke, their disease rate is significantly higher than that of pipe and cigar smokers.

Since the end of World War I, the cigarette has dominated the tobacco market. The modern cigarette was invented in Spain where cigar clippings were stuffed into performed paper tubes and sold very cheaply. These "little cigars" became popular during the Crimean War of 1845 when soldiers learned to wrap Turkish tobacco in nitrate-impregnated cannon fuse paper which burned very evenly. The J. B. Duke Company of Durham, North Carolina, started producing machine-made cigarettes in 1883, but plug tobacco (peaking at three pounds per capita per year in 1890) and the masculine stogy (peaking at 86 cigars per capita per year in 1907) dominated the market for many years. Victorian gentlemen gathered after dinner for cigars and brandy, and their ladies were left to compare complaints about nursemaids. A farm woman could smoke a pipe, a European woman could smoke a thin cigar, but it was generally agreed that a cigarette-smoking woman was obviously of loose morals, for example, Bizet's Carmen or Puccini's Mimi in *La Boheme*. After World War I, chawin' tobacco fell into disfavor, and the brass spitoon became a collector's item. With the impetus of the most expensive (and successful) advertising campaign ever mounted, both men and women were urged to use the cigarette as an entry to wealth, health, and sexual magnetism, and cigar-chomping villians flickered on the silver screen. Seven-eighths of all tobacco consumed in North America today is in the form of cigarettes, the bulk being filter-tipped which reduces the amount of tobacco per unit and increases industry's profits; filters are cheaper than tobacco.

Since tobacco has a high value per pound of leaf, many growers are small landholders with less than 10 hectares. Using, in many cases, marginal farm lands unsuitable for growing any other cash crop, their livelihoods are absolutely dependent on a continuing market. All in all, 250,000 hectares are planted each year by over 600,000 farmers in the United States alone. North Carolina and Kentucky accounts for 300,000 tobacco-farming families and half of the U.S. production. Since the lands are generally infertile and have been drained of nutrients by almost 200 years of tobacco cropping, extensive fertilization and labor-intensive cultivation are economically justified by the relatively high dollar value of the crop. If the land were converted to raise food, dollar productivity would be reduced below that capable of supporting these farmers. Because of these economic and social considerations and the powerful financial and political pressures exerted through the manufacturing sector, producers of pesticides and fertilizers and other subsidiary industries, tobacco remains on the list of "basic crops" of the

United States Department of Agriculture and considerable research effort of the USDA is devoted to tobacco. The U.S. Commodity Credit Corporation finances price supports for wheat, peanuts, rice, soybeans, corn . . . and tobacco.

ADDITIONAL READINGS

Brooks, J. F. *The Mighty Leaf.* Boston: Little Brown, 1952.
Corti, E. *A History of Smoking.* New York: Harcourt Brace Jovanovich, 1932.
Goodspeed, T. H. *The Genus* Nicotiana. Waltham, Mass.: Chronica Botanica, 1954.
Heimann, R. K. *Tobacco and Americans.* New York: McGraw-Hill, 1960.
Hewitt, E. J., and T. A. Smith. *Plant Mineral Nutrition.* New York; Wiley, 1975.
Lévi-Strauss, C. *Introduction to a Science of Mythology,* vol. 2. *From Honey to Ashes.* New York: Harper & Row, 1973.
Robert, J. C. *The Story of Tobacco in America.* New York: Knopf, 1960.
Tso, T. C. *Physiology and Biochemistry of Tobacco Plants.* Stroudsbury, Pa.: Dowden, Hutchinson & Ross, 1972.
Wagner, S. *Cigarette Country: Tobacco in American History and Politics.* New York: Praeger, 1971.

CHAPTER 9
Plants in the Environment

Floods

℘ For behold, I will bring a flood of waters upon the earth
  GEN. 16:17

The announcement of the flood in Florence on November 4, 1966, aroused the immediate reaction of horror felt by most people when they hear of a natural catastrophe. Fragmentary stories on the wire services were followed by detailed descriptions of specific paintings damaged or destroyed, books soaked with water and oil, and accounts of death, hardship, and personal heroism. The word pictures of the odors of sewage, rotting animal and human flesh, mold, mud, and oil were graphic. In the succeeding days, as is usual in our overcommunicated world, the stories receded into the back pages of the newspapers and fresh stories of other tragedies took over the world's attention.

Yet, there remained a nagging feeling that this flood should never have happened at the level of intensity and with such severe loss to the treasures of Renaissance Italy—indeed of Western civilization. As we read more about the history of the Florentine region, it becomes clear that the 1966 flood is an almost classic example of how to ensure that a flood will recur.

GEOLOGY

The area known as Tuscany is a well-marked and fairly compact geologic unit which rises gradually from the western coast of Italy to the heights of the Apennine Mountains. It is a harsh land of hills and valleys, of winter rains and summer droughts. The valley of the Arno lies on the northern margin of the African trade winds and on the southern margin of the prevailing westerlies that bring rain from the Atlantic Ocean. In winter, the warm air above the sea forms a low-pressure cap that draws winds over the water. As these warm, moisture-laden winds flow up the hills to the Casentino mountain range, they cool, and rain may fall furiously for days in succession. Of the 25 to 40 inches of rainfall that Tuscany receives annually, over 80 percent falls between October and January—much of this from late October through the middle of December. In the summer, there may be no rain for several months, because the hot, rapidly moving winds from Algeria do not pick up enough moisture from the sea.

Today, the hills surrounding Florence can scarcely support olives and grapes. There are occasional patches of herbs, including lavender, myrtle, rosemary, and thyme; aromatic, romantic, lovely to look at in flower, and completely useless as ground cover. The soil is leached-out

Adapted from "The Florence Floods," Natural History 78(1969):46–55. Copyright American Museum of Natural History.

clay with pockets of stone and sand. The sun beats down on impoverished and badly worn soil, changing it into a cementlike hardpan. In spite of, or perhaps because of, its aridity, the land above the city of Florence is very beautiful. From the surrounding hills, one can take in the peaks of the Casentino mountain range, bared and gleaming in the summer sun. First in abrupt descents and then more gently, the highlands fall away to the foothills. Sere, brown, and rocky, the hillsides catch the early morning or late afternoon sun, leaving the hollows in deep shadow.

The present city of Florence stands on land that has been a part of the civilized world for more than 3000 years. Excavations have unearthed artifacts dating back to the Bronze Age, and there is some evidence that Neolithic cultures flourished there even earlier. And well they might have, for the Tuscan valleys and hills were fertile and wooded, the climate was generally mild, with just enough seasonal variation to stimulate, but not debilitate, the people, and there was reasonable protection from marauders from the north. The major river, the Arno, was a freshwater stream, abounding with fish, flowing throughout the year, and was broad enough to permit travel, and hence trade, with the coastal peoples in the delta that is now Pisa.

ETRUSCA

The earliest complex civilization about which we have information is that of the Etruscans. They settled in what is now Tuscany about 800–600 B.C., coming from northwest Asia Minor via the Caucasus. In addition to being farming peoples, they were experienced merchants, and much of our knowledge of them is derived from their commercial activities. The Etruscans found, settled, and exploited a land of natural beauty. The Arno and its tributaries were clean, clear, fishable, and navigable. The hills were completely forested. In the valleys were found the Aleppo pine and two evergreen oaks, the Holm oak and the Valonia oak, forming open forests, with understories of small trees and shrubs. The midmountains contained excellent stands of beech, Spanish chestnut, and deciduous oaks, and the upper mountains were clothed with black pine and several species of fir. There was an abundance of animal life, for figures on Etruscan pottery depict deer and bear hunts. In the fourth century B.C., Theophrastus mentioned the beech groves of Tuscany as being of such excellent quality that the wood was bought from the Etruscans for ships' keels. Even by 300 B.C. the forests were so dense that Roman legions hesitated to invade. Plant cover was still notable several hundred years later when Strabo reported that the central Apennines and the hills of Etruria provided wood for the construction of Rome. The Sila hill range yielded "fine trees and pitch" for shipbuilding activities centered near Pisa at the mouth of the Arno.

Etruscans were a vigorous people who used their land for many activities. They planted olive trees and grapevines on hillsides and farmed valleys for wheat and barley. They mined some copper, zinc, and tin to make bronze for tools and weapons, and they smelted mercury and some iron. Sheep were grazed on the lower hills to supply white, black, and brown wools, much prized for their softness, fineness, and natural colors. Thus, some deforestation and grazing injured the land, but the populations were too small to damage severely the fragile hills. By the end of the Etruscan hegemony, the Romans were lavish in their praise of this still-wooded, fertile, and well-watered land.

THE FIRST FLOODS

The first adequate records of floods date from the twelfth century A.D., actually 1117, when the Ponte Vecchio, then the only bridge over the Arno, was swept away. Records show clearly that the Tuscan hills were denuded, the Arno was a seasonal stream, and the land was plagued by drought. If we date the demise of Roman rule at about A.D 300, and we know that the land was ruined by 1117, we are left with approximately 800 years to account for in terms of physical geography. What happened to the forests, the clear streams, the good soil, and the fine climate?

The main city of the Etruscans was not modern Florence, but the now-small hill town of Fiesole. Present-day Florence was a suburb, about 15 miles downstream on the Arno. Florence itself was situated where the Arno issued from the mountains at about 180 feet above sea level, a few miles upstream from its present location. The upriver location was determined by a simple fact—the site of the present city was a marsh. Indeed, when winter snows melted in February, the river was not fordable. Livy noted that in the late winter of 217 B.C., Hannibal and his armies floundered for four days in the muck of the valley floor. Yet the river bottomland was very fertile, producing an excellent crop of hard wheat, some millet, and good barley. Snowmelt served to supply nutrients, and the water moved out to sea by March, leaving time to plant. There was a second reason for the lowly political position of Florence, one related to the defense of the area. The Etruscan military wanted its citadel on the heights, not on the nearly indefensible valley floor.

Fiesole, as the political and defense center, gathered the greater share of the population, but it soon became apparent that this small, hilly area could not provide an adequate living for the populace. The hills about Fiesole had to be intensively cultivated, and the only practical way to do this was by terracing. The technique was certainly not invented by the Etruscans; the Greeks and other peoples of Asia Minor had been cutting horizontal ridges into their hills for eons. In the Tuscan context, terracing was necessary more as a means of fully utilizing all land within the protected reach of the fort than as a conservation measure (although it certainly served well in this respect). Cereal crops were not planted on the terraces but were grown on the Arno valley floor, and the crop was stored in the granaries of Fiesole. Vines were planted on the terraces, particularly on the drier, stony, south-facing slopes. Initially, the trees were native species, but these were replaced by the olive prior to the Roman invasions. Thus, even before the rule of Rome, some of the plant cover on the hills had been removed. Since the Romans left no records of flooding, we can assume that the major portion of the Tuscan hills was still covered with plants that sopped up the water and prevented the destructive run-off that resulted in floods.

ROMAN RULE

Tuscany, being on the route from the lands of the Franks and the Goths to that of the Romans, served as a staging ground for many armies. The mountains were no longer an effective barrier from the north as disciplined armies replaced wandering bands of military plunderers. Records are hazy and contradictory, but it seems that the Romans defeated the Etruscans several times between 300 and 200 B.C. and that by about 200 or 150 B.C. Roman rule was firmly established. The Roman road system was extended through Tuscany, and during its construction the engineers tried to stay in the piedmont except where there was a necessary valley river crossing. The Via Julia Augusta crossed the Arno at the site of the present Ponte Vecchio to connect with the Via Cassia going north. Even then, the Roman legions knew it as a bad crossing, because it was muddy and tended to flood in the spring.

Under the protection of the legions and the Roman governor Sulla, the population of Fiesole began to move down the Arno to where life was easier. When Sulla was defeated by invading Franks, Florence was leveled and burned, and again the people retreated to the hills. By 59 B.C., Florence was, under the auspices of Julius Caesar, built at its present location and given the name Julia Augusta Florentina, the latter word apparently being a good omen rather than having any botanical allusions. The city was a typical walled Roman city-fort with temples, baths, aqueducts, and an efficient sewage-disposal system. There was a central marketplace, now the Piazza Vittorio Emanuele. Apparently there was some attempt to drain the swampy area then outside the city walls. Although flooding in the winter did not assume sufficient importance to record, the Romans were well aware that snowmelt was a danger, for they erected levees on the banks of the now-contained, but still free-flowing Arno.

The ancients knew the value of woodland as flood protection. Plato deplored the reckless cutting of Greek forests and pointed out the cause-and-effect relationship between forest destruction and rapid run-off of rain from bared hills (Figure 9.1). Pliny quoted Plato when he

admonished the Romans to maintain and to replenish their forests, and while we have no direct evidence, it seems likely that some form of forest conservation was practiced in the Tuscan hills.

MALARIA

With the population increasing, terracing also increased, but this stopped abruptly. Malaria, which had been introduced into Sicily from North Africa during the fourth century B.C. moved onto the continent during the Second Punic War of 218–204 B.C. By the beginning of the Christian era, it reached Tuscany, where it became endemic in the lower valleys and in the city. Within less than a century, Florence and the hill towns were abandoned. Terraced farms, no longer cultivated, were subjected to erosion that washed away precious soil. It is easy to visualize the stripping of the soil from the hills, the mud slides, and the consequent silting of the tributary rivers and, eventually, of the Arno itself. Although no records exist, we are fairly safe in saying that minor flooding occurred in the Arno basin from about the middle of the first century A.D.

During the subsequent 300 years, the town of Florence continued to decline, local trade decreased, and this once relatively important

Roman city became little more than a village. As the Roman Empire began to fall apart during the third and fourth centuries, tilling declined and plowed land and terraced hillsides were subjected to direct erosion. The upper hills, however, began to restore themselves, and flooding probably decreased.

The fortunes of the city took a turn for the better by the end of the fourth century as malaria began to disappear. The Lombards made Florence into a military center, and when the Franks replaced the Lombards in 774, the entire region assumed an increased role as a trade center. By then, Fiesole had declined to the level of a town, a condition that has never been reversed.

As populations grew, there was a greater demand for wood to be used for construction and for fuel. The hills were rapidly stripped, so that by the end of the seventh century, wood had become so scarce and expensive that stone became the primary construction material. From the perspective of the art historian, this had some importance. The artisans of the time were both good craftsmen and good architects. These early stone buildings served as models for the Renaissance masterbuilders, who made Florence a place of enduring beauty and, incidentally, laid out a city especially suited for flooding. As the city began to assume its present shape, virtual bathtubs were created there—piazzas and squares with a single inlet street and one or a few outlets. Water rushing into one of these squares could not easily pour out the other side, and the whole area filled up. Most of the listed highwater marks of the 1966 and earlier floods occurred in these piazzas.

THE WOOL TRADE

Although the Etruscans and the Romans engaged in the wool trade, and Florence had been a weaving center for many centuries, increased use of wool cloth by peoples less exalted than the aristocracy resulted in a great increase in woolen mill activity. As woodlands were decimated, the cleared land that resulted became pasture for goats and sheep. Both these animals nibble forage to the root level, effectively killing individual plants. Overgrazing became so severe that, by the eighth century, Florentine mills could no longer depend on local wool but were forced to import it from England and Spain. Parenthetically, it is likely that the denuded Spanish hills are partially the result of overgrazing by sheep to supply wool to Florence.

The effect of land clearing on the climate of Florence is not easy to document. Modern ecological and ecophysical studies show that forested areas serve as natural air conditioners. Heat is collected by trees during the day and released slowly at night. Modified by the cooling power of evaporation, heat never builds up as it does when captured by stone and paving; a walk from a city street into a park at dusk illustrates this dramatically. The presence of nothing but baked clay on the hills above Florence served to perpetuate its summer heat and drought. Forests served, as Etruscans and Romans knew, to ameliorate the heat of the summer. Not incidentally, the absence of the forests with their water-holding capacity resulted in loss of scarce summer moisture, and rivers and brooks ceased flowing in the summer.

By the beginning of the ninth century Florence had about 30,000 people and the old walls had been breached and built farther out several times. The region had many different rulers, none particularly firm or effective. Weak rulers gradually let power slip from their hands, and authority, power, and ownership of the land became divided among Roman Catholic bishops and a few wealthy landowners. These interlocked concentrations of land ownership led steadily toward the development of a feudal system in which large areas were tilled by serfs for the production of a few economically important crops. Goats and sheep continued to graze, and meadowlands were cut from the forests farther up the hills. With the removal of the sparse grass, the summer droughts baked and caked the clay soil to the consistency of rock, and natural seeding in it became virtually impossible. Bound to squeeze the last florin of profit from the country, the Church and the landowners made no attempt to replenish their land. Tuscany, a land noted for good water, forests, and fine agricultural bottomland, inexorably declined into the arid, bare, and impoverished country that we see today. The greed of the Church, the avarice of the aristocracy, and the ignorance of the people have given us this legacy.

FLOOD RECORDS

The remainder of this sad tale can be told rather quickly. The first complete record of a major flood was provided by Giovanni Villani for the flood of November, 1333. There was four feet of water, the city walls collapsed (never to be rebuilt), and three of the then four bridges over the Arno were destroyed; the Ponte Vecchio stood firm. Quoting Villani:

> Wherefore everyone was filled with great fear and all the church bells throughout the City were rung continuously as an invocation to heaven that the water rise no farther. And in the houses, they beat the kettles and brass basins raising loud cries to God of "misericordia, misericordia," the while those in peril fled from roof to roof and house to house on impoverished bridges. And so great was the human din and tumult that it almost drowned out the crash of the thunder.

Not unexpectedly, the flooding was worse in the low-lying districts nearest the Arno. Three hundred people died in the hard-hit districts of San Piero Scheraggio and the Porta San Piero. On the Via de Neri the flood crest was 13 feet 10 inches, only 14 inches lower than during the 1966 flood. Villani carefully listed the losses of grain and casks of wine, and in quite modern fashion he noted the cost of bridge repair, as well as municipal outlays for cleaning the streets and for minor compensations to citizens.

Villani noted the extraordinary rainfall over the Arno watershed, but then wrote: "On May 15 there was an eclipse of the moon in the sign of Taurus . . . and then at the beginning of July there followed a conjunction of Saturn with Mars at the end of the sign of the Virgin." The Church had the last word, stating that "by means of the laws of nature, God pronounced judgment on us for our outrageous sins." There was slight consolation from the King of Naples, who told the Florentines that "whom God loveth, He chasteneth."

A tiny note of biological intelligence was, however, struck by Gianbattista Vico del Cilento, who recommended a government sponsored reforestation plan. Two hundred years later—the plan never having gotten off the ground—Guistino Fortunato dismissed Vico's plan, stating that trees were ugly and useless.

We need not detail the economic and biotic history of Florence beyond 1333. In spite of accurate information on causes of floods and adequate technical information on flood control, nothing was done. Vico was probably not the first, and he certainly was not the last, to chastise the Florentines for their laxity. Giovanni Batista Adriana wrote that the severe flood of 1547 badly disturbed the people who demanded that something be done. They were told that God wanted, with this act, to signify that worse calamities would occur, so they decided to wait. In 1545, Michelangelo cautioned a nephew about buying a house in the Santa Croce district, "every year there the cellars flood." His warning was apt, for the flood of 1547 was followed by another in mid-September of 1557, when the entire Santa Croce district was under several feet of water. Lorenzo de' Medici warned the population about the Arno: "Its arrogant anger spits and beats away at the weak banks." But in spite of his authority to order new construction, he did nothing, and there is no evidence that the people heeded the warning.

Leonardo da Vinci reported that the streams of Tuscany were muddier when they passed through populated districts and suggested that the mud was washed-out soil that clogged the streams and leached the soil of the hills. He conducted surveys and designed projects to develop water impoundments in the hills, to dredge the tributaries of the Arno and the Sieve, and to develop chambers under the city to hold excess floodwaters. These, too, were ignored.

The Tuscan flood situation had only one useful effect. In 1861, President Lincoln appointed the Vermonter George Perkins Marsh as American minister to the kingdom of Italy. Marsh's travels in Tuscany and his reports on the relatively minor floods of 1861 and 1863 undoubtedly contributed to his thinking in the preparation of *Man and Nature or Physical Geography as Modified by Human Action*, published in 1864. As a result of interest (indeed the controversy) created by this book, Marsh was asked to help compile the irrigation laws of Italy, laws never adequately enforced, which could have alleviated the floods that have occurred during the past 100 years. Marsh's book so stimulated Gifford Pinchot that he persuaded President Theodore Roosevelt to set in motion the legislation leading to the conservation policies of the United States.

PRESENT PROSPECTS

For what it is worth, the statistical picture is gloomy. Since 1333, there has been a moderate flood in Florence every 24 years, a major flood every 26 years, and a massive flood every 100 years. Of the massive floods, those of November, 1333; November, 1844; and November, 1966, have been the worst. Since 1500 there have been more than 50 floods qualifying as moderate and 54 qualifying as major.

In 1797, a French engineer, Fabre, announced that alpine torrents which flooded the lowlands were a direct result of deforestation of the heights. One hundred years later, in 1890, the French government undertook the job of reforestation in the Hautes-Alpes. Engineer Surell reported that as reforestation continued, floods decreased in number and severity. This report was widely distributed in Europe, was acted upon in Germany and Switzerland, but was ignored in Italy.

Apparently no one in Florence could believe that water could cover their city in the twentieth century. But they, themselves, place marble markers on their homes and churches "commemorating" previous floods. They, themselves, proudly indicate the high-water marks of floods dating back 500 years. They, themselves, speak of their dry, bare hills which "shimmer in the summer sun," and they cannot fail to see that their "ribbon of shining silver," their Arno, is a muddy, silted-in, malodorous stream that almost stops flowing during the summer.

The *New York Times* of January 2, 1969, reported that during the week preceding Christmas of 1968—two years after the most damaging flood in Florentine history—the Arno crested toward the flood mark. "Only after the river began to subside at 9:00 P.M. did the crowds move homeward." The bed of the Arno is still filled with the debris of the 1966 flood and has never been dredged. It is estimated that the silt from 1966 alone is 2 m thick.

There are two dams on the Arno, the Levane dam at Livorno, 35 miles upstream from the city, and the Penna dam, yet higher into the hills. These are power dams and do not provide for flood control. No money has been (or apparently ever was) appropriated for reforestation as recommended in 1334 by Vico, for the construction of water impoundments and for dredging the Arno as recommended by Leonardo, or for caring for the land as recommended by Plato.

It is beyond the scope of this book to discuss the storage of irreplaceable books and art treasures in the basements of Santa Croce, noted by Michelangelo as the district where the cellars flood every year. It is beyond the scope of this book to discuss the reasons for the lack of coordination between the two damsites. It is beyond the scope of this book to discuss the lack of retaining walls bordering the Arno; even

Lorenzo did nothing about them. As George Santayana warned in his *Reason in Common Sense*, "when experience is not retained . . . infancy is perpetual. Those who cannot remember the past are condemned to repeat it" (Figure 9.2).

Dust

At 10 A.M. the dust in the air was so dense
that objects could not be distinguished 100 yards off.
No one who could possibly remain indoors was on the street.
REPORT OF U.S. WEATHER BUREAU,
DODGE CITY, KANSAS, APRIL 8, 1890

It was no accident that great civilizations of the ancient world developed in areas where the soil was renewed by flooding of rivers whose precious mud, high in organic matter annually inundated the land. Egypt's dynasties were dependent upon the Nile, India's on the Indus, Mesopotamia's on the Tigris and Euphrates, and China's on the floodplains of the Yangtse and Yellow rivers. The Tigris and Euphrates valleys and the valley of the Indus have long since become arid, and Egypt utilizes a good deal of the electric power from its ill-conceived

Aswan Dam to make fertilizers which the untrammeled river provided free for thousands of years. The great rivers of China, by conscious effort, are still serving people's needs. As the world's populations began to increase, less fertile and less desirable territory was farmed, including dry uplands and the semiarid and arid lands that constitute more than half of the world's arable area. It is these lands, called marginal by agricultural economists, which must be considered, for it is just these lands upon which the bulk of the people in the world must depend for food.

ARID LANDS

Among our most fragile lands are those which meteorologists classify as semiarid or arid. Here, rainfall is the limiting factor in plant development. For our purposes, we can call any area receiving less than 25 cm of precipitation per year an arid zone, and any area receiving less than 50 cm per year as semiarid; lands receiving less than 20 cm per year are deserts. Arid and semiarid land usually show extreme fluctuations in rainfall patterns, and the 25 or 50 cm limits are based on very long-term data. Rainfall is rarely "normal." The statistical average is based on a minimum of 30 years, but individual yearly totals, monthly averages, or daily precipitation recordings may differ markedly from this average, normal, or expected figure. Even using a 30-year average can be dangerous because of variations among successive periods. The midwesterner speaks of "spells" of rainy or dry weather, knowing that short-term changes in expected weather patterns are matters of chance and hence not predictable with any accuracy. Rainfall does, however, tend to come and go in cycles, perhaps with cyclic patterns of 11 or 22 years corresponding to sunspot activity, but deviations from such cyclic distribution patterns are frequently seen.

Semiarid and arid lands are further characterized by the plants which grow on them. With few exceptions, grasses dominate the landscape; early explorers of North America's grasslands spoke with awe of "seas of grass" with waves produced by the winds which swept over the land. Grasses are dominant because they possess characteristics which permit them to survive where other plants cannot. Most arid land grasses are perennial, with new leaves formed each year from shoot buds on the rootstock. Reproduction is mostly vegetative with extension of rootstocks from a center, producing mounds or bunches which may extend for several meters. Individual plant mounds may become confluent, with kilometers covered by the same species. The extensive root systems spread laterally as well as vertically, binding the soil into a tough sod. Mats of dead leaves and stems insulate the soil from heat, prevent wind and water erosion, reduce evaporation of moisture from the soil surface, and trap snow. Trees in grasslands are restricted to watercourses; most tree species cannot survive with the limited rainfall, and various sod-forming grass species tend to cover the ground so completely that tree seeds cannot get established. In contrast to many broad-leaved plants, grass leaves have the ability to curl, with the stomata-bearing lower epidermis on the inside to reduce water loss by transpiration. When undisturbed by man, grasslands can support huge herds of grazing animals—bison and prong-horn antelopes in North America and many species of antelope in Africa and Asia—which are preyed upon by carnivores like the wolf and the big cats. Mice, moles, birds, and their predators form interlocking food chains, all dependent upon the grass.

GRASSLANDS

Undisturbed grassland soils are usually very fertile, dark brown to almost black, and topsoil depths may exceed 2 m. All soils are mixtures of mineral particles derived from rocks plus living and nonliving organic material, the humus, which provides the texture and structure of soil, its water-holding capacity, its biochemical and biophysical properties, and its ability to support the growth of plants. Without humus, plants will not, unless carefully and individually tended, develop properly. Over eons of time, as grass roots and leaves decayed or were digested by soil organisms, the life-sustaining humus developed. It provides suitable habitat for the animals whose tunnels and runs aerate the soils, with water-holding capacities which allow a build-up of a ground water supply which sustains the grass roots during periods of low rainfall.

Based on vegetation types, grasslands may be divided into *prairie communities* characterized by tall-grass species and *short-grass plains* or *steppes*. Various species of grass are adapted to the climatic differences between Alberta and Texas, both of which have prairie and steppe communities. The tall grass prairies, with 50 cm of rain per year, are dominated by the bluestems (*Andropogon spp.*) and Indian grasses (*Sorghastrum spp.*), with grama (*Boutiloua spp.*) and buffalo grass (*Buchloe spp.*) found on the steppes where rainfall averages less than 25 cm per year. The best wheat lands—those of Kansas, Nebraska, the eastern Dakotas, and the prairie Provinces of Saskatchewan and Manitoba—are tall grass prairies.

All semiarid and arid lands are drought-prone because of the deviations from normal rainfall patterns. Aridity carries the connotation of a permanent condition of low average rainfall, but drought is a temporary condition which occurs aperiodically in climatic zones where precipitation is usually adequate for the dominant plant cover. Drought is a relative term, depending upon life styles and land-use practices of the inhabitants. Conditions which a vegetable grower might consider drought would be fine growing weather for a wheat farmer; and the sheep herder would be happy with rainfalls which would not permit wheat to develop. Prairie and steppe grasses are drought-enduring plants, selected over eons of time for their capacity to survive during those periods when precipitation is less than average. Leaves will wither and die, but root systems—protected from desiccation by the leaves, tapping ground water supplies, and utilizing the moisture held by the spongy humus—can survive for several years, putting forth new leaves when rain again falls.

Survival and regrowth of plants is abruptly changed when the natural plant cover is replaced by crops or is destroyed by overgrazing of domestic animals. Semiarid lands, because of their fine soils, can be excellent agricultural lands, yielding reasonably well when the human populations dependent upon them are small, and when only small areas of grasslands are put into production. But when people possessing high levels of technology intrude upon virgin plant communities, they possess the tools to impose their wills upon the land. Land, in the modern view, is a commodity to be exploited, to be mined, and the natural plant cover must be completely removed to make way for crop plants. Opened to the onslaught of the wind, thunderstorms, and hot sun, humus levels, no longer build from interlocking grass plants begin to decrease, earthworms no longer open water channels, and the soil deteriorates. In Paul Sears' words, "Deserts are on the march."

THE AMERICAN DUST BOWL

From the Mississippi to the Rockies, the Great Plains flow in undulating waves north, from Texas to the tundra of Canada. The climate is continental, with hot summers and very cold winters. The long-term average rainfall is less than 50 cm per year, and much of the area receives less than 30 cm per year—in a good year.

In 1821, Stephen F. Austin led a group of immigrants from U.S. territory into the Mexican province of Texas. First occupying the wooded river valleys of east Texas, settlers slowly moved into that portion of west Texas which was included in the Great American Desert. In spite of danger from the Comanche and from Mexicans who realized that their country was being illegally taken over, settlers found that the semiarid grasslands supported the growth of long-horn cattle. These were fattened on the grass and driven to Abilene, Dodge City, and other railheads to the eastern markets. The Indian Territory to the north contained few whites until the end of the Civil War, but since the upper plains states had already become settled, covetous eyes were cast on these potentially valuable lands. Although the Territory was semiarid at best, it was covered with short grasses and the soil was fertile.

In 1889, the Oklahoma Territory was formally opened to settlement. In March and early April, thousands of people with horses, carriages, and wagons lined up on its boundaries. At noon on April 22, cannons were fired and the settlers rushed into the area to stake out homesteads. Those who jumped the gun were called "sooner's," and Oklahoma, the "Sooner State," is proud of its law-breaking founders. Within a month the population increased from fewer than 5000 people—mostly despised Indians—to over 60,000 farmers. The sod was broken immediately and a wheat crop planted. Yields of the 1889 crop compared favorably with those of Kansas, but as a sign of what was to come, the rains of 1890 failed to materialize and the crop was almost a complete failure. Rainfall was about 75 percent of average, and, although the plants started out well, a hot summer and dry winds from New Mexico reduced the growth and seed production of the plants.

AN ESSAY ON WATER STRESS

A student leaves the dormitory on Friday afternoon, comes back from a wonderful weekend late Sunday night, and finds the geranium on the windowsill looking as bad as the student feels. Leaves are flaccid, shriveled, and some are beginning to turn yellow brown along their margins. Stems are drooping and flowers have fallen. Rushing to the rescue, water is poured onto the bone-dry soil until it flows from the bottom of the pot—usually all over the floor. If the plant has not been wilted too long and if the room hasn't been too hot, the plant will begin to recover within a few hours. Leaves will again become succulent and turgid, and the stems will again become upright. Since many houseplants are fairly rugged, recovery is frequently complete, although the death toll of dorm plants is high.

Wilting is a cellular phenomenon, the loss of water from the vacuoles of living cells. Vacuolar water has moved through the cytoplasm, cell membrane, and cell wall into the spaces between the cells from where it is pulled up to the stomata in the leaves and thence out into the atmosphere. As water pressure within the cells decreases, vacuoles collapse and cells become flaccid, much like a balloon with most of the air let out. The flaccid cells are no longer capable of maintaining the rigidity of stem and leaf tissues, and the plant wilts. Wilting is a biological example of the economist's Law of Supply and Demand; transpirational losses of water from the leaves exceed the supply of water available to the roots. As with economic supply and demand, the consequences reverberate in the plant long after the plant is visually recovered.

Water is a necessary condition for all life. Its role in plant and animal cells is varied and irreplaceable. As the closest substance to a universal solvent, inorganic and organic substances are dissolved in water for transport and for reactivity. A dry enzyme protein, meat tenderizer for example, cannot act to soften a tough steak until the protein assumes its three-dimensional, hydrated form, and functions as a proteolytic enzyme. Table sugar cannot be split or lysed into its component hexoses without the addition of a molecule of water, and the synthesis and degradation of all complex biological molecules also involve water addition or removal as part of the reaction system. The passage of molecules and ions into and out of cells occurs in aqueous solutions. In living rooms, greenhouses, and in the field, solar energy can accumulate in plant leaves, raising temperatures to lethal levels unless there is vaporization of water from leaf surfaces with the "cooling power of evaporation."

Wilting is one extreme of a continuum between the fully water-saturated or turgid cellular state to the dehydrated cell. Although the wilted state can result in cellular or plant death, most plants under field or natural conditions are rarely at either end of the continuum, but are usually somewhere in between; plants live perpetually in a state of water stress. Those adapted to dry (*xeric*) habitats have been selected for one or more responses which lead to survival. A drought-resistant species such as white pine (*Pinus strobus*) will maintain cellular turgidity and exhibit little evidence of stress under conditions of available soil moisture that would cause the permanent wilting and death of most house plants. The pine survives because its needles have a smaller surface area and fewer stomata per unit area than does a geranium and hence loses water at a proportionally slower rate. Most cacti do not have leaves, and their green, photosynthetic stems have few, if any, stomata. Some desert plants lose their leaves during the dry season, growing a new set only when water becomes available during the rainy season. Other plants, especially the desert annual flowers, are capable of going through seed germination, vegetative growth, flowering, fruiting, and seed maturation in short periods of time when spring rains wet the soil. These are drought-avoiding adaptations, and, like drought resistance, are programmed into the cellular hereditary material. Knowing this, geneticists can choose individuals or races of economically important plants possessing drought-resistance or drought-avoidance characteristics as crop plants. They may use these plants as breeding stock to improve existing crops and extend their growing range. Green Revolution cereal grains (Chapter 2) now include drought-resistant characters and were specifically engineered for drought-prone climates.

In order to develop such improved crops, an understanding of the cellular consequences of water stress is necessary. When asked how severe water stress must be in order to reduce growth and yield, the plant physiologist gives a typical scientific answer: "it depends." It depends on what cellular function or cellular system is limited for growth and yield, on the species under study, on the expected severity of water deficit, and on when during the growing season this deficit is expected. With appropriate instruments, and under controlled conditions, test plants are placed under conditions of zero water stress and the level of stress can be raised to any point up to permanent wilting. Various components of the physiological, biochemical, and growth activity of the stressed plants can then be evaluated to determine whether they are, relative to appropriate controls, functioning efficiently. Different cell types are not injured to the same extent by a given level of stress nor does injury start at the same stress level for all species. In general, cell division and cell enlargement are the most sensitive; this is why that windowsill geranium grows poorly for a while after it has been severely water-stressed.

Protein and chlorophyll synthesis, activities of growth-regulating chemicals, photosynthesis, respiration, and the accumulation of metabolites within cells are injured as stress becomes more and more severe, but the sequence is variable with species, stage of development, and even the previous water status of the plant. Seedlings and young sprouts are usually very sensitive to even slight reductions in water supply, for they are in a period of rapid cell division and cell enlargement. Plants grown with high levels of water—most houseplants are in this category—are ill adapted to any but short-term water stress. A tree forming its winter buds and preparing for winter dormancy suffers little from water stress, and, indeed, high levels of available water may suppress normal winter-hardiness. Corn and soybeans, both adapted to hot, dry summers, may wilt at the end of a hot, midwest "scorcher," but will suffer little permanent damage, while a rice plant at the same stage of development will show reduced yields after just a few evening wiltings.

With much of the world's available agricultural land subjected to drought, investigation of water stress in the laboratory and in the field are basic to food production. Physiologists, ecologists, taxonomists, anatomists, cell biochemists, geobotanists, and horticultural and agricultural scientists are cooperating in this vitally important study.

DUST STORMS

In spite of an occasional poor harvest, wheat farmers in Oklahoma prospered. The fertility of the soil and the water stored in underground aquifers and in the humus produced bumper crops. These were sold in 1914–1918 to Europeans so busily engaged in killing each other that they couldn't plant their own crops. Wheat more than doubled in price, and the areas cultivated increased rapidly. In the 1920s there were a few more bad years, each a bit more severe than the one before. The land was "dry-farmed," the farmers disk-harrowing a dust mulch on their fields to prevent soil-water evaporation and there was still some humus left. Much of the land was either tenant-farmed or farmed on lease, and neither the owners nor the farmers gave any thought to the possibility that fertility would run out, that ground water supplies would be used up, or that the humus would disappear. Wheat was still in short supply, and the government was encouraging cultivation of all available land. Loans for machinery and land purchase could easily be financed by banks whose officers were themselves speculating in land and whose financial stability was unregulated by the states or the federal government. American ebullience, boosterism, and an unquenchable optimism glossed over the few bad years; more sod was removed and everyone was mining the soil for all it was worth.

Figure 9.3
A dust storm in Texas in the 1930s. Courtesy U.S. Department of Agriculture.

The inevitable could not have occurred at a worse time. Europe's agricultural recovery was complete, and American wheat prices were falling. The stock market crash of 1929 and the following depression dried up the money supply that was keeping the expansionist economy buoyant. The bank reorganization ordered by the New Deal administration of Franklin D. Roosevelt uncovered and made public an inadequate debt financing structure. Those banks that had not already succumbed to the panic withdrawal of funds began to call in their loans from farmers who could not pay them. The crushing blow was drought. From 1930 to 1937, the area that was to be known as the dust bowl received less than 65 percent of the expected, "normal" rainfall. This would have been enough for survival of tough, resilient prairie and steppe grasses, but it was not enough to allow wheat to grow. The prairie grasses were gone, and so was the humus, the ground water, and the soil fertility. As the hot, dry, bare soil radiated heat from cloudless skies, winds arose from the west and southwest and swept unhindered across the land. The first dust storm was reported on 12–13 November 1933 and was followed by so many others that complete records have not been kept (Figure 9.3). Custodians wiped Oklahoma topsoil from the desks of senators in the Capitol building, and homemakers in Boston and Atlanta complained about the dirt on freshly laundered sheets hanging on their clotheslines. At its peak, the dust storms affected 50 million acres centered in Oklahoma and the Texas panhandle, but drought and dust extended beyond the Canadian border. Dust filtered into feed stacks, it covered pastures, and filled in waterholes. Livestock

Figure 9.4
Aftermath of a 1938 dust
storm. Courtesy U.S. Department of Agriculture.

died of starvation, suffocation, and thirst (Figure 9.4). Static electricity
rendered automobile ignition systems inoperable, and countless stalled
cars and trucks were abandoned. Trains were cancelled because the engineers couldn't see the rails, and, if they did move, they spun their
wheels on rails made slick and greasy by the crushed bodies of millions
upon millions of grasshoppers who filled the air, eating anything that
was still green. Driving sand scoured the paint from houses and sand-
blasted windowpanes. Jackrabbits were blinded by driving particles of
sand and children choked to death. When light rains did fall, they came
down as mud showers.

Wheat yields averaged 2.5 bushels per acre, and, because of the
depression, sold for eight cents per bushel, less than the cost of the
seed. Bank loans could not be extended and many farms were sold to
pay the interest on these loans. Bread-lines in the cities were paralleled
by the long lines of cars and trucks carrying people away from their
land. The "Okies" left for the golden dream of California, and Route 66
was the highway to a state where, it was thought, farming could again
be a way of life. One jalopy had a whitewashed sign reading:

1930—Frozen Out
1931—Dried Out
1932—Hailed Out
1933—Grasshoppered Out
1934—Dried Out
1935—Rusted Out
1936—Blown Out
1937—Moving Out

Figure 9.5
An Oklahoma migrant family
in a camp in San Jose, California, in the late 1930s.
Photo from Resettlement Administration, U.S. Department of Agriculture. (The
Granger Collection)

California, beset by reductions in available irrigation water, increased salt accumulation in its irrigated fields, and its own depression,
was in no mood to welcome the tattered immigrants to towns with 30
percent of their work force on relief (Figure 9.5). The immigrants were
jailed for vagrancy, harassed, shot at, denied entrance to the state, and
otherwise set upon, cheated, and hated as potential rivals for jobs. The
record of the Okie migration is nowhere as movingly put as in John
Steinbeck's *The Grapes of Wrath.* Today's migrant farm workers, still inadequately protected by minimum wage laws and lacking adequate
housing and education, are the children and grandchildren of the
Okies.

And the land? In 1935, the U.S. Department of Agriculture mobilized soil scientists, agronomists, and botanists into the Soil Conservation Service to restore the ravaged countryside. Irrigation and
water-conservation practices were introduced (Figure 9.6), new
methods of tilling the soil were developed, shelterbelts of trees were
planted to break the force of the winds, and arid soil was covered with a
mantle of prairie and steppe grasses seeded from trucks and planes.
New crops were introduced to bind the soil, legumes were recommended to replace lost nitrogen and humus, and silted-in ponds and
streams were cleaned out. The more marginal lands were purchased by

state and federal authorities and declared to be immune to further agricultural exploitation. The prosperity of the war years of 1941–1946 reestablished credit for the purchase of machinery and allowed those farmers who had not left the land to again become self-supporting.

The one thing which could not be done was to ensure more adequate and more dependable rainfall patterns. The former dust bowl is still, geologically, geographically, and biologically, semiarid to arid land, subject to the vagaries of weather and climate. A reduction of 25 percent in average precipitation in any given year has a reasonably high statistical probability, and, although much less predictable, there is absolute assurance that an extended period of drought can be expected. In 1954, after two years with subnormal rain, dust storms occurred in the former dust bowl and in western Kansas and eastern Colorado. Some dust blowing occurred in Nebraska and the Dakotas. Floods in Bangladesh, droughts in India, the Sahel, and other parts of North Africa, crop failures due to disease in Central America, frost in the Soviet Union, and excessively hot weather in China caused a rise in the value and price of wheat to the point that the restored grasslands of Oklahoma, Texas, and parts of New Mexico are again being plowed to plant wheat. And who is to say nay? In the United States, Canada, and other countries, land use is determined by the owner of the land. He or she is not its steward, but its exploiter. Government cannot tell the individual what use of the land is best for the future of that land. And everyone knows that there will again be years when hot winds will suck the moisture from the soils and when the rains will not fall and when the topsoil, deprived of its grass, will blow and blow and blow.

ADDITIONAL READINGS

Carter, V. G., and T. Dale. *Topsoil and Civilization,* rev. ed. University of Oklahoma Press, 1974.

Hyams, E. *Soil and Civilization,* rev. ed. New York: Harper & Row, 1976.

Hylander, C. J. *Wildlife Communities.* Boston: Houghton Mifflin, 1966.

Lang, O. L., L. Kappen, and E. D. Schulze (eds.). *Water and Plant Life: Reports of the Wurzberg Conference in 1974 on Plant-Water Relations.* New York: Springer-Verlag, 1976.

Sears, P.B. *Deserts on the March.* rev. ed. University of Oklahoma Press, 1974.

Steinbeck, J. *The Grapes of Wrath.* New York: Viking Press, 1939.

Tannehill, I. R. *Drought.* Princeton University Press, 1947.

Yearbook of Agriculture. Soils and Man. Washington, D.C.: U.S. Department of Agriculture, 1938.

Shade

❧ I made me gardens and orchards
And I planted the trees in them
ECCLES. 2:5

A set of before-and-after photographs of roads leading out of cities can be called ecological pornography. What used to be a tree-lined, shaded highway has often become a maze of transmission lines, interspersed with hamburger joints, motels, and gas stations. Even residential streets and college campuses have had trees removed because it became "necessary" to put up an apartment building or a tower to house administrative functionaries whose role in education is unknown to faculty or students.

THE JOY OF TREES

All cultures have had a feeling of awe and reverence for trees (Chapter 3), and while the awe remains, reverence has been replaced by aesthetic appreciation. Explorers and early settlers of North America were both amazed and annoyed by the dense forests along the Atlantic seaboard and set forth with a will to remove the trees and bring the land into agricultural production. Trees became commodities, sources of fuel and timber needed by the community, and they were again worshipped, this time as sources of wealth. Paul Bunyon is a bigger folk hero than Johnnie Appleseed. By the middle of the nineteenth century, most seaboard forests were gone, and the axe and saw were chewing their way through Ontario and the American Middle West. After the American Civil War, lumber barons cleared large parts of New England, upper Michigan, Wisconsin, and adjacent Canada, leaving the land naked to wind and rain and subject to erosion and flooding. From being barriers in the road to agricultural and urban expansion, trees suddenly became precious reminders of a paradise lost. As cities expanded, demand of their inhabitants for a piece of nature led to expansion of public parks—thus repeating Europe's experience—and tree planting became a social good. Establishment of national parks, national and state forests, and wilderness areas and demands for a sense of social responsibility from the lumber companies has given some assurance that trees will remain part of the North American environment. Activist groups such as the Audubon Society, Wilderness Society, Nature Conservancy, and the Sierra Club, provide a vigilant and effective voice in the give and take of economics and conservation. It is now easier to mobilize a community endangered by tree removal than it is to obtain consensus for sewage systems or even public education.

With the increased urbanization of the past quarter century, a new assault on trees has become apparent, and new political and social issues have arisen. Private profit from public resources has intensified, and tract developers and their allies, the multinational lumber corporations, possess great political and economic influence with regulatory agencies and government. In addition, we are now seeing the decline and death of trees on a scale which has raised the consciousness of urban and suburban populations to unprecedented heights. Local governments, perpetually short of money, are seeing funds desperately needed for social services diverted to removal of dead trees. Funds are rarely available for replacement, and a skinny sapling, whose total cost may exceed $100, is no substitute for a mature specimen.

In North America, relatively few of the hundreds of tree species have been used extensively as shade trees. Some conifers, particularly Douglas fir (*Pseudotsuga taxifolia*), true firs (*Abies spp.*), spruces (*Picea spp.*), and the smog-resistant *Ginkgo* have been used. Deciduous trees and broad-leaved evergreens are more common. Elm (*Ulmus americana*), maple (*Acer spp.*), ash (*Fraxinus spp.*), willow (*Salix spp.*), the London plane (a hybrid: *Platanus x P. acerifolia*), some magnolias, and palms are extensively grown. All of these trees are in trouble. Since it is impossible here to discuss the full range of problems, conceptual understanding of the shade tree problem can be obtained by the case-history method in which a few examples are given.

CHESTNUT BLIGHT

The American chestnut (*Castenea dentata*) once covered nine million acres in North America, with dense stands found from northern New England and adjacent Canada down to the Gulf of Mexico and west to the Mississippi River. As late as 1910, chestnuts were growing on 40 percent of the forested land of Connecticut. Valued for its decay-resistant wood used for fence rails, railroad ties, utility poles, and house foundation timbers, one out of every ten hardwood trees that went through the sawmills was a chestnut. Its nuts were staple food for wild turkeys, grouse, and squirrels. Roasted, ground into flour, or fed to hogs, the nuts were important carbohydrate and protein sources on farms. When woodlots were cut out and roads widened, chestnut trees were left standing, to become beautiful and useful shade trees. In 1904, a shipment of Japanese chestnut (*C. crenata*) trees was imported into New York City. Since the trees looked healthy, they were outplanted in the Bronx zoological park, but no one was aware that they contained the spores of a fungus, *Endothia parasitica*, against which the American species had no resistance. In less than 25 years, most of the mature chestnuts east of the Mississippi were dead. The fungus, an ascomycete, is exquisitely adapted to its parasitic role. Asexual spores, shed as jellylike yellow masses in wet weather, are easily carried in rain and by insects or birds to other chestnuts. The sexual ascospores are shed in dry weather and, being very light, are carried by the wind. Spores can germinate in wounds or in cracks in the bark, and the fungal mycelial threads grow rapidly, killing phloem and cambium cells, until the twig or trunk is girdled and dies.

By 1930, the American chestnut had regretfully been written off; attempts to discover a cure had failed, and the blight had spread inexorably throughout North America. It was, however, noticed that the fungus does not kill the root collar or the roots, and trees began to coppice, to regenerate new sprouts. Hopes that these new trees, sometimes 10–15 m tall and occasionally bearing nuts, were resistant were dashed when, as bark began to furrow, they became infected and the cycle was repeated. To "bring back the chestnut," a nation-wide search for seed was started, and nurseries of seedlings were planted with the thought of finding individual trees resistant to the fungus. Since this is a very long-range project, final assessment of its potential has not been made, but chances for success are low. In 1956, chestnut seeds, irradiated in the cobalt-gamma ray reactor at Brookhaven National Laboratory were planted on the theory that mutations induced by the ionizing radiation would confer blight resistance. Over 10,000 trees were planted in five states, and, in 1973, seeds from these trees were germinated to see if any resistant strains were produced. The chances for success are slight.

It has long been known that the oriental chestnut trees are blight resistant. The Chinese chestnut (*C. mollissima*) and the Japanese species were introduced into North America as substitutes, but they are less vigorous, require considerable maintenance, are not as handsome, and their nuts are not as tasty as those from the American species. They cannot compete with native trees and most forest plantings have died out. Hybrids of native and oriental trees have been made, but none to date are fine specimens. Since the fundamental difference between a resistant and a susceptible tree is unknown, genetic engineering cannot proceed on a rational basis. A fungicide, Lignosan, can protect healthy trees and arrest the spread of the fungus, but it is expensive and hence impractical for any but individual specimens.

In the 1930s, *Endothia* was introduced into Europe, and it seemed that this same unhappy story was going to be repeated. About 1950, it was noticed in Italy and France that diseased trees of the European chestnut (*C. sativa*) was beginning to recover. French plant pathologists isolated a strain of the fungus which differed from the expected strain. Cultures of this hypovirulent or "H" strain were sent to the Connecticut Agricultural Experiment Station, center for research on chestnut blight, and researchers confirmed the European reports that inoculation of diseased trees with H-strain caused regression of canker development and recovery of the trees. A cross country skier in Michigan also noticed recovery of coppice sprouts near Ann Arbor, and other H-strains were obtained from samples of bark and wood. Although the evidence is incomplete, it appears that H-strains carry a virus that can invade and destroy virulent strains of *Endothia* whose toxins are responsible for the death of phloem and cambial cells. H-strains are less vigorous than virulent fungi, and their spores are disseminated less widely; in France, H-strains spread only 10 m per year, a fraction of the dissemination of virulent spores. There is, however, now hope for the return of the American chestnut.

DUTCH ELM DISEASE

The elm (Ulmaceae) family is a relatively small taxon whose trees are among our most important shade plantings. In Europe and in North America, elms grace streets and parks. Ulmaceous species of *Zelkova* are planted in Crete, the Caucasus, and Japan, and hackberry (*Celtis spp.*) cultivars are used as shrubby decorative accents. There are several North American species of elm, but the American elm is the most planted shade tree.

America's love affair with the elm is of long standing. In 1646, citizens of Boston planted American elms in the Boston Common for "the relief of travelers." A century later, on August 14, 1765, the "Free Men and the Sons of Liberty" met under one of these trees to hang an effigy of Lord Butte, the British parliamentarian who conceived the hated Stamp Act. A rallying place for rebels, the Liberty Tree was cut down by British soldiers in 1775, a less than futile gesture since hundreds of elms, all over the colonies, were designated as Liberty Trees. A major centennial project in the United States was the planting of millions of American elms along roads and in parks across the country.

The elm is not a commercially valuable tree. Its wood is occasionally used for furniture, but any owner of a wood-burning stove will tell you how difficult it is to cut and split. It is, however, unexcelled as a shade tree. Rows of elms on each side of a street can grow so that their umbrella-shaped crowns meet to form a cool, green arch, and their thick, gray trunks line miles of town and city roadways. Ohio, very proud of its elms, was in 1930 the first state to experience the Dutch Elm disease. Leaves of 60-year-old trees began to wilt, turn yellow, then

Figure 9.7
Removal of an American elm
dying of Dutch elm disease.
Courtesy U.S. Department of
Agriculture.

brown, and fall. Usually first noticed in hot midsummer, it looked as if
one side or one main branch of a tree had been scorched by fire. The
disease was progressive, moving from twigs to larger branches until the
entire crown was devoid of leaves, buds were shriveled, and the tree
died within a year or two (Figure 9.7). New York was next, with the city
being hardest hit, and dead elms were reported from Ontario and
Quebec down to Louisiana and as far west as Nebraska by 1935. In
1975, the disease reached California. The disease is caused by a fungus,
Ceratocystis ulmi, which grows rapidly in the vessels (Figure 9.8). The
fungal cells bud, multiplying much like yeast cells, quickly plug up the
water-conducting xylem cells. Leaves, deprived of water and minerals,
soon wilt, lose the ability to carry on photosynthesis, and the death of
twigs, branches, and the stem becomes only a matter of time. There is
some evidence that *Ceratocystis* also produces a toxin which kills tree
cells.

The spread of Dutch Elm disease from tree to tree is due, at least
partly, to two insects. Both the European bark beetle (*Scolytus multi-
striatus*) and a native bark beetle (*Hylugopinus rufipes*) bore into cracks in
elm bark where they tunnel out extensive galleries in which to lay their
eggs. The eggs hatch into larvae which feed on young bark tissues,
opening up additional galleries. Although these tunnels and galleries
injure trees, most healthy elms can withstand the damage. When the
beetles mature into flying insects, they emerge from the tree. If this tree

Figure 9.8
Disease cycle of Dutch elm
disease caused by *Ceratocystis
ulmi*. Redrawn from George
N. Agrios, *Plant Pathology*
(New York: Academic Press,
1969).

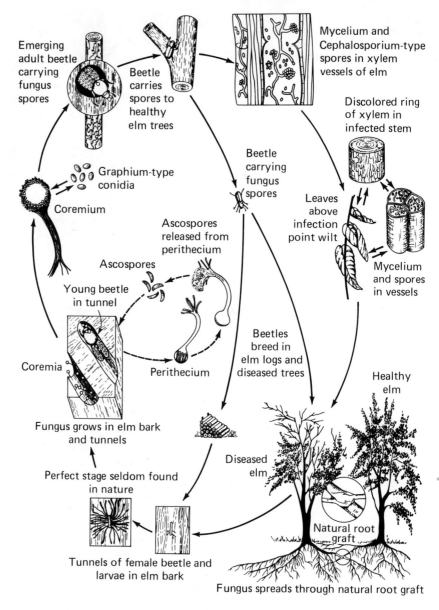

Emerging
adult beetle
carrying
fungus
spores

Beetle
carries
spores to
healthy
elm trees

Mycelium and
Cephalosporium-type
spores in xylem
vessels of elm

Discolored ring
of xylem in
infected stem

Graphium-type
conidia

Coremium

Beetle
carrying
fungus
spores

Leaves
above
infection
point wilt

Mycelium
and spores
in vessels

Ascospores
released from
perithecium

Ascospores

Young beetle
in tunnel

Perithecium

Beetles
breed in
elm logs and
diseased trees

Healthy
elm

Coremia

Fungus grows in elm bark
and tunnels

Perfect stage seldom found
in nature

Diseased
elm

Natural root
graft

Tunnels of female beetle and
larvae in elm bark

Fungus spreads through natural root graft

was infected with *Ceratocystis* fungi, the adult insect would carry both
spores and mycelium to another tree where the fungus would be introduced when tunnelling begins.

For many years it was believed that these insect vectors were
solely responsible for transmission of Dutch Elm disease, but more recent studies have demonstrated that root systems of adjacent trees form
natural grafts, allowing infection to move directly from tree to tree.
Until this phenomenon was discovered, control of the disease consisted

of attempts to kill bark beetles with insecticides and the removal of dead trees which served as sources of infective fungi. Although these methods were partly successful, the ease with which the beetles can carry the fungus and the development of resistance to insecticides like DDT and methoxyclor merely slowed down the spread of the disease. Nevertheless, the ravages of the disease are essentially a man-made problem. By our extensive planting of elms, we developed a monoculture of *Ulmus americana,* in every way comparable to a field of wheat or corn, through which a disease or pest can sweep unhindered. A magnificent line of elms is a chain, with the weak link being one infected tree.

Demands for the elimination of Dutch Elm disease have not been met. To date, no resistant individuals have been found, and attempts to develop genetic resistance are not far advanced. The American elm has twice the chromosome number of the resistant European species and direct hybridization is impossible. It is possible to induce chromosome doubling in the European elm to make it genetically compatible with the American species. Once this is accomplished, another 15–20 years are needed before the tetraploid (doubled chromosome numbered) tree will mature and flower, and an additional 10–15 years are needed before the hybrids can be evaluated. A cross between an English species (*U. glabra*) and *U. carpinifolia* from France was made in Holland in 1963 and proved to be very resistant. Unfortunately, this tree rarely reaches 10 m in height, far too short to become a useful shade tree. Asiatic elms are resistant to the fungus, but are not particularly handsome trees and do not excite the imagination of urban planners and park departments.

Both species of bark beetle were found in North America in 1904, but the introduction of the fungus did not occur until 1933, probably on a shipment of logs sent from England. In 1935 the fungus was isolated and identified by Dutch plant pathologists, and Holland was inadvertently and incorrectly blamed for being the place of origin of the disease. There is little doubt that Dutch Elm disease was present for centuries in much of Europe where it was a minor problem because European elms are relatively resistant. All this doesn't matter any more. Quarantine restrictions have little meaning in an age when jet planes carry people and cargo around the world in a matter of hours. What does matter is whether the American elm, as a shade tree or even as a part of the forest environment, will be able to survive.

There are two current lines of attack on the problem. Instead of using insecticides which are less and less effective as beetles are selected for resistance and as evidence accumulates on the potential and manifest insults of chemicals on the environment, new methods of reducing the beetle population are being developed. Most hopeful is the use of pheromones, insect scents by which beetles can be lured to traps. Although not tested with bark beetles, it is also possible to sterilize male beetles of some species with radiation and release them to mate, unsuccessfully, with female beetles. These potentially useful techniques cannot help infected trees or trees root-grafted to infected trees. Direct chemical assault on the fungus in situ was started in 1972 when the fungicide Benomyl (Lignosan) was developed for infected elms, and possibly as a prophylactic. Still experimental, Benomyl seemed to be fairly effective, but it is expensive and its protective action is only moderately long-lasting. Its proponents correctly point out that use of Benomyl, prompt removal of dead trees, pruning, chemical control of insects, and other practices are "buying time"—time for genetic research to be completed, time to develop other treatments and preventatives, and time to permit the planting and establishment of other tree species to break the monocultural lockstep that resulted in the devastation of the elm.

MAPLE WILT

Native and imported maples, although not as common as elms, are basic landscaping and shade trees. North American sugar maple (*Acer saccharum*), red maple (*A. rubrum*) and silver maple (*A. saccharinum*), the European or Norway maple (*A. platanoides*), the Japanese red maple (*A. palmatum*), and the Chinese Amur River maple (*A. ginnala*) are handsome shade trees whose spring and fall colors delight the eye. Box elder (*A. negundo*) is used as a foundation planting. Unfortunately, members of the genus are susceptible to a wide range of leaf, stem, and root diseases, attacks by insects, and abnormal reactions to environmental conditions. Leaf spottings and discolorations caused by various fungi (*Verticillium, Phyllosticta, Rhytisma*) or bacteria (*Pseudomonas aceris*) and insect depredations are common, but rarely fatal to the plant. Stem and branch diseases are more severe. Cankers (open, nonhealing wounds) progressively destroy the wood and weaken the trunk or may kill cambial or phloem cells to effectively girdle and thus kill the tree. Sapsuckers chip holes in the trunks, the injured bark dies, and the holes serve as ports of entry for fungi, bacteria, and insects. Roots may rot following invasion of fungi, may swell and produce galls when attacked by nematodes, or have their meristems injured by insects. All in all, maples are delicate trees whose susceptibility is at variance with our image of a strong tree. Like the elm, the rapid dissemination of many diseases is due to the practice of planting these trees in rows.

℘ AN ESSAY ON CAUSATION

Diseases of shade tree maples are, like those of other plants, "caused" in most cases by definable, isolatable organisms. By suitable techniques, the virus, bacterium, or fungus can be obtained in pure culture, inoculated back into healthy specimens where it induces the typical disease symptoms, reisolated, and used again to infect the host.

This lockstep of demonstrating cause and effect in disease was developed by the German bacteriologist Robert Koch in his studies on human disease. Koch's postulates provide excellent experimental evidence for the causal nature of disease. But cause and effect in the biological world are rarely as simple as the postulates would suggest. A sapsucker hole will not kill a sugar maple tree, but if the wound is infected with a wilt fungus, the resulting maple wilt may result in the tree's death. The sapsucker, trying to get a meal, is not the cause of the disease, but without its activity, the fungus would not have been able to invade the tree. The concept of predisposition is not a new one, but its relevance to disease of ornamental and urban trees is becoming more important. Maple decline illustrates this very well.

Consider the stress imposed on trees in an urban environment. In large cities, trees are planted in small holes chopped out of cement sidewalks to a depth determined by the enthusiasm of the employees of the street department. Filled with poor soil, a 3-m sapling is planted, irrigated primarily with contaminated, alkaline water running off the pavement, and fertilized by dogs on leashes. Immediately adjacent to the tree is a busy city street where automobiles emit clouds of exhaust containing oxides of nitrogen, sulfur dioxide, and carbon monoxide. These block the respiratory process, and exhaust particulates clog the stomata and interfere with normal transpiration of water and uptake of carbon dioxide. The asphalt and the cement reirradiate heat in the summer time, raising ground temperatures to Death Valley levels, and hold cold in the winter, lowering temperatures to the arctic permafrost level. When it rains, sulfur compounds emitted from factories and power-generating plants are converted to sulfuric acid which accumulates in the minimal amount of soil but cannot leech out because conduits, pipes, and foundations are just below ground level. The city street is no place for a plant. Photosynthesis, water relations, and respiration are far from normal, root growth is restricted, and bruises, cuts, and scrapes form openings through which disease-causing organisms can enter. A direct connection between low vigor and lowering of disease resistance is difficult to establish because the term vigor is difficult to define or measure quantitatively, but vigor, however defined, is certainly a component in predisposition to disease. And yet, trees are necessary, indeed vital, for city living

---

Maples in suburban or even rural environments are not exempt from man-made predisposing factors. Economic considerations—the tourist industry, long-distance trucks which carry the goods needed for people, and inadequate public transportation—dictate that roads shall be ice free. Salts, usually sodium chloride or calcium chloride, have been substituted for sand. Although most salt will leech down below the root growing zone of shallow-rooted maples, the salts can cause some damage and will accumulate in ground water supplies and injure plants at some distance from the point of application. By altering electrical charges on soil particles, the character of soils may be altered. Automobile exhausts near heavily travelled highways cause leaf damage; sugar maples rarely reach their potential fall glory along such roads, and in areas where winters are severe, one can see a zone of ice on roadside plantings from the water vapor in automobile exhausts.

MAPLE DECLINE

Construction work can cause severe damage to trees, and maples are among the most susceptible of our shade trees. Their shallow root systems spread out several meters from the trunk, usually beyond the circle of the outer twigs, gathering water and minerals for plant growth and development. Like all living cells, those of the roots metabolize, respiring to produce energy, synthesizing new cells, and growing in length. Oxygen requirements for roots are high, and a friable, well-structured soil is a necessity. Construction of a new road or modification of an existing road to carry more traffic usually results in direct injury to tree root systems, as well as indirect damage caused by compaction of the soil and alteration in the pattern of water percolation. Mature specimens along newly constructed roads frequently show very high mortality.

The seemingly inexorable march of "progress" can be measured by conversion of forests or meadowlands—usually of excellent quality—into housing developments. It is standard practice, because it is more "economical," to chainsaw down all trees in the area, bulldoze the topsoil off for resale, and bring in heavy machinery which packs down the existing subsoil. Once the buildings are up, a two-to-four cm layer of topsoil is spread, planted with short-lived annual ryegrass (*Lolium spp.*), and a few poor quality saplings are inserted into small holes. Lacking adequate mineral supplies, inadequately supplied with water, and attempting to expand a root system into compacted subsoil, many saplings die as soon as the house is sold (never before), while others grow poorly and are susceptible to disease. Builders with some feeling for trees, whether aesthetic or economic, may leave mature specimens. Careless construction crews use the trees as guy standards or alkalize the soil with bits of building debris, resulting in decline and death. The more common cause of lowered vigor is the practice of back-filling or raising the ground level to promote drainage. As with compacting soil, this lowers the capacity of the soil to hold water and lengthens the path for oxygen penetration to the root zone.

Although first noted in the Northeast, maple decline has been reported in many states and provinces. The first symptoms include a premature yellowing or reddening of the leaves accompanied by browning of leaf margins. Leaves may fall in midsummer. The crown of the tree begins to show dieback of branch tips and fine twigs which continues toward the center of the crown and down toward the trunk. Bare branches with a tuft of leaves at the tip are seen the following spring. Growth is severely reduced, and, all too frequently, large portions of the tree deteriorates to the point that half or all of the crown is dead. Attempts to isolate a causal organism have not failed; indeed, at least six pathogenic fungi have been found in dead or dying trees. None, however, fulfill Koch's postulates, although each can induce some of the observed symptoms. Maple decline is a *disease syndrome,* a complex of biotic and nonbiotic causes, each dependent and related to the other. Salt injury, inadequate water, insect damage, compaction, back-filling, noxious gases, and other insults predispose the tree to reduced vigor which, in turn, facilitates the depredations of viruses, fungi, bacteria, insects, nematodes, and hence disease, decline, and eventual death. Obviously, there is no single or even multiple cure or control for maple decline. One cannot move a 20 m maple back from a road or reroute traffic or individually water and fertilize a thousand maples along a 20-mile stretch of superhighway.

Maintaining shade and ornamental trees as part of our inhuman environment is, to date, an unsolved problem. The American Horticultural Society has noted that man-made environmental stresses are increasing and pose threats to current and future installations of trees. Where a specific condition exists, it is possible to plant trees known to have the capacity to resist the condition. *Ginkgo* is resistant to automobile exhaust fumes; Russian olive (*Elaeagnus angustifolia*) is salt tolerant whereas pines are not; and many conifers are drought resistant and capable of growing on sites where water patterns have been upset. The experience with the American elm has taught us that diversity of species provides more safety than monoculture. Our experience with the chestnut should suggest caution against introduction of foreign species without long, intensive study. Our experience with maple decline has shown the need for interaction between the builder and the botanist. May we profit from these experiences!

ADDITIONAL READINGS

Agrios, G. N. *Plant Pathology.* New York: Academic Press, 1969.
Daubenmire, R. F. *Plants and Environment,* 2nd ed. New York: Wiley, 1959.
Environmentally Tolerant Trees, Shrubs and Growth Covers. Mt. Vernon, Va.: American Horticultural Society, 1975.
Feininger, A. *Trees.* New York: Viking Press, 1968.
Li, H. L. *The Origin and Cultivation of Shade and Ornamental Trees.* University of Pennsylvania Press, 1963.
Menninger, E. A. *Fantastic Trees.* New York: Viking Press, 1967.
Pirone, P. P. *Maintenance of Shade and Ornamental Trees.* New York: Oxford University Press, 1941.
Symposium on Trees and Forests in an Urbanizing Environment, August 1975. Amherst: Extension Service, University of Massachusetts.

Index